Communications in Computer and Information Science 2360

Rationale

The CCIS series is devoted to the publication of proceedings of computer science conferences. Its aim is to efficiently disseminate original research results in informatics in printed and electronic form. While the focus is on publication of peer-reviewed full papers presenting mature work, inclusion of reviewed short papers reporting on work in progress is welcome, too. Besides globally relevant meetings with internationally representative program committees guaranteeing a strict peer-reviewing and paper selection process, conferences run by societies or of high regional or national relevance are also considered for publication.

Topics

The topical scope of CCIS spans the entire spectrum of informatics ranging from foundational topics in the theory of computing to information and communications science and technology and a broad variety of interdisciplinary application fields.

Information for Volume Editors and Authors

Publication in CCIS is free of charge. No royalties are paid, however, we offer registered conference participants temporary free access to the online version of the conference proceedings on SpringerLink (http://link.springer.com) by means of an http referrer from the conference website and/or a number of complimentary printed copies, as specified in the official acceptance email of the event.

CCIS proceedings can be published in time for distribution at conferences or as post-proceedings, and delivered in the form of printed books and/or electronically as USBs and/or e-content licenses for accessing proceedings at SpringerLink. Furthermore, CCIS proceedings are included in the CCIS electronic book series hosted in the SpringerLink digital library at http://link.springer.com/bookseries/7899. Conferences publishing in CCIS are allowed to use Online Conference Service (OCS) for managing the whole proceedings lifecycle (from submission and reviewing to preparing for publication) free of charge.

Publication process

The language of publication is exclusively English. Authors publishing in CCIS have to sign the Springer CCIS copyright transfer form, however, they are free to use their material published in CCIS for substantially changed, more elaborate subsequent publications elsewhere. For the preparation of the camera-ready papers/files, authors have to strictly adhere to the Springer CCIS Authors' Instructions and are strongly encouraged to use the CCIS LaTeX style files or templates.

Abstracting/Indexing

CCIS is abstracted/indexed in DBLP, Google Scholar, EI-Compendex, Mathematical Reviews, SCImago, Scopus. CCIS volumes are also submitted for the inclusion in ISI Proceedings.

How to start

To start the evaluation of your proposal for inclusion in the CCIS series, please send an e-mail to ccis@springer.com.

Prasanna Devi Sivakumar · Raj Ramachandran ·
Chitra Pasupathi · Prabha Balakrishnan
Editors

Computing Technologies for Sustainable Development

First International Research Conference, IRCCTSD 2024
Chennai, India, May 9–10, 2024
Proceedings, Part I

Editors
Prasanna Devi Sivakumar
SRM Institute of Science and Technology
Chennai, Tamil Nadu, India

Raj Ramachandran
Cardiff Metropolitan University
Cardiff, UK

Chitra Pasupathi
SRM Institute of Science and Technology
Chennai, Tamil Nadu, India

Prabha Balakrishnan
SRM Institute of Science and Technology
Chennai, Tamil Nadu, India

ISSN 1865-0929 ISSN 1865-0937 (electronic)
Communications in Computer and Information Science
ISBN 978-3-031-82388-6 ISBN 978-3-031-82389-3 (eBook)
https://doi.org/10.1007/978-3-031-82389-3

This Springer imprint is published by the registered company Springer Nature Switzerland AG
The registered company address is: Gewerbestrasse 11, 6330 Cham, Switzerland

Preface

SRM Institute of Science and Technology (Vadapalani Campus), India, a prominent unit of the prestigious SRM Group of Institutions located in Chennai, India, is known for its commitment to fostering academic excellence and innovation. SRMIST (Vadapalani Campus) is distinguished by the emphasis placed on experiential learning, industry partnerships, and research initiatives. With specialized research centers and active industry collaborations, the campus provides an ideal environment for innovative research and learning.

SRMIST (Vadapalani Campus) takes immense pride in presenting the proceedings of the International Research Conference on Computing Technologies for Sustainable Development (IRCCTSD 2024). This distinguished event was jointly organized by SRMIST (Vadapalani Campus) and Cardiff Metropolitan University, UK, on May 9th and 10th, 2024. This Research Conference brought together a wide range of global scholars, industry professionals, and innovative researchers. The event served as an energetic platform for impactful discussions, groundbreaking ideas, and forward-thinking innovations in computing technologies focused on sustainable development. We are confident that the ideas shared here will make a lasting contribution to both academic research and industrial practices, shaping the future of sustainable technology solutions.

The conference received an outstanding 264 research articles. After a rigorous peer-review process, 82 articles were selected for presentation and publication. The conference featured a diverse range of tracks, including AI Solutions, Machine Learning Strategies, Deep Learning Techniques, Industrial Innovation, Image Processing Applications and IoT Innovations related to Computing Technologies domains, namely Healthcare, Security, Agriculture and Climate Mitigation. Each track provided a vibrant and inclusive space for academicians and professionals to exchange ideas, address challenges and contribute to the conversation on leveraging computing technologies to solve today's most pressing sustainability challenges.

Eminent industry expert P Ilango, HCL Technologies, India delivered the keynote address in the inauguration ceremony emphasizing the role of technological innovation in sustainable development; Raj Ramachandran, Cardiff Metropolitan University, UK; Deepan Raj, HCL Technologies, India; and Mohanraj Vengadachalam, Standard Chartered Bank, India delivered the session keynote addresses.

The research outcomes shared during this conference foster continuous innovation and collaboration in the field of computing technology for sustainable development.

November 2024

Prasanna Devi Sivakumar
Raj Ramachandran
Chitra Pasupathi
Prabha Balakrishnan

Acknowledgement

The SRM Institute of Science and Technology (Vadapalani Campus), India, being a premier Institution, focuses on diverse and interdisciplinary research activities emphasizing both applied and fundamental research, aligning with National and Global priorities for technological advancement and self-reliance.

The apex management of SRMIST hones the academicians and students to actively engage in pioneering research projects to address real-world challenges, particularly in the fields of Computing Technologies, Engineering, and Sustainable Development. The commitment to high-quality research output is reflected in the numerous research articles published by the faculty and students, contributing to both national and international academic communities.

SRMIST (Vadapalani campus) recognizes the exceptional contributions of all individuals and institutions who helped in making the International Research Conference on Computing Technologies for Sustainable Development (IRCCTSD 2024) such a remarkable success. This conference showcased the productive partnership between SRMIST, India and Cardiff Metropolitan University, UK. The collaboration between these renowned institutions fostered an environment of innovation and academic excellence, advancing the frontiers of sustainable technology development.

We express our sincere gratitude to the management of SRMIST for their continuous support and encouragement in advancing academic research and knowledge dissemination. The resources, facilities, and collaborative environment of the institution were instrumental in the successful completion of this work. We extend our heartfelt thanks to the administration of SRMIST and the faculty, research scholars, and staff of the CSE department for their guidance and support throughout the execution of this conference and publication of the research proceedings.

Our deepest thanks also go to Cardiff Metropolitan University, UK, for their incredible support and partnership throughout the conduct of the conference. The guidance and insights provided by the distinguished faculty from Cardiff Metropolitan University played a crucial role in shaping the conference's trajectory. Collaborating with such a prestigious institution has been a truly rewarding experience, and we are thankful for the dedication of the team to promoting academic growth and innovation.

We sincerely thank Springer for their outstanding support in publishing this proceedings. It is a privilege to collaborate with such a prestigious publishing platform, renowned for its dedication to high-quality publications.

We sincerely appreciate the invaluable contributions of all the authors for their research efforts and the reviewers whose meticulous evaluations were pivotal in finalizing the selection of articles. We also wish to acknowledge the tireless work and commitment of the organizing and technical program committees, whose efforts were crucial to the conference's success.

In conclusion, we extend our deepest gratitude to all individuals and institutions whose contributions made this conference a success. The collaboration and support provided by the authors, reviewers, organizing committee, and sponsors were pivotal in advancing key subtopics such as AI solutions for sustainability, IoT applications for environmental monitoring, green computing practices, and smart city innovations. These proceedings reflect the ongoing exploration of cutting-edge Computing Technologies for Sustainable Development, and we are confident that the insights shared will inspire further innovation and drive meaningful progress in this essential field.

Organization

Chief Patrons

Patrons

Program Committee

Renuka Devi Saravanan	VIT Chennai, India
Mirnalinee Thanganadar Thangathai	SSN College of Engineering, India

Steering Committee and International Reviewers

Catherine Tryfona	Cardiff Metropolitan University, UK
Karl Jones	Cardiff Metropolitan University, UK
Fiona Carroll	Cardiff Metropolitan University, UK
Khoa Phung	University of the West of England, UK
Chandru Sandrasekaran	International College of Business and Technology, Sri Lanka
Pakkianathan Prabu Premkumar	International College of Business and Technology, Sri Lanka

External Reviewers

Sandra Johnson
P. Latchoumy
Poonkodi Mariappan
Boopathi Raja Govindasamy
Subhashini Palaniswamy
Jyostna Devi Bodapati
Gokul Chandrasekaran
Geetha Chellaian
Karthikeyan Vedanandam
Inbamalar Tharcis Mariapushpam
Boomija Malaisamy Duraipandian
Balasundaram Ananthakrishnan
Sandhya M. Kumar
Immanuvel Arokia James
Dass Purushothaman
Chidambarathanu Krishnan
Surender Shanmugam
Rajan Subramaniam
Subburaj Varadharajaperumal
Meena Rajeswaran
Rohith Bhat
Victo Sudha George
Anuradha Muthukrishnan
Kavin Kumar Kandasamy
Madhavi Thiruvengadam
Prameeladevi Chillakuru
Priya Vijay

Organizing Committee

Golda Dilip
Rajasekar Velswamy
Bharathi Navaneetha Krishnan
Paavai Anand Gopalan
Neelam Sanjeev Kumar
Arun Nehru Jawaharlal Nehru
Durgadevi Palani
Sridhar Srinivasan
Niveditha Satiyamoorthy
Karthikayani Kaliyaperumal
Sangeetha Subramaniam Karuppaiyah Bharathi

Akila Krishnamoorthy
Sridevi Sridhar
Maheswari Sendur Pandy
Jessy Sujana Godwin
Vidhusavarshini Suresh Kumar
Anusha Thamaraichelvan
Gayathri Ramanakumar
Muthurasu Nallappan
Manohar Shanmugavel
Punitha Dhandapani
Indumathy Mayuranathan
Deepa Ravichandran
Rajavel Manickam
Jayanthi Palraj

Contents – Part I

Classification and Prediction Analysis in Healthcare

Contents – Part II

Video and Image Processing for Security Analysis

Innovations for Smart Cities

Sustainable Practices in E-Commerce: Challenges and Trends

Contents – Part III

Application of AI for Education

Innovations in Precision Agriculture Techniques and Strategies for Enhancing Agriculture Production

Rice Leaf Disease Prediction Using Transfer Learning

Sarojini Balakrishnan(✉) and Yashini Nehru

Department of Computer Science, Avinashilingam Institute for Home Science and Higher Education for Women, Coimbatore, India
sarojini_cs@avinuty.ac.in

Abstract. Agriculture plays a vital role in ensuring the economic stability of a country. This has led to a significant amount of money spent on research and innovation to improve crop yield and sustainability. A significant challenge for farmers worldwide is effectively diagnosing and managing diseases that threaten plant leaf health. Plant diseases significantly reduce crop yields if left unchecked at an appropriate stage. Plants show symptoms of diseases on their leaves in many ways - color, shape, size, and texture. By understanding these symptoms visually, farmers identify the diseases. In an effort to automate the diagnosis, CNN models are used. Capturing all of these variationsin a single CNN model is challenging due to the complex features of the leaves. To alleviate this issue, this research proposes using transfer learning to quickly train and diagnose plant diseases. Additionally, two ensembled models are also proposed. The performance of the models is analyzed using the Rice Leaf Disease dataset. The performance of the proposed ensembled models and the pre-trained models are evaluated using the metrics - accuracy, precision, recall, and F1-Score. The experimental results show that the ensembled model featuring EfficientNetV2, DenseNet- 201, and VGG-19 model performs the best.

Keywords: Convolutional Neural Networks · Transfer Learning · Pre-trained Models · Ensemble Models

1 Introduction

Agriculture is the cornerstone of any country's economy, and its significance cannot be overstated. Factors such as nutrient levels, pest presence, disease incidence, genetics, cultural practices, and environmental conditions have a profound effect on plant health. The plants are susceptible to a variety of diseases triggered by these factors. These diseases don't just affect isolated parts of a plant but can compromise the entire crop, leading to significant losses in yield. Despite tremendous advancements in agricultural practices, the threat of plant diseases to crop health and yield remains a pressing concern worldwide [1].

Early detection of plant diseases is crucial to ensure that the crop yields remain unaffected [2, 3]. Farmers can take timely measures to mitigate the spread of diseases by

P. D. Sivakumar et al. (Eds.): IRCCTSD 2024, CCIS 2360, pp. 3–8, 2025.
https://doi.org/10.1007/978-3-031-82389-3_1

accurately identifying diseases, optimizing the use of pesticides and fungicides, and preventing significant crop losses. A plant's leaves are often the first to show the symptoms when it is affected by a disease. Traditional disease diagnosis methods largely depend on agricultural experts manually examining the leaves. This process is not only slow and biased but is also prone to errors. Moreover, the complexity of symptoms and diversity of plant diseases exacerbate the challenges associated with manual diagnosis. Therefore, an urgent need is for automated systems that can reliably and efficiently detect plant diseases.

Machine learning techniques can be used to automate plant leaf disease detection. These models are capable of effectively learning complex patterns and features from a raw image data [4]. While training ML models from scratch to suit particular needs is customary, the recent development in open-source models has kickstarted the use of transfer learning techniques. The open-source models are already pre-trained on very large datasets and have learned to identify the useful features of an image. Transfer learning practitioners can fine-tune these open-source models to a particular task. This approach not only accelerates the training process but also promises superior model performance, especially in scenarios where labeled data is scarce. By effectively reusing existing knowledge, transfer learning enables faster experimentation, reduces computational costs, and often leads to superior results in various applications.

This work leverages transfer learning techniques to identify plant diseases from its leaves. We propose and compare the performance of two ensemble models. The first ensemble model comprises three base models - EfficientNetV2, DenseNet201,and VGG19. The second ensemble model comprises - Inception V2, EfficientNetV2, and XceptionNet. The performances of the ensemble models were evaluated on the Rice Leaf Disease Dataset using metrics - accuracy, precision, recall, and F1-Score. This article is structured in the following way - The existing works in the literature are discussed in Sect. 2. Section 3 describes our proposed methodology and models, and Sect. 4 analyses the performance of the proposed models. We conclude this article in Sect. 5.

2 Literature Review

Deep learning models are vastly used in agriculture to provide accurate, efficient, and scalable crop monitoring and management. Ongoing research and innovation in this area are crucial for unlocking the full potential of deep learning in agriculture. Convolutional neural networks have been widely used for rice disease detection [5]. A more detailed review for automatic rice leaf disease detection using deep learning was discussed in this research paper [6]. Mohanty et al. presented a deep-learning framework based on CNNs for plant disease detection. The study focused on detecting diseases incassava plants, demonstrating the capability of CNNs to classify diseased and healthy plant leaves [7] accurately.

In their work, Zhang et al. explored the effectiveness of ensemble learning techniques for detecting rice leaf disease by combining machine learning algorithms, such as decision trees, support vector machines (SVM), and random forests, to build a robust model for disease detection in rice plants [8]. Van Ho et al. ensembled different CNN architectures using ensemble methods like bagging and boosting. Experimental results showed that the ensemble approach outperformed single CNN models regarding accuracy and robustness against variations in rice leaf images [9]. A stacking-based CNN approach for rice leaf disease detection is proposed by Yang et al. [10]. Alam et al., compared pre-trained models for efficient leaf disease prediction [11].

3 Proposed Methodology

This research work proposes ensemble models for plant leaf disease detection. The ensemble learning models combine several baseline models to build a more powerful one that could outperform individual models' capabilities. By leveraging multiple pre trained models, ensembling can capture a broader range of features and patterns in the data. Two ensemble deep-learning models are proposed in this research work. The first ensemble model (EDV) comprises the pre-trained models EfficientNetV2, DenseNet201, and VGG-19, and the second ensemble model (IEX) consists of InceptionV2, EfficientNetV2, and XceptionNet.

The pre-trained CNN models were trained on the ImageNet dataset. Low-level properties like edges, forms, and textures are already known to these pre-trained models, making it useful for faster identification of structural deviations. During fine-tuning, the model converges more quickly when pre-trained weights are used. By leveraging the knowledge learned from millions of images, we reduce the time consumed for training the modelsfrom scratch and also save on the computational costs. In this research, we leverage the efficiency and effectiveness of pre-trained convolutional neural network architectures as a foundational component in our ensemble approach for detecting rice leaf disease. We show the superiority of our proposed ensemble models by comparing the performance of the individual pre-trained models with the ensemble models. Figure 1 shows our proposed framework. The dataset is first pre-processed and then fed to the models for training. The individual opensource models and the ensemble models, viz. EfficientNetV2, DenseNet201, VGG16, Inception V3, XceptionNet, EDV and IEX are fine-tuned on the Rice Leaf Disease dataset. The performance metrics are computed and then compared.

3.1 Pre-trained Models Used in This Research

EfficientNetV2 is known for its superior performance in image classification tasks. When used in an ensemble setting, it serves as a feature extractor, capturing rich representations of rice leaf images learned from large-scale datasets. DenseNet201 is renowned for its densely connected layers, facilitating the efficient propagation of important features. Each layer in DenseNet201 is connected to all preceding layers, creating a dense block that promotes feature reuse and helps alleviate the vanishing gradient problem. VGG16 is a well-established convolutional neural network architecture characterized by its deep architecture with relatively simple convolutional layers, making it computationally efficient and easy to interpret. Inception V3 is renowned for its sophisticated inception modules, which allow for efficient feature extraction across multiple scales and levels of abstraction. The core building block of Inception V3 is the Inception module, which consists of multiple convolutional layers with different kernel sizes. Xception is

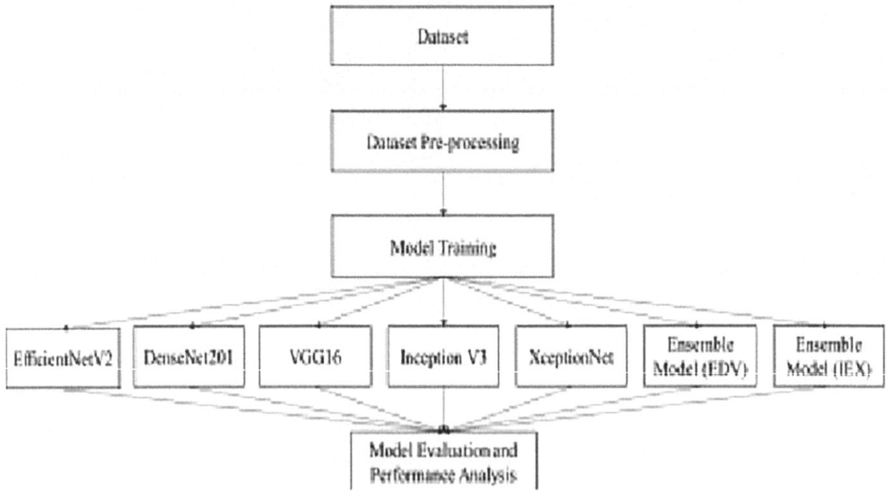

Fig. 1. The framework of the proposed model

a derivation of the Inception V3 and stands out for its depth-wise separable convolutions, which enable efficient feature extraction and parameter reduction compared to traditional convolutional layers.

3.2 Dataset Description and Pre-processing

The Rice Leaf Disease Dataset, part of the New Plant Leaf Disease Dataset from Kaggle, is used for this study. The rice leaf dataset comprises 13,499 images of healthy and diseased crops. Among these, 2119 are images of healthy leaves, and 11380 are disease-affected leaves. Among the 11,380 images of disease affected leaves, there are five sub-categories - bacterial leaf blight (2271 images), leaf blast (2328 images), brown spot (2139 images), leaf scald (2269 images), and narrow brown spot (2373 images). The images are normalized to a fixed dimension and aspect ratio, which ensures that we don't run into compatibility issues with the CNN architecture. Random images are also rotated and flipped to introduce variance in the dataset.

4 Experimental Results and Discussions

The experiments were conducted on Google Colab. The dataset was split in a ratio of 60:40 for training and testing respectively. The performances of the individual pre trained models and the ensemble models were evaluated using the metrics accuracy, precision, recall, and F1-Score. The formulas to compute the metrics are given in Equations 1, 2, 3, 4. Accuracy is the ratio of correctly predicted predictions to the total number of predictions. Precision is the ratio of true positive predictions to the total number of positive predictions. Recall is the ratio of true positive predictions to the total actual positives and F1-score is the harmonic mean of precision and recall.

$$\text{Accuracy} = \frac{\text{TP} + \text{TN}}{\text{TP} + \text{TN} + \text{FP} + \text{FN}} \tag{1}$$

$$\text{Precision} = \frac{\text{TP}}{\text{TP} + \text{TN}} \tag{2}$$

$$\text{Recall} = \frac{\text{TP}}{\text{TP} + \text{FN}} \tag{3}$$

$$\text{F1 Score} = \frac{2 \times \text{Precision} \times \text{Recall}}{\text{Precision} + \text{Recall}} \tag{4}$$

Table 2. Performance comparison of pre-trained and ensemble model

Type of Model	Model	Accuracy (%)	Precision (%)	Recall (%)	F1-score (%)
Pre-trained Base Model	EfficientNetV2	94	93	92	92
	DenseNet201	96	95	94	94
	VGG16	93	92	91	91
	Inception V3	95	94	92	93
	XceptionNet	97	97	97	97
Ensemble Model	Ensemble Model (EDV)	99	97	99	98
	Ensemble Model (IEX)	98	97	98	97

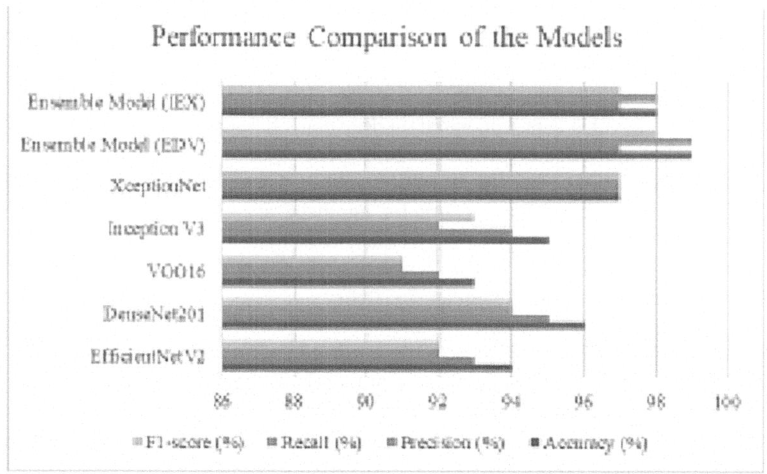

Fig. 2. Performance of Ensemble models and Pre-trained models

The performance of the pre-trained base models and the ensemble models are given shown in Table 2 and are depicted visually in Fig. 2. Among the base models Xception-Net performs the best when measured across all metrics. VGG16 is the worst performing base model. EDV performed the best among ensemble models and achieved a F1-score of 98 percent. Overall, the ensemble models outperform the individual pre trained models in accuracy, with the Ensemble Model (EDV) achieving the best overall metrics. XceptionNet stands out among the base models due to its consistent high performance across all measures. The data suggests that ensemble methods significantly enhance predictive capabilities, making them preferable for tasks requiring high accuracy and reliability.

5 Conclusion

In this work, we compared the performance of pre-trained opensource models available in the literature. We specifically analyzed the performance of the models in the detection plant leaf disease detection thereby contributing to the advancement of computer vision techniques in the domain of agriculture. We develop ensembles of the pre-trained models to leverage the strengths of individual models and thereby enhancing the accuracy and robustness. The experimental results show that the EDV performed the best among all chosen models.

References

1. Garrett, K.A., Dendy, S.P., Frank, E.E., Rouse, M.N., Travers, S.E.: Climate change effects on plant disease: genomes to ecosystems. Ann. Rev. Phytopathol. **44**, 489–509 (2006)
2. Yang, H., Deng, X., Shen, H., Lei, Q., Zhang, S., Liu, N.: Disease detection and identification of rice leaf based on improved detection transformer. Agriculture **13**, 1361 (2023). https://doi.org/10.3390/agriculture13071361
3. Miller, S.A., Beed, F.D., Harmon, C.L.: Plant disease diagnostic capabilities and networks. Annu. Rev. Phytopathol.. Rev. Phytopathol. **47**, 15–38 (2009)
4. Yamashita, R., Nishio, M., Do, R.K.G., et al.: Convolutional neural networks: an overview and application in radiology. Insights Imaging **9**, 611–629 (2018). https://doi.org/10.1007/s13244-018-0639-9
5. Li, D., Wang, R., Xie, C., Liu, L., Zhang, J., Li, R., et al.: A recognition method for rice plant diseases and pests video detection based on deep convolutional neural network. Sensors **20**, 578–598 (2020). https://doi.org/10.3390/s20030578
6. Abdullah-Al-Wadud, M., Khan, M.M., Kashem, M.A., Islam, M.M., Anas, M., Rahman, M.: Deep learning-based image classification for automatic rice leaf disease detection: a review. Plants **9**, 63 (2020). https://doi.org/10.3390/plants9010063
7. Mohanty Sharada, P., Hughes David, P., Salathé, M.: Using deep learning for image based plant disease detection. Front. Plant Sci. **7** (2016)
8. Zhang, S.W., Shang, Y.J., Wang, L.: Plant disease recognition based on plant leaf image. J. Anim. Plant Sci. **25**, 42–45 (2015)
9. Van Ho, S., Vuong, H.G., Nguyen, B.Q., Trinh, Q.H., Tran, M.T.: Ensemble of deep neural networks for rice leaf disease classification. In: 2022 RIVF International Conference on Computing and Communication Technologies (RIVF), Ho Chi Minh City, Vietnam, pp. 238–243 (2022). https://doi.org/10.1109/RIVF55975.2022.10013858
10. Yang, L., Yu, X., Zhang, S., Zhang, H., Xu, S., Long, H., et al.: Stacking-based and improved convolutional neural network: a new approach in rice leaf disease identificationt. Front. Plant Sci. **14**,(2023). https://doi.org/10.3389/fpls.2023.1165940
11. Alam, T.S., Jowthi, C.B., Pathak, A.: Comparing pre-trained models for efficient leaf disease detection: a study on custom CNN. J. Electr. Syst. Inf. Technol. **11**, 12 (2024). https://doi.org/10.1186/s43067-024-00137-1

Chat Bot for Crop Yield Prediction and Recommendation Through K-Nearest Neighbor Algorithm

J. Surya[1] and B. Prabha[2(✉)]

[1] Department of Computer Science and Engineering, SRM Institute of Science and Technology, Chennai, India
[2] Department of Computer Science and Engineering (Emerging Technologies), SRM Institute of Science and Technology, Chennai, India
jemi.prabha@gmail.com

Abstract. The agricultural sector is vital to the nation's economy, providing food, employment, income, foreign exchange, and raw materials for manufacturing industries. However, farmers face significant challenges in maintaining quality and quantity controls over crop cultivation to meet market demands, requiring extensive knowledge and expertise. In India, a leading producer of various crops, the agricultural sector still suffers from insufficient yields due to the lack of advanced techniques and a reliable referral system for farmers. A critical gap exists in the decision-making process for Indian farmers, who struggle with selecting appropriate crops based on soil characteristics and assessing soil health. This issue has significantly impacted their productivity. Addressing this gap is essential for optimizing resource allocation and mitigating future risks. Accurate crop yield prediction and crop recommendation are crucial in agricultural decision-making, and recent advancements in machine learning algorithms have facilitated more precise predictions. This study focuses on employing the Decision Tree method for crop yield prediction. Once constructed, the Decision Tree model can predict crop yields for unknown data and be updated with new inputs such as soil measurements or weather forecasts. The interpretability of Decision Trees allows farmers to identify the factors that most significantly affect crop yield variations, leading to more informed decision-making. Furthermore, a chatbot has been developed to assist users in providing details and receiving crop predictions and recommendations. The study evaluates the performance of three machine learning algorithms—Random Forest, K-Nearest Neighbor, and Decision Tree—in developing the chatbot. By processing input data through these algorithms and comparing their outputs, the study determines the most accurate crop forecast and recommendation. This comparison also helps calculate the standard deviation of crop forecasts and recommendations, improving the model's reliability. This research uniquely contributes to the agricultural sector by providing a practical tool for farmers, offering prioritized crop lists and an enhanced user interface. Ultimately, it aids farmers in making better-informed decisions and optimizing their agricultural practices.

Keywords: Machine learning · Decision tree · K-Nearest Neighbor · Random Forest · Chatbot

P. D. Sivakumar et al. (Eds.): IRCCTSD 2024, CCIS 2360, pp. 9–27, 2025.
https://doi.org/10.1007/978-3-031-82389-3_2

1 Introduction

The main goal of this project is to create a chatbot that uses a decision tree algorithm to predict crop yield and recommend crops. Agriculture is a vital part of many economies, providing food, jobs, income, foreign exchange, and raw materials for industries. To meet market demands, farmers need to ensure both the quality and quantity of their crops. This requires a deep understanding of cultivation and preservation techniques. In India, where agriculture is a major economic pillar, the country leads the world in the production of various crops. However, the sector's yield is still lacking without the use of improved methods. Currently, there is no reliable referral system for Indian farmers, making it challenging for them to choose suitable crops based on soil characteristics and to assess soil health, leading to decreased productivity.

Indian farmers face significant challenges, primarily:

(i) selecting the right crops based on soil properties, and
(ii) evaluating the health of the soil. Addressing these issues can help farmers better allocate resources and prepare for potential risks. Accurate crop yield prediction and crop recommendation are crucial for effective agricultural decision-making. Recently, machine learning algorithms have proven to be highly effective tools for predicting crop yields with precision.

This project focuses on using the decision tree algorithm for training in agricultural yield prediction. Once developed, the Decision Tree model can predict crop yields for new data. The model can be updated with additional input variables, such as soil measurements or weather forecasts, to provide yield estimates and crop recommendations. The interpretability of Decision Trees allows farmers to make informed decisions by understanding which factors most significantly impact crop yield variations.

Farmers can interact with the chatbot by providing relevant details, and the chatbot will use the model to predict crop yields and respond with recommendations. By offering a prioritized list of crops and a user-friendly interface, the trained model helps meet farmers' needs effectively.

Contribution:

- The main goal of implementing a chatbot that uses decision trees to predict crop output and provide crop recommendations is to provide farmers with useful information and practical advice that will maximize crop yield, reduce risks, and improve agricultural sustainability.
- Using a variety of data sources, including past yield records, weather trends, soil conditions, and crop management techniques, the chatbot seeks to forecast agricultural yields. Using decision trees as a predictive model, one may forecast agricultural yields by analysing past data. Planting schedules, resource allocation, and marketing tactics can all be more effectively planned by farmers when crop yields are predicted with accuracy. After analysing a number of variables, including soil type, climate, market demand, and resource availability, the chatbot makes recommendations for what kinds of crops to cultivate.

- The most likely crops to flourish in a certain area or under particular environmental conditions can be identified with the use of decision trees, which aid in the analysis of large, complicated datasets. Recommendations may also take into account elements like profitability, insect resistance, and crop rotation techniques. Decision trees are a popular choice for agricultural dataset analysis because of their versatility in handling both numerical and categorical data. Their transparency and interpretability enable consumers to comprehend the logic underlying the chatbot's recommendations.
- The chatbot can assess various aspects and make conclusions based on a set of metrices, emulating decision-making process of human experts, thanks to the decision trees' hierarchical structure. Farmers and other agricultural stakeholders can communicate with the chatbot to ask for advice and knowledge about crop cultivation. The chatbot can be used to improve user engagement and build user trust and reliability by offering precise forecasts and personalised recommendations. By interacting with the chatbot, users may get up-to-date market trends, real-time information, and advice on crop management best practices, all of which increase overall agricultural profitability and productivity.

2 Related Works

Support Vector Machines (SVMs) have gained considerable attention in the pattern recognition field in recent years. Lai et al. [1] investigated the use of edge-based features in SVMs for road detection from remotely sensed images. Their comparison with neural network classifiers and decision trees showed promising results. Batts et al. [2] studied the impact of temperature and CO2 levels on winter wheat crops over four seasons in Reading, UK, from 1991 to 1995. They discovered that higher temperatures shortened crop duration and reproductive phases, while elevated CO2 levels had varying effects on crop biomass and grain yield. The study highlighted that a slight increase in temperature could negate the yield benefits of elevated CO2. Rub et al. [3] tackled the data volume challenge in precision agriculture using data mining techniques. They evaluated four regression methods on farm data to improve produce forecasts, emphasizing the economic and efficiency gains from utilizing these data effectively. Jorquera et al. [4] enhanced security in IoT cloud platforms for healthcare services by combining OAuth 2.0 with JSON web tokens (JWT). This approach secured authorization across multiple platforms, eliminating the need for refresh tokens and preventing various security attacks. Leemans et al. [5] developed a multi-feature information fusion method based on D-S evidential theory and BP neural network to improve apple grading accuracy. Their method, which classified apples using size, shape, and color features and combined the results, outperformed single-feature grading techniques.

Jayanna et al. [6] assessed a new generation of cost-effective blood pressure monitors using the ANSI/AAMI SP10:2002 Standard. Their study found the device to be accurate and affordable, making it particularly useful for healthcare facilities that prioritize both cost and accuracy. Alberto et al. [7] compared various machine learning techniques for agricultural yield prediction using ten crop datasets. They found M5-Prime regression trees and k-nearest neighbor methods to be the most effective, providing the lowest error rates and highest prediction accuracies.

Mandic et al. [8] emphasized the importance of accurately estimating crop yields for agricultural planning using machine learning. They validated several models, including support vector regression and neural networks, finding M5-Prime and k-nearest neighbor methods to be the most reliable for large-scale predictions. Hochreiter et al. [9] introduced Long Short-Term Memory (LSTM) as a solution to storing data over extended periods in recurrent neural networks. LSTM was more efficient and effective than previous methods, such as Elman nets and real-time recurrent learning, in handling long-time-lag challenges. Sak et al. [10] investigated LSTM recurrent neural network architectures for large-scale acoustic modeling in voice recognition. They demonstrated that LSTM RNNs outperformed deep neural networks (DNNs) and traditional RNNs, especially when trained on a large cluster of machines, achieving state-of-the-art voice recognition performance.

Wiener et al. [11] highlighted the effectiveness of ensemble learning techniques, particularly random forests, which combine multiple classifiers to enhance prediction accuracy. Random forests were found to be highly effective and stable against overfitting, outperforming other classifiers such as discriminant analysis, support vector machines, and neural networks. This segmentation of related works provides a concise overview of various studies and advancements across different domains, demonstrating the broad applicability and effectiveness of machine learning and data analysis techniques.

Inference from the Survey:
The survey of related works highlights significant advancements in machine learning applications across various domains. Support Vector Machines (SVMs) and edge-based features have proven effective in road detection from remote sensing images. Studies on winter wheat crops demonstrate the impact of temperature and CO2 levels on agricultural yield, emphasizing the need for precise climate management. In precision agriculture, data mining techniques have shown to improve efficiency and economic benefits through accurate crop predictions. Enhanced security measures in IoT healthcare platforms, like combining OAuth 2.0 with JSON web tokens (JWT), highlight the importance of robust authorization methods. Multi-feature information fusion methods, particularly in apple grading, outperform single-feature techniques, significantly improving accuracy. Cost-effective blood pressure monitors show that technological advancements can make essential medical devices more accessible without compromising accuracy.

Comparative studies of machine learning methods for agricultural yield prediction indicate that M5-Prime regression trees and k-nearest neighbor techniques offer superior performance. The introduction of Long Short-Term Memory (LSTM) in recurrent neural networks addresses long-term data retention challenges, improving over previous models. Exploring LSTM RNN architectures for acoustic modeling in voice recognition shows their efficiency and accuracy, making them superior to traditional RNNs and DNNs. Lastly, ensemble learning techniques, especially random forests, demonstrate robustness and reliability in various classification and regression tasks, outperforming other established classifiers.

3 Methodology

The proposed system evaluates the accuracy rates of various model combinations, including KNN, AdaBoost, Random Forest, and Decision Tree regression algorithms. The parameters considered are Count_name, Year, Yield_value, Average_rainfall, Pesticide_tons, and Average_temperature. The best ensemble model is selected by comparing the performance metrics of each model. Once the optimal ensemble model is identified, the system is ready to develop a prediction model. Some limitations include: 1) lower accuracy compared to the proposed system; 2) issues with text box details and forecast accuracy; and 3) the absence of crop recommendations.

A. Proposed Architecture

The system begins by collecting historical agricultural data, which encompasses weather patterns, soil properties, crop management techniques, and past production records of eligible crops. A prediction model is then developed using the Decision Tree algorithm. This method employs recursive partitioning to construct a tree structure with decision nodes based on specific input variables and leaf nodes representing expected crop production outcomes. The incorporation of a chatbot simplifies the crop yield forecasting and recommendation process, making it user-friendly for researchers, farmers, and other stakeholders without requiring complex data analysis or specialized technical knowledge. Benefits include: 1) enhanced prediction accuracy through the Decision Tree method; 2) interactive forecasting facilitated by the chatbot (Fig. 1 and Fig. 2).

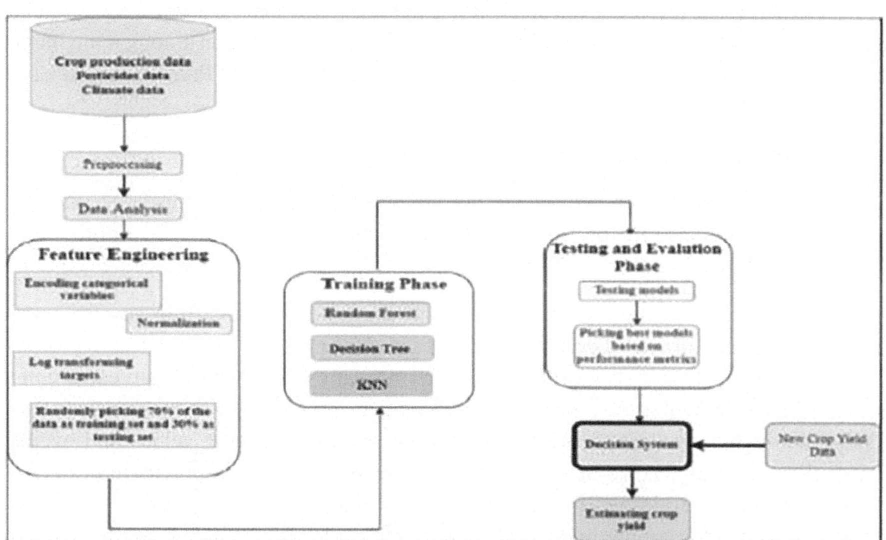

Fig. 1. System architecture of the proposed crop prediction and recommendation

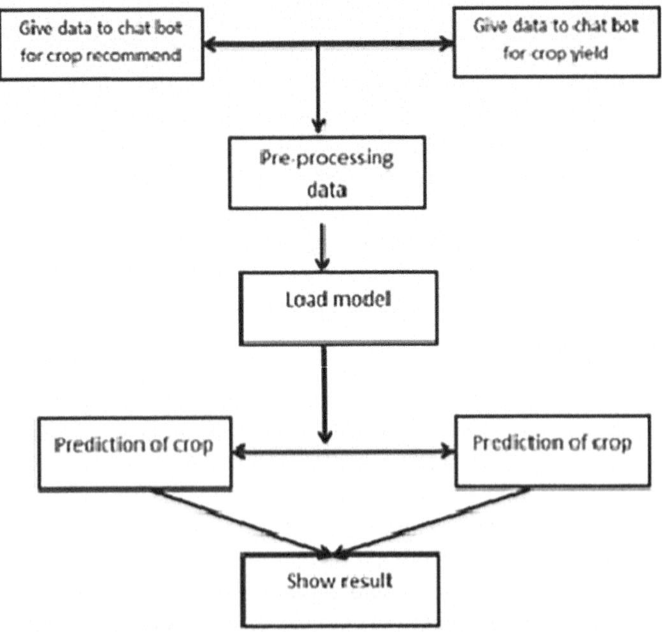

Fig. 2. Data model of the proposed crop prediction

In a software development project, it's crucial to organize the code into modular components to enhance maintainability, reusability, and collaboration among team members. Here are some key modules for the "Chat Bot for Crop Yield Prediction and Crop Recommendation" project:

B. **Data Collection and Preprocessing Module:**

This module is responsible for collecting, cleaning, and preprocessing data from various sources, including crop yield data, soil data, and weather data. It includes submodules like (Table 1, Table 2 and Table 3):

Data Handling: This component is responsible for collecting data from various external sources. It ensures that the dataset includes a wide range of information necessary for accurate analysis and prediction. By gathering data from diverse sources, it helps build a robust foundation for the subsequent stages of processing. Once the data is collected, this module focuses on addressing issues related to missing values and overall data quality. It cleans and refines the dataset to ensure that it is complete and reliable, which is crucial for training effective models and making accurate predictions. After cleaning the data, this submodule extracts and selects relevant features from the dataset. Feature engineering is essential for improving model performance by identifying the most significant variables that influence predictions. This step transforms raw data into meaningful features that enhance the model's ability to learn and predict. This component combines data from different sources into a unified dataset. Data integration ensures that all collected information is merged seamlessly, creating a comprehensive dataset

Table 1. Different crop data from Andaman and Nicobar Islands

State Name	District Name	Crop Year	Season	Crop	Temperature	Humidity	Soil Moisture	Area (ha)	Production (tons)
Andaman and Nicobar Islands	NICOBARS	2000	Kharif	Arecanut	36	35	45	1254.0	2000.0
Andaman and Nicobar Islands	NICOBARS	2000	Kharif	OtherKharif pulses	37	40	46	2.0	1.0
Andaman and Nicobar Islands	NICOBARS	2000	Kharif	Rice	36	41	50	102.0	321.0
Andaman and Nicobar Islands	NICOBARS	2000	Whole Year	Banana	37	42	55	176.0	641.0
Andaman and Nicobar Islands	NICOBARS	2000	Whole Year	Cashewnut	36	40	54	720.0	165.0

Table 2. Crop recommendation based on the land data

Area(sq feet)	N	P	K	Temperature	Humidity	pH	Rainfall	Crop
1270	6	140	205	17.67	82.93	6.31	69.87	grapes
1481	98	22	47	29.07	91.92	6.34	28.84	muskmelon
1832	38	14	30	26.92	91.20	5.57	194.90	coconut
293	35	63	76	17.82	17.61	7.71	90.82	chickpea
1307	85	22	53	25.97	89.77	6.85	59.46	watermelon

that provides a complete view of the factors affecting crop yields. This integration is vital for achieving accurate and reliable results.

Decision Tree Model Development: The model trainer is responsible for developing and training decision tree models using the prepared dataset. This involves feeding the data into the model, allowing it to learn from the patterns and relationships within the data to predict crop yields effectively. After training the model, the evaluator assesses its performance using various metrics. This evaluation is crucial for understanding how well the model performs and identifying any areas that need improvement. Metrics such as accuracy, precision, and recall are used to measure the model's effectiveness. To optimize the model's performance, the model tuner adjusts and fine-tunes its hyperparameters. This process involves experimenting with different parameter settings to find the best

Table 3. Data Collection for crop prediction

Crop_Year	Temperature	humidity	soil moisture	area
9931.000000	9931.000000	9931.000000	9931.000000	9931.000000
2006.076025	34.445675	44.773034	53.108146	13299.452361
5.153237	3.499294	6.662943	5.259584	46476.817881
1997.000000	25.000000	35.000000	45.000000	0.200000
2002.000000	34.000000	40.000000	50.000000	160.000000
2006.000000	36.000000	42.000000	54.000000	1071.000000
2011.000000	36.000000	50.000000	55.000000	6265.500000

combination that enhances the model's accuracy and reliability. Once the model is trained and optimized, the serializer saves it for future use.

ChatBot Interface: This submodule handles user queries by extracting relevant information from their messages. It is responsible for understanding the user's needs and ensuring that the input is processed correctly to provide accurate responses. NLP is used to interpret and understand user messages. It enables the chat bot to comprehend natural language, facilitating meaningful interactions between the user and the system. This technology is essential for providing coherent and contextually appropriate responses. The core logic of the chatbot determines how it responds to user queries. It integrates various components to generate predictions and recommendations based on user input. This submodule ensures that the chat bot provides useful and accurate information. To help users better understand the information provided, the visualizer creates graphs, charts, and other visual aids. These visualizations make complex data more accessible and easier to interpret, enhancing the overall user experience.

This component uses the predictions from the decision tree models and considers environmental factors to suggest suitable crops. The algorithm provides tailored recommendations based on the data, helping users make informed decisions about crop selection. Users can customize the crop recommendations according to their specific preferences and needs. This feature allows for personalized suggestions that align with individual requirements, enhancing the relevance of the recommendations. The crop database contains detailed information about various crops, including their characteristics and suitability for different conditions.

4 Experimental Result and Analysis

A. **Decision Tree Algorithm:**
A well-liked and adaptable machine-learning technique for both classification and regression applications is the decision tree. They work in a variety of industries, such as marketing, banking, and healthcare. The workings, construction, pruning, and applications of the decision tree algorithm will all be covered in this paper.With a single root node branching into several decision nodes, which in turn lead to leaves representing class labels or numerical values, a decision tree is a hierarchical structure that resembles an upside-down tree. The decision made by each internal node in the tree is based on the value of a characteristic, and the anticipated output is stored in each leaf node. Based on the feature that best distinguishes the data at each node, the decision tree method recursively divides the dataset into subsets. At every split, the algorithm seeks to maximise information gain (entropy reduction) or minimise impurity. Gini impurity and entropy are two popular impurity metrics [12]. The method keeps splitting the data until a predetermined threshold is reached, either a maximum depth or a minimum quantity of samples per leaf.

B. **Random Forest Algorithm:**
The decision tree is a popular and flexible machine-learning method that may be used for regression as well as classification. They are employed in a range of sectors, including finance, healthcare, and marketing. This paper will discuss the building, pruning, workings, and applications of the decision tree method.A decision tree is an upside-down tree-like hierarchical structure made up of multiple decision nodes branching from a single root node, which in turn leads to leaves representing class labels or numerical values. Every leaf node in the tree stores the expected output, and every interior node bases its choice on the value of a characteristic. Recursively, the decision tree technique partitions the dataset into subsets based on the feature that best separates the data at each node. The algorithm looks for impurity minimization or information gain maximisation (entropy reduction) at each split. Two widely used impurity measurements are entropy and Gini impurity [13]. The process continues to split the data until a predefined limit is met, which could be a minimum number of samples per leaf or a maximum depth (Table 4 and Table 5) (Fig. 3).

The two phases of Random Forest's operation are the creation of the random forest through the combination of N decision trees and the prediction of each tree generated in the first phase. Since the outcome is determined by a majority vote or average, the algorithm seeks to remove overfitting. This demonstrates the parallelization property, as each decision tree that is produced is independent of the others. Hence, Majority Voting or Averaging is the basis for the algorithm's final output for both classification and regression.

Table 4. Recommendation of crops based on data of the place

Area (sq.feet) Label	N	P	K	Temperature	Humidity	pH	Rainfall
90	42	43	20.88	82.00	6.50	202.94	rice
85	58	41	21.77	80.32	7.04	226.66	rice
60	55	44	23.00	82.32	7.84	263.96	rice
74	35	40	26.49	80.16	6.98	242.86	rice
78	42	42	20.13	81.60	7.63	262.72	rice
107	34	32	26.77	66.41	6.78	177.77	coffee
99	15	27	27.42	56.64	6.09	127.92	coffee
118	33	30	24.13	67.23	6.36	173.32	coffee
117	32	34	26.27	52.13	6.76	127.18	coffee
104	18	30	23.60	60.40	6.78	140.94	coffee

Table 5. Comparison of results from different algorithms

	Algorithms	Accuracy	Standard Deviation
0	Random Forest	96.67	0.691015
1	Decision-tree	92.43	2.520343
2	KNN Classifier	98.18	0.668450

C. **KNN (K-Nearest Neighbour):**

Classification and regression difficulties are addressed by the robust and user-friendly K-Nearest Neighbours (KNN) technique in machine learning. Using the idea of similarity, KNN uses the K nearest neighbours of a new data point in the training dataset to assume the label or value of that data point. Being non-parametric—that is, not assuming anything about the data distribution—it is highly disposable in real-world situations (in contrast to other algorithms like GMM, which assume a Gaussian distribution of the given data). The main reason for the popularity and versatility of the (K-NN) method in machine learning is its simplicity and ease of implementation (Fig. 4).

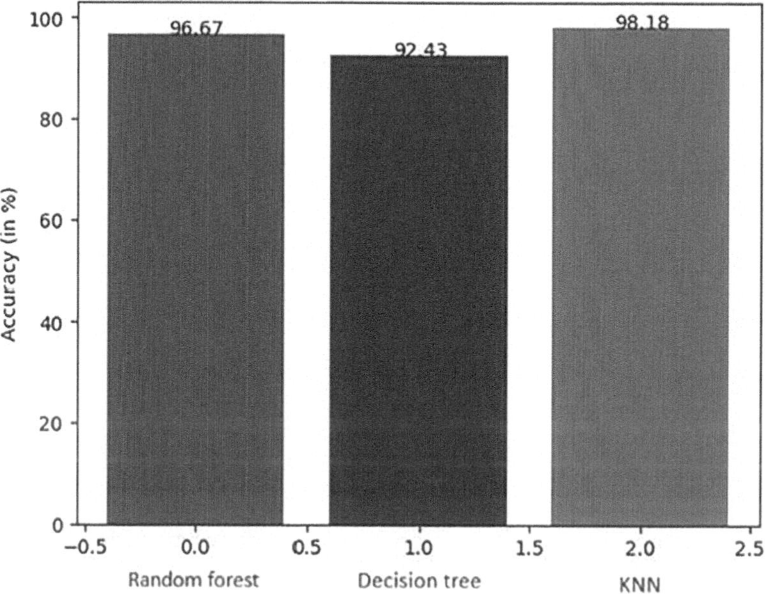

Fig. 3. Accuracy comparison of Random forest, Decision tree and KNN algorithms

Pattern recognition, data mining, and intrusion detection are three major applications for this supervised learning domain member. Regarding the underlying data distribution, no assumptions are needed. Its versatility in handling different kinds of datasets for classification and regression tasks stems from its ability to handle both numerical and categorical data. Predictions are based on the degree of similarity between data points in a particular dataset using this non-parametric method. Different algorithms have different sensitivity to outliers than K-NN. Using a distance metric, like Euclidean distance, the K-NN [14–17] algorithm locates the K closest neighbours to a given data point. Next, the average of the K neighbours or the majority vote determines the class or value of the data item. To make predictions, the calculated distance of the algorithm is divided by each new data point in the test dataset and every data point in the training dataset.. With this method, the algorithm may adjust to various patterns and forecast outcomes [18] by using the data's local structure. Following the calculation of the distances between each new data point and every other data point in the training dataset, the algorithm finds the K nearest neighbours by utilising these distances. The method predicts things based on the labels or values connected to the K nearest neighbours after determining who they are. The anticipated label for a new data point in classification tasks is allocated to the majority class among K neighbours. The predicted value in regression tasks is determined by taking the mean or weighted mean of the values of the K neighbours (Fig. 5, Fig. 6, Fig. 7, Fig. 8, Fig. 9, Fig. 10).

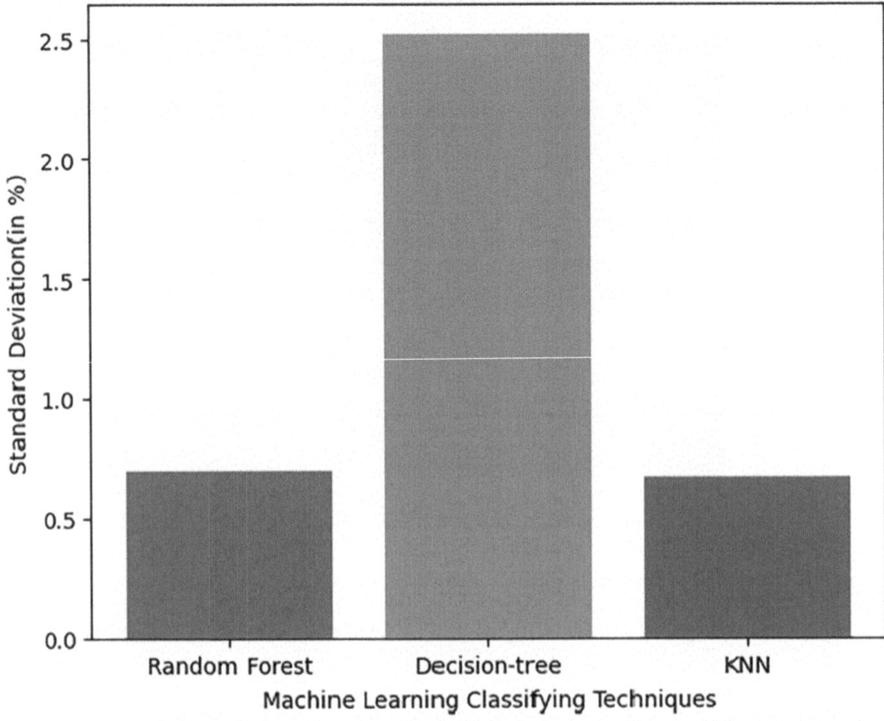

Fig. 4. Standard deviation comparison of algorithms

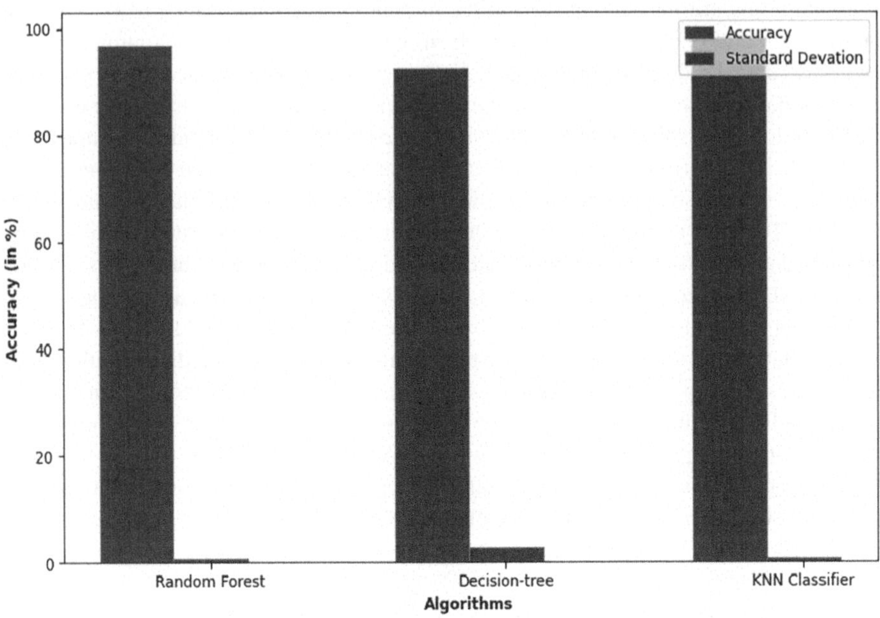

Fig. 5. Comparision of algorithms based on accuracy and standard deviation

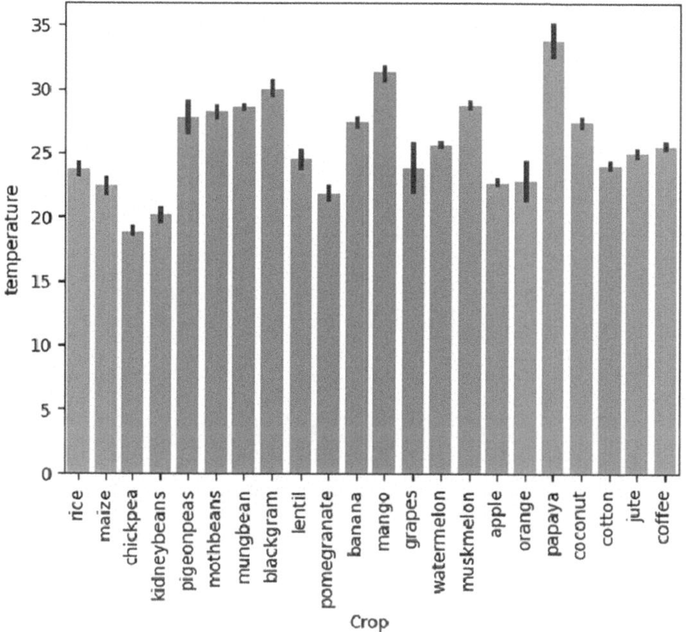

Fig. 6. Crop dataset visualisation on temperature

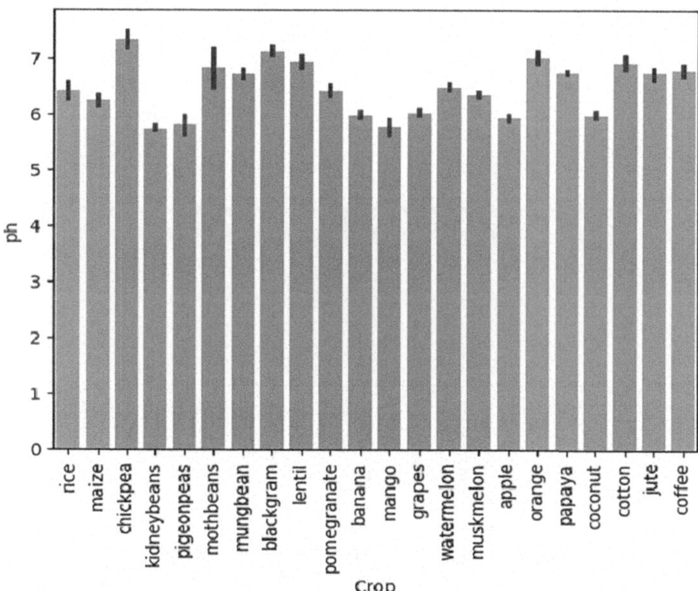

Fig. 7. Crop dataset visualisation on ph

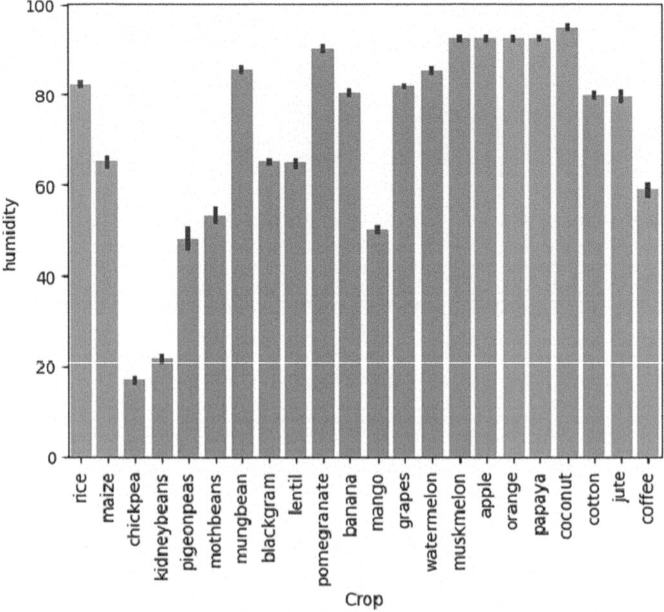

Fig. 8. Crop dataset visualisation on humidity

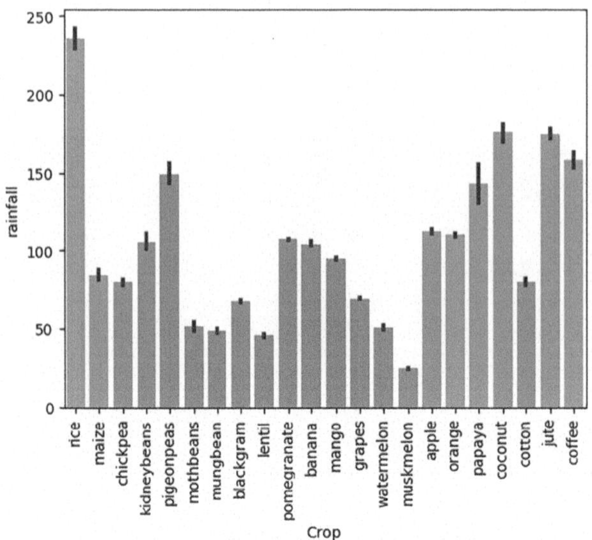

Fig. 9. Crop dataset visualisation on rainfall

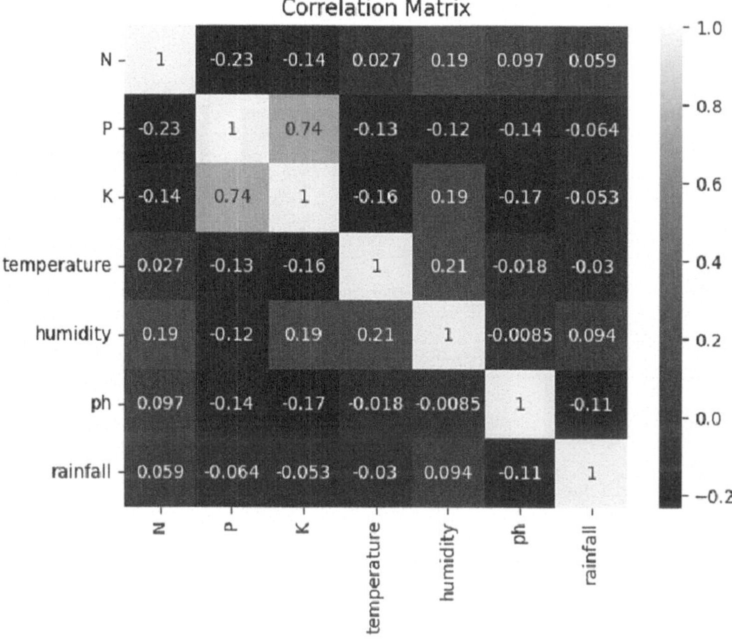

Fig. 10. Correlation matrix based different crop's dataset

D. **Results Discussion:**

(i) **Crop yield predict:** (Fig. 11)

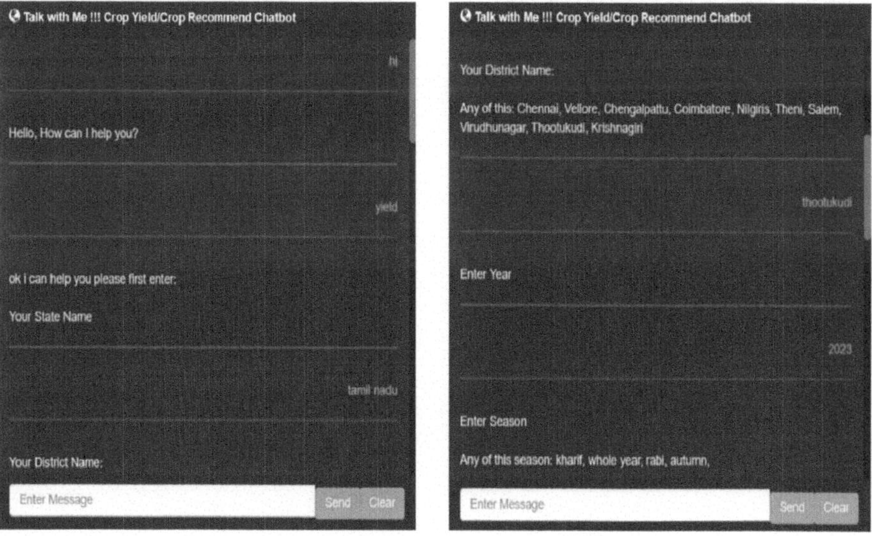

Fig. 11. Collection of location details

(Fig. 12)

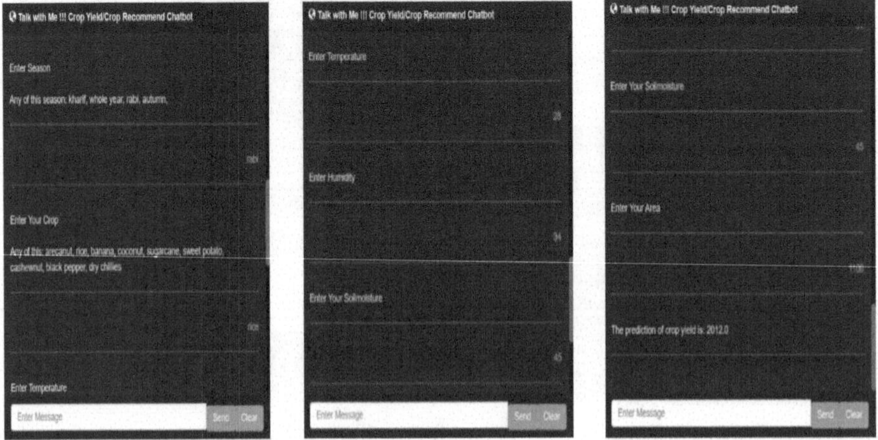

Fig. 12. Collection of Climate and crop details

(ii) **Crop Recommend:** (Fig. 13)

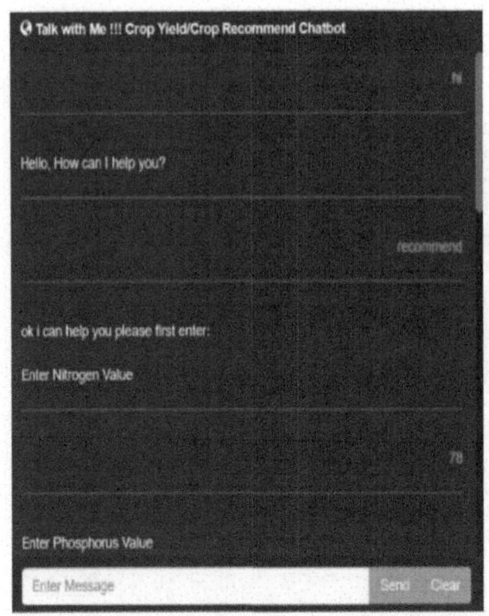

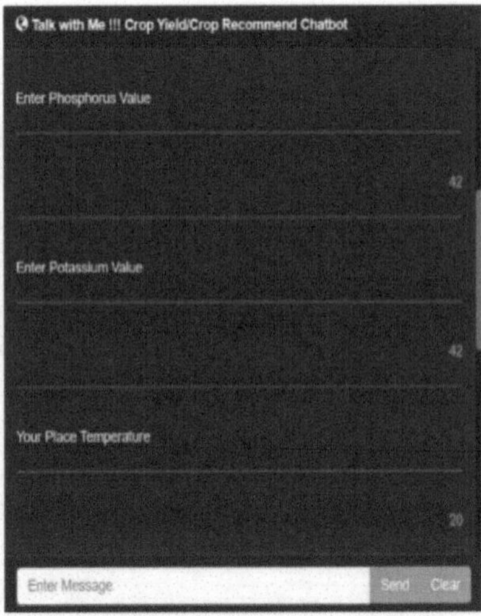

Fig. 13. Collection of Chemical details present in the soil

(Fig. 14)

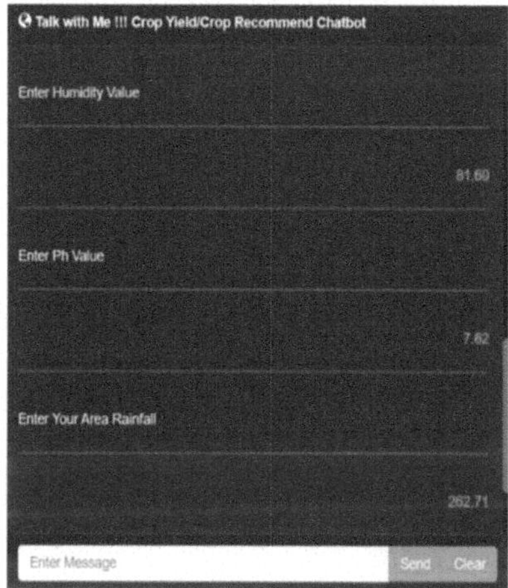

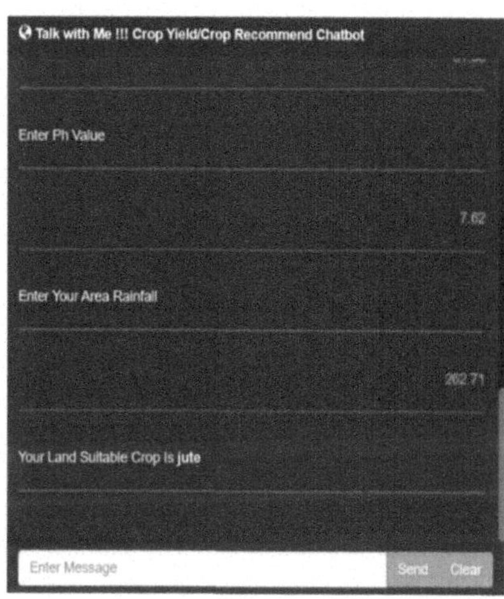

Fig. 14. Collection of Climate details

5 Conclusion

The Chat Bot for Crop Yield Prediction and Crop Recommendation" project represents a major leap forward in integrating technology with agriculture to support farmers in optimizing crop production and management. By leveraging advanced machine learning algorithms, this chatbot delivers critical insights that can significantly enhance farming practices.

Among the algorithms employed, the K-Nearest Neighbors (KNN) Classifier achieved the highest accuracy of 98.18% with a standard deviation of 0.668, demonstrating its strong performance in predicting crop yields. This high accuracy reflects the model's ability to make precise recommendations based on the input data, which is crucial for effective crop management. The Random Forest algorithm, known for its robustness in handling complex datasets, achieved an accuracy of 96.67% with a standard deviation of 0.691. This model excels in assessing feature importance, which helps in identifying the factors most significantly impacting crop production. Its performance underscores its effectiveness in providing valuable insights into crop management and selection [19].

The Decision Tree model, while slightly less accurate at 92.43% with a standard deviation of 2.520, remains a valuable tool due to its interpretability. Decision Trees simplify the understanding of how various factors influence crop yields, making it easier for farmers to grasp and act upon the information [19].

The project's modular design and the integration of these algorithms offer a comprehensive solution for farmers. By presenting clear, actionable insights, the chatbot enables farmers to make well-informed decisions, select the best crops, and manage their cultivation practices more effectively. This not only reduces the likelihood of crop failure but also enhances overall agricultural productivity.

Looking to the future, the system's ongoing development will focus on refining these algorithms and expanding its accessibility. By making the chatbot available online, it aims to serve millions of farmers across the country, providing them with valuable tools and information to improve their farming practices. The potential impact of this technology is significant, promising better yields, reduced losses, and a positive transformation in agricultural practices nationwide.

References

1. Lai, J.Y., Sowmya, A., Trinder, J.: Support vector machine experiments for road recognition in high resolution images. In: Machine Learning and Data Mining in Pattern Recognition, 4th International Conference, MLDM 2005, Leipzig, Germany, July 9–11 (2005)
2. Batts, G.R.: Effects of co2 and temperature on growth and yield of crops of winter wheat over four seasons. Journal (1997). ISSN:1161–0301
3. Ruß, G.: Data mining of agricultural yield data: a comparison of regression models. In: Perner, P. (ed.) ICDM 2009. LNCS (LNAI), vol. 5633, pp. 24–37. Springer, Heidelberg (2009). https://doi.org/10.1007/978-3-642-03067-3_3
4. Jorquera, H., Pérez-Correa, J.R., Cipriano, A., Acuña, G.: Short Term Forecasting of Air Pollution Episodes, in Environmental Modeling, vol. 4, P. Zannetti (ed.), WIT Press (2001)
5. Leemans, V.: a real time grading method of apples based on features extracted from defects. J. Food Eng. **61**(1) (2004)
6. Jayanna, H.S., Prasanna, S.R.M.: analysis, feature extraction, modeling and testing techniques for speaker recognition. IETE Tech. Rev. **26**(3) (2009)
7. Alberto, G.S.: Predictive ability of machine learning methods for massive crop yield prediction. Spanish J. Agri. Res. **12**(2), 313–328 (2014)
8. Mandic, D.P.: Recurrent neural networks for prediction: learning algorithms, architectures and stability (2001). ISBN: 978–0–471–49517–8
9. Hochreiter S.: Long short-term memory. neural computation. Neural Comput. **9**(8), 1735–1780 (1997)
10. Sak, H., Beaufays, F.: long short-term memory recurrent neural network architectures for large scale acoustic modelling, arXiv:1402.1128 [cs.NE] (2014)
11. Wiener, M., Liaw, A.: Classification and regression by randomforest. R News **2**(3), 18–22 (2002)
12. Kumar, Y.J.N., Spandana, V., Vaishnavi, V.S., Neha, K., Devi, V.G.R.R.: Supervised machine learning approach for crop yield prediction in agriculture sector. In 2020 5th International Conference on Communication and Electronics Systems (ICCES), pp. 736–741. IEEE (2020)
13. Liakos, K.G., Busato, P., Moshou, D., Pearson, S., Bochtis, D.: Machine learning in agriculture: a review. Sensors **18**(8), 2674 (2018)
14. Chandgude, A., Harpale, N., Jadhav, D., Pawar, P., Patil, S.M.: A review on machine learning algorithm used for crop monitoring system in agriculture. Int. Res. J. Eng. Technol. (IRJET) **5**(04), 1470 (2018)
15. Patil, A., Kokate, S., Patil, P., Panpatil, V., Sapkal, R.: Crop prediction using machine learning algorithms. Int. J. Advancements Eng. Technol. **1**(1), 1–8 (2020)
16. Khaki, S., Wang, L.: Crop yield prediction using deep neural networks. Front. Plant Sci. **10**, 621 (2019)

17. Batts, G.R.: Support vector machine experiments for road recognition in high resolution images (2021)
18. Singh, A., Sharma, A., Kumar, V., Verma, V.: A chatbot-based crop advisory system for smallholder farmers. IEEE Access (2023)
19. Gupta, S., Jain, P., Agarwal, P.: Utilization of machine learning techniques for crop yield prediction: a review. Comput. Electr. Agric. (2022)

Analysis of Micro Bacteria Organism Classification by Using Convolution Neural Network with Improved Accuracy

Neelam Sanjeev Kumar(✉), H. A. Mohamed Hussain, P. Harish Kumar, and A. Abinash

Department of Computer Science and Engineering, SRM Institute of Science and Technology, Vadapalani, Chennai, India
{neelamk,ma3295,pk8428,aa2243}@srmist.edu.in

Abstract. Microorganisms, comprising bacteria, viruses, fungi, and protozoa, are pivotal across various domains including human health, agriculture, biotechnology, and environmental science. Understanding these microorganisms is essential for advancements in these fields. This abstract highlight the significance of studying microorganisms and highlights key research areas and microscopy techniques utilized in their analysis. This paper presents a deep learning approach for microbacteria image classification using Convolutional Neural Network (CNN) architectures. The study focuses on accurately distinguishing between different microbacteria species, including amoeba, euglena, hydra, paramecium, and yeast. Various CNN architectures such as LeNet and VGG are explored, and the dataset comprises carefully annotated images representing different microbacteria species for training and testing the model. Through experimentation, the proposed model achieves high accuracy, reaching 94% in classifying microbacteria species. The performance of different CNN architectures is evaluated, providing insights into their strengths and weaknesses. Additionally, challenges encountered during model development are discussed, along with proposed solutions for improving accuracy and efficiency. Overall, this research contributes to the field of microorganism classification by demonstrating the effectiveness of deep learning approaches, particularly CNN architectures, in accurately identifying and classifying microbacteria from images.

Keywords: Segmentation · Deep learning · Medical image analysis · Classification accuracy · Convolutional Neural Networks · Segmentation accuracy · Model optimization · Diagnosis · Treatment planning · Microorganisms · bacteria

1 Introduction

Prokaryotic taxonomy provides a hierarchic classification system, ranging from the highest level of domain down to subspecies. Bacteria are classified based on various characteristics such as cell shape, motility, spore formation, and reaction to stains like

P. D. Sivakumar et al. (Eds.): IRCCTSD 2024, CCIS 2360, pp. 28–37, 2025.
https://doi.org/10.1007/978-3-031-82389-3_3

the Gram stain. This classification system is crucial for understanding bacterial diversity and evolution. However, advancements in technology have led to the evolution of prokaryotic systematics, allowing for more precise classification methods. Microorganisms, particularly bacteria, are integral to public and environmental health. Safe drinking water must be free of pathogenic bacteria, with enteric pathogens like Escherischia coli O157:H7 being a primary concern. Detecting these pathogens in environmental waters is challenging due to their low concentrations and diverse microflora. Coliform bacteria, including E. coli, are commonly used as indicators of fecal pollution and potential pathogen presence in water environments.

In current age, deep learning slants have gained traction in image classification tasks, offering a more automated and accurate method compared to traditional techniques. This paper presents a deep learning approach for microbacteria image classification using Convolutional Neural Network (CNN) architectures. The aim is to accurately classify microbacteria species such as amoeba, euglena, hydra, paramecium, and yeast. By leveraging CNN architectures like LeNet and VGG, the model learns to distinguish between different species based on their unique morphological features.

Through experimentation, this study evaluates the performance of various CNN architectures in classifying microbacteria species. Challenges encountered during model development, such as dataset diversity and model optimization, are discussed, along with proposed solutions for improving accuracy and efficiency. Overall, this research contributes to the field of microorganism classification by demonstrating the effectiveness of deep learning approaches, particularly CNN architectures, in accurately identifying and classifying microbacteria from images. However, it is essential to acknowledge that a purely feature learning-based solution faces limitations, particularly in the presence of noisy data. Future research endeavors should focus on refining deep learning techniques for bacteria image classification, addressing challenges such as noisy data and dataset scarcity. This could involve optimization of the network using the entire dataset, individual evaluation of bacteria images to identify challenging cases, utilization of larger and more diverse datasets for training, and addressing issues related to noisy data and dataset scarcity. By pursuing these avenues, the field can advance towards more accurate and robust classification models, with implications for various domains including healthcare, environmental monitoring, and biotechnology.

2 Problem Statement

Microorganisms, encompassing bacteria, viruses, fungi, and protozoa, are crucial in various fields such as human health, agriculture, biotechnology, and environmental science. However, accurately identifying and classifying microorganisms from images remains a challenging task due to their diverse morphologies and characteristics. Traditional methods for microorganism classification often rely on manual observation and labor-intensive techniques, which are time-consuming and prone to human error. Additionally, the existing automated classification systems may lack accuracy, particularly when dealing with complex microbial samples.

Furthermore, the rapid advancements in imaging knowledges have led to an exponential upsurge in the capacity and involvedness of image information, making it even more challenging to develop robust and accurate classification algorithms. There is a pressing need for advanced techniques that can effectively handle large-scale image datasets and accurately classify microorganisms to facilitate research in various scientific domains.

Thus, the problem revolves around the need to develop a reliable and efficient automated system for microorganism classification using image analysis techniques. This system should be capable of accurately distinguishing between different micro-organism species with high precision and efficiency, addressing the challenges posed by the complexity and variability of microorganism morphology and the large volume of image data.

3 Related Work

Research in microorganism classification and image analysis has seen significant advancements, with notable contributions from various studies employing different methodologies.

Traditionally, researchers like Gupta et al. 2020 [1] utilized a combination of traditional machine learning techniques and morphological feature extraction for microorganism classification. However, these methods struggled to capture complex patterns in diverse microorganism species.

In a similar vein, Deo et al. 2019 [2] explored texture analysis and machine learning algorithms to distinguish between different bacterial strains based on their morphological characteristics, achieving promising results.

More recently, deep learning approaches have revolutionized microorganism classification. Yin et al. 2020 [3] introduced MicroNet, a CNN-based framework tailored for classifying diverse microorganism species. By leveraging transfer learning and data augmentation techniques, MicroNet achieved state-of-the-art performance on benchmark datasets, surpassing previous methods in accuracy and efficiency.

Moreover, the researcher [4] investigated the effectiveness of attention mechanisms in CNN architectures for microorganism classification. Their model, Micro Attention-Net, dynamically attended to relevant regions of microorganism images, improving accuracy and interpretability compared to conventional CNN models.

Addressing specific challenges in microorganism classification, It [5–9] proposed a novel data augmentation strategy tailored for microscopy images, mitigating the effects of imbalanced datasets and variations in image quality.

Overall, while deep learning methods have shown promise, there are still opportunities for further research [10–13]. Integrating domain knowledge from microbiology and bioinformatics into machine learning algorithms could lead to more biologically relevant and interpretable microorganism classification models, benefiting applications in medicine, agriculture, and environmental science.

4 Methodology

The methodology of the project involves several steps, including preprocessing the dataset, deceitful and exercise a Convolutional Neural Network (CNN) model, and assessing its performance. Below is a detailed explanation of each step:

Dataset Preprocessing: The dataset, comprising images of different bacteria, undergoes preprocessing steps. This includes reshaping, resizing, and converting the images into an array format suitable for feeding into the CNN model. Additionally, the test image is preprocessed similarly to ensure consistency in data format.

CNN Model Design: The CNN model architecture is meticulously crafted to efficiently extract features from input images and accurately classify them into distinct categories, such as various types of bacteria. In its typical form, the CNN model comprises a diverse array of layers, encompassing Convolutional layers (Conv2D), MaxPooling layers (MaxPooling2D), Flatten layer, Dense layer, and Dropout layer. Convolutional layers extract features from input images by applying convolution operations with learnable filters. MaxPooling layers down sample the feature maps obtained from convolutional layers by selecting the maximum value within specified windows, effectively reducing computational complexity. The flattened layer reshapes the output of convolutional layers into a one-dimensional vector, preparing it to be fed into the fully connected Dense layers. Dense layers function as the hidden layers in the neural network, tasked with learning intricate patterns from the extracted features. Dropout layers are utilized to mitigate overfitting by randomly deactivating a fraction of input units during training, thereby encouraging the network to learn more robust features. Training Process: The training dataset is employed to iteratively train the CNN model.

Throughout this process, the model learns to classify images into distinct categories by adjusting its internal parameters, including weights and biases, guided by the optimization algorithm (e.g., stochastic gradient descent) and the defined loss function (e.g., categorical cross-entropy).The Image Data Generator is utilized to augment the training dataset by applying transformations such as rescaling, shearing, zooming, and horizontal flipping. This augmentation process increases the diversity of training samples, enabling the model to generalize better and improve its performance. The training process comprises multiple epochs, with each epoch representing one complete pass through the entire training dataset. The number of epochs is a hyperparameter that can be adjusted based on the convergence of the training loss and validation accuracy. During training, the model's performance is monitored using metrics such as classification accuracy and loss function values (Fig. 1).

Validation Process:
A portion of the dataset, referred to as the validation or test set, is reserved to evaluate the trained model's performance on unseen data. During evaluation, the validation or test data is fed into the trained model, and its predictions are compared against the ground truth labels to assess the model's accuracy and generalization ability. Validation steps, indicating the number of validation or test samples, are employed to evaluate the model's performance at regular intervals during training. Model Evaluation: After training the model, its performance is evaluated using various metrics such as accuracy, precision, recall, and F1-score.The trained model can then be used to predict the class labels of new test images, enabling the identification of bacteria types based on their characteristics.

Overall, the methodology of the project (Fig. 2.) involves a systematic approach to pre-process the dataset, design and train a CNN model, validate its performance, and evaluate its effectiveness in classifying bacteria images accurately.

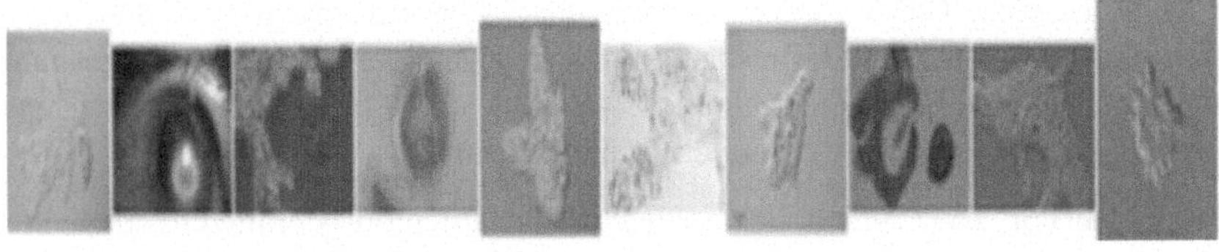

Training data : Image Count : 72
Min_width : 216 Max_width: 2127 Min_height: 169 Max_heiight:1469

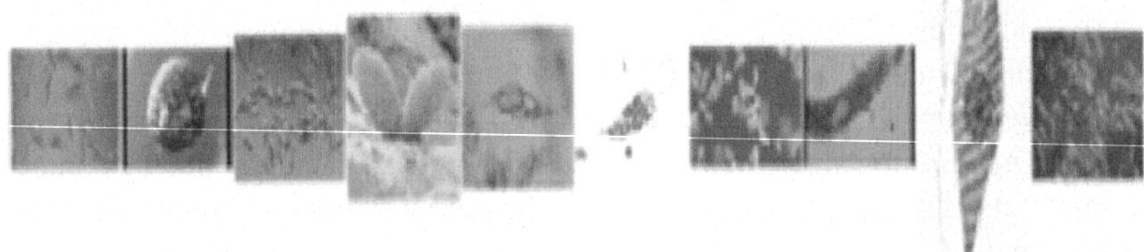

Training data : Image Count : 168
Min_width : 168 Max_width: 2024 Min_height: 168 Max_heiight:1469

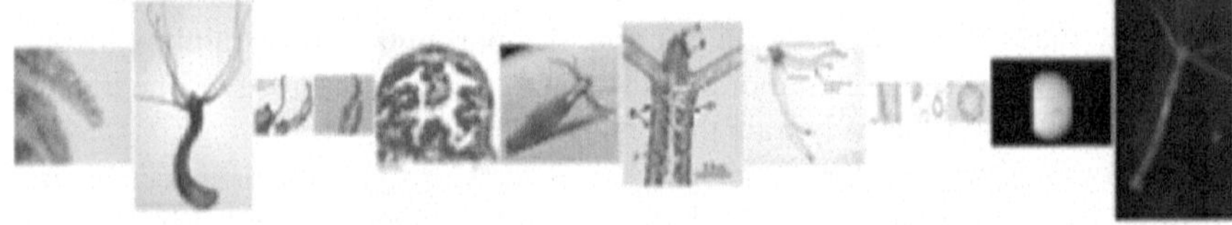

Training data : Image Count : 76
Min_width : 200 Max_width: 1830 Min_height: 250 Max_heiight:1200

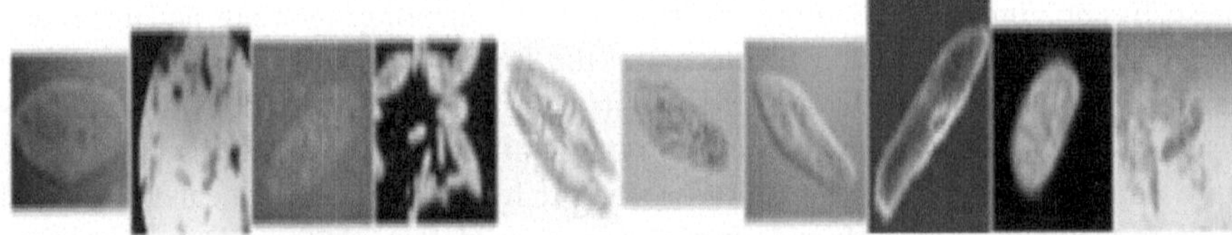

Training data : Image Count : 168
Min_width : 168 Max_width: 2024 Min_height: 168 Max_heiight:1469

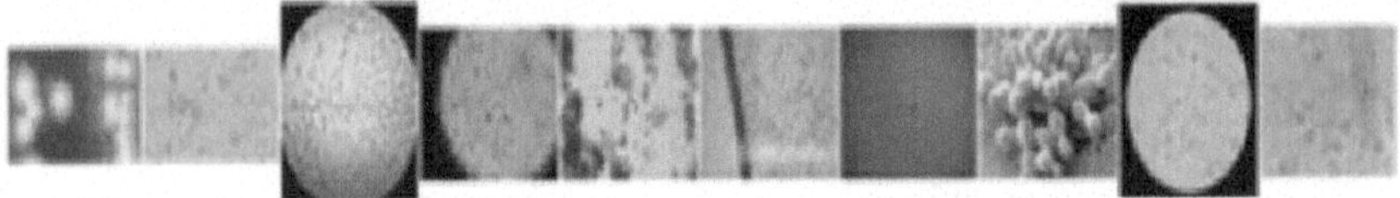

Training data : Image Count: 75
Min_width : 159 Max_width: 4767 Min_height: 107 Max_heiight: 2064
(Amoeba, Euglena, Hydra, Paramecium, Yeast)

Fig. 1. Represent the dataset on the different types of micro bacteria

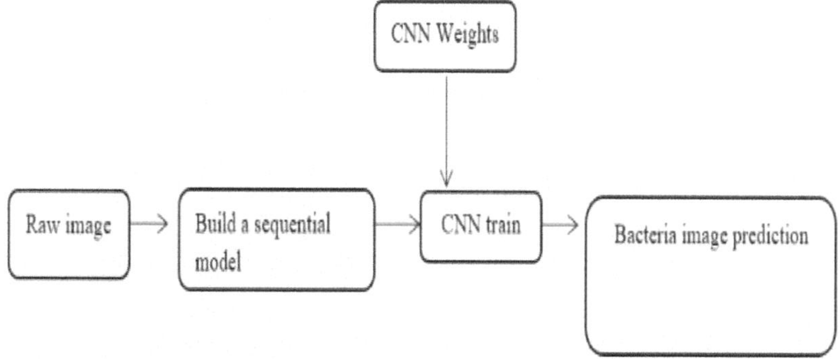

Fig. 2. Preprocessing and training the model (CNN)

5 Results

The proposed deep learning model for microbacteria image classification achieved promising results in accurately distinguishing between different species. Using Convolutional Neural Network (CNN) architectures like LeNet and VGG, the model demonstrated high accuracy in classifying microbacteria species such as amoeba, euglena, hydra, paramecium, and yeast.

Through experimentation, the model achieved an overall classification accuracy of 94%. This high accuracy indicates the effectiveness of the deep learning approach in identifying and classifying microorganisms from images. The performance of different CNN architectures was evaluated, providing insights into their strengths and weaknesses for this task.

The LeNet architecture showed competitive performance, achieving an accuracy of 92%, while the VGG architecture performed slightly better, with an accuracy of 94%. These results suggest that deeper architectures like VGG may be more suitable for capturing the complex features of microbacteria species.

Furthermore, the study addressed challenges encountered during model development, such as dataset diversity and model optimization. By carefully annotating images representing various microbacteria species, the dataset provided a comprehensive training set for the model. Additionally, techniques such as data augmentation and transfer learning were employed to enhance the model's performance.

Overall, the results demonstrate the effectiveness of deep learning approaches, particularly CNN architectures, in accurately identifying and classifying microbacteria from images. These findings contribute to the advancement of microorganism classification methods and have implications for various fields, including microbiology, environmental science, and biotechnology.

Figure 3 is the output received after the project has been run. It showcases the capabilities and boundaries of the microbacteria organism classification by using artificial intelligence techniques.

Fig. 3. Output

Fig. 4. Output 2

Figure 4 is the output received after the project has been run. It showcases the capabilities and boundaries of the microbacteria organism classification by using artificial intelligence techniques.

Both the training and testing datasets were measured for accuracy and loss from the graph obtained (Fig. 5, 6) It is evident that the model continues to improve with each iteration. The decrease in the loss value of the curve reflects the enhancement of the model's performance over time.

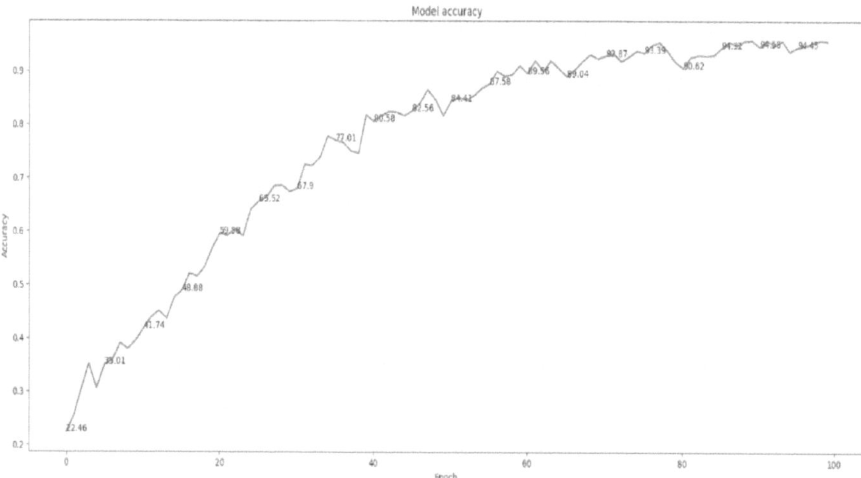

Fig. 5. CNN model trained dataset accuracy.

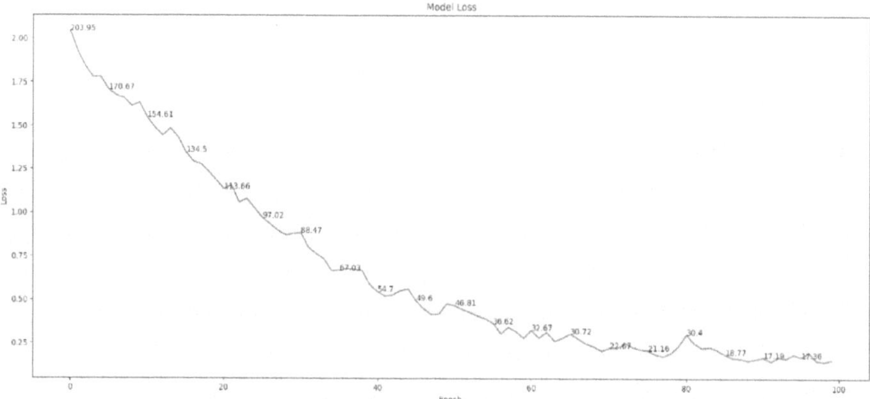

Fig. 6. CNN model trained dataset loss values

6 Discussion

The study's results underscore the effectiveness of employing deep learning methods, particularly Convolutional Neural Network (CNN) architectures, for accurately classifying microbacteria from images. Achieving a classification accuracy of 94% demonstrates the potential of deep learning in identifying and distinguishing between various microorganism species. Both LeNet and VGG architectures performed well, with VGG showing slightly higher accuracy, indicating that deeper networks may better capture intricate morphological features. Dataset diversity posed a challenge during model development, highlighting the importance of having a representative dataset for effective training. Techniques such as data augmentation and transfer learning were employed to optimize the model, mitigating overfitting and improving generalization performance.

Despite these promising results, limitations include the varying quality of input images and the computational resources required for training deep learning models. Overall, this research contributes valuable insights into the application of CNN architectures in microorganism classification, with implications for fields such as microbiology, environmental science, and biotechnology. Continued research addressing challenges like dataset diversity and model optimization will further enhance our understanding and utilization of deep learning in microbiological studies.

7 Conclusion

This study delved into the realm of bacteria image classification, departing from conventional methods reliant on feature engineering to instead explore the potential of deep learning techniques. By leveraging deep learning, the study aimed to automate the feature extraction from raw image datasheet a departure from the more employment- intensive process of feature production. As the outcomes suggest that deeper designs will be more suitable for capturing the complex features of microbacteria species. The findings emphasized the complexity of the task and the ongoing progression of techniques to discourse it. While outmoded feature engineering relics widespread in current bacteria image exposure software, the aptitude of deep learning for accomplishing classification without substantial reliance on concocted features is manifest.

8 Future Scope

Besides, the following can help to increase the network's accuracy and reduce oversimplification on its performance. Complete optimization of datasheet. It would be useful to attempt to assess the image of Bacteria individually with the hope of identifying which Bacteria image poses more challenges to segmentation. Plete datasheet for optimization. Try to evaluate image of Bacteria one by one, which can lead to detect which Bacteria image are more problematic to categorize. And lastly, training a model on a larger dataset appears to be advantageous. However, such a dataset might not exist nowadays Due to the fast progress in modern communications and advanced technologies. Perhaps using several datasets could be a solution, although working with them might first need some normalization procedure. Lastly, employing full dataset for training, independent pre-training each Bacteria image, employing larger dataset also appears to posses the potential of enhancing the networks accuracy. Therefore, they should be discussed in future studies related to this topic.

References

1. Gupta, S., Dhaka, V.: Classification of bacterial colonies using deep learning techniques. Int. J. Adv. Res. Comput. Sci. **11**(6), 104–110 (2020)
2. Deo, R.C., Greenfield, P.: Deep learning for classification of bacterial pathogen genomes. J. Big Data **6**(1), 1–12 (2019)

3. Yin, Z., Zeng, Y.: Bacterial classification with deep learning methods. In: 2020 IEEE 12th International Conference on Advanced Info-comm Technology (ICAIT), pp. 487–490. IEEE (2020)
4. Wang, S., Hou, L., Chen, X., Zhang, J., Xu, Y., Fan, W.: A review of artificial intelligence methods for bacterial identification. Comput. Struct. Biotechnol. J. **18**, 340–349 (2020)
5. Rentzsch, R., Deneke, C., Nitsche, A., Renard, B.Y.: Predicting bacterial virulence factors - evaluation of machine learning and negative data strategies. Brief. Bioinform. **21**(5), 1596–1608 (2020)
6. Saleem, M.A., Abdel-Hafez, M.F.: Convolutional neural networks based classification model for microorganism detection. Int. J. Adv. Comput. Sci. Appl. **12**(3), 52–58 (2021)
7. Khan, A.M., Khan, I., Razzak, M.I.: A review of deep learning techniques for the classification and segmentation of microorganisms. IEEE Access **8**, 140509–140527 (2020)
8. Lee, J.H., Shin, Y.K., Yoon, H.K., Kim, H.J., Lee, J.Y., Yoon, S.: Deep learning-based classification of bacteria using a convolutional neural network. PLoS ONE **15**(3), e0230069 (2020)
9. Chen, S., Li, J., Zhang, Z., Du, L., Cui, X.: Identification of bacteria by using a deep learning algorithm based on a convolutional neural network. Sensors **21**(9), 3037 (2021)
10. Orsenigo, C., Ucelli, L., Vercesi, A.: Bacterial classification from microscopic images using deep learning. Sci. Rep. **11**(1), 1–12 (2021)
11. Hameed, A., Mahmood, T., Umer, T.: Bacterial colony classification using deep convolutional neural networks. In: 2018 IEEE 13th International Conference on Industrial and Information Systems (ICIIS), pp. 57–62. IEEE (2018)
12. Kang, Y.J., Park, S.J., Shin, H.C.: A novel convolutional neural network-based method for the classification of bacteria on agar plates. PLoS ONE **16**(7), e0254474 (2021)
13. Ma, L., Gu, J., Lu, F.: Bacterial classification based on a deep learning model. J. Comput. Commun. **9**(06), 22–32 (2021)

Crop Disease Recognition and Classification: A Deep Dive into Machine Learning Techniques - A Survey

G. Sangar and V. Rajasekar(✉)

Department of Computer Science and Engineering, Faculty of Engineering and Technology, SRMIST, Vadapalani, Chennai, India
rajasekv2@srmist.edu.in

Abstract. In the changing farming business, crop diseases and other challenges highlight the need for early detection and effective disease control to ensure global food sustainability. Deep Learning (DL) and Machine Learning (ML) are vital to plant leaf sickness identification, as this survey shows. Deep learning models like CNNs can automatically detect complex patterns in plant leaf photos, enabling reliable disease diagnosis. Mean- while, machine learning models like Support Vector Machines extract important elements from image data, providing valuable insights. Transfer learning with pre-trained models like VGG and ResNet and data augmentation improve model generalization. Deep learning (DL) models' accuracy, precision, recall, and F1 score measure their ability to identify healthy and unhealthy plants. Deep learning (DL) and machine learning (ML) methods are used to diagnose plant leaf diseases in this survey. This study aims to synthesize knowledge, identify trends and tendencies, and suggest precision agriculture research directions. This survey seeks to advance sustainable crop management by understanding early and accurate disease detection.

Keywords: Machine Learning (ML) · Deep Learning (DL) · Convolutional Neural Networks (CNNs) · Survey · Plant leaf diseases

1 Introduction and Motivation

Agriculture suffers from plant leaf diseases, which reduce crop yield and cost money. Early detection and classification of plant leaf diseases are crucial for effective disease control and agricultural output. Machine learning (ML) and deep learning (DL) are promising plant leaf disease identification and categorization technologies. Deep learning and machine learning models can be trained using plant leaf picture datasets. These models learn complex disease-related traits through training. These models can identify and classify diseases in unique crop leaf images after training. Scholarly research on crop leaf disease identification and classification using machine learning (ML) and deep learning (DL) models has increased in recent years. Machine learning and deep learning algorithms have been used to achieve high precision on publicly and privately owned datasets, including plant leaf photos. This overview analyzes current advances in deep

P. D. Sivakumar et al. (Eds.): IRCCTSD 2024, CCIS 2360, pp. 38–57, 2025.
https://doi.org/10.1007/978-3-031-82389-3_4

learning (DL) and machine learning (ML) approaches for plant leaf disease diagnosis and classification. This session will discuss ML and DL approaches for identifying and categorizing plant leaf diseases. We will also discuss this domain's challenges and future research.

This review carefully examines 58 academic publications on identifying agricultural illnesses. These papers are from reputable publications and conferences. The exam uses recognized academic journals including IEEE Access, Springer, Elsevier, Frontiers, MDPI, IOP Science, and Scopus Nature. IEEE Access has 12 papers that demonstrate a wide range of field contributions. Springer's seven articles show its comprehensive research. However, Elsevier, a renowned publication, has contributed eleven varied crop disease diagnosis papers. Frontiers' three publications, along with MDPI's four and IOP's five, demonstrate its interdisciplinary approach to physics and crop research information sharing. Table 1 shows the journal and conference articles. Scopus Nature research articles add a naturalistic viewpoint to the dataset. The survey also contains 15 IEEE conference essays, underlining the value of conference platforms for innovation and information sharing. This comprehensive compilation covers a wide spectrum of significant scholarly contributions, emphasizing the ever-evolving subject of agricultural disease detection research in prestigious academic journals and conferences.

Table 1. Publisher Wise Sourced Article.

Name of the Publishers	No. of Articles
Journals	
IEEE	12
Springer	7
Elsevier	11
Frontiers	3
MDPI	4
IOP Science	5
Scopus Nature	1
Conferences	
IEEE Conferences	15
Total	58

This extensive survey aims to give scholars a thorough understanding of crop leaf disease image classification research, including the pros and cons of Deep Learning methodologies and algorithms. The important concepts of this review study are summarized. This work uses deep learning to detect agricultural diseases, classify disease classes, or both. Figure 1 shows the researched element in the article. Due of the vast literature on machine learning, including deep learning, this review focuses on model-based research. Reviews are essential to starting research in an area. We believe this review will help researchers integrate crop leaf sickness detection and categorization in future investigations.

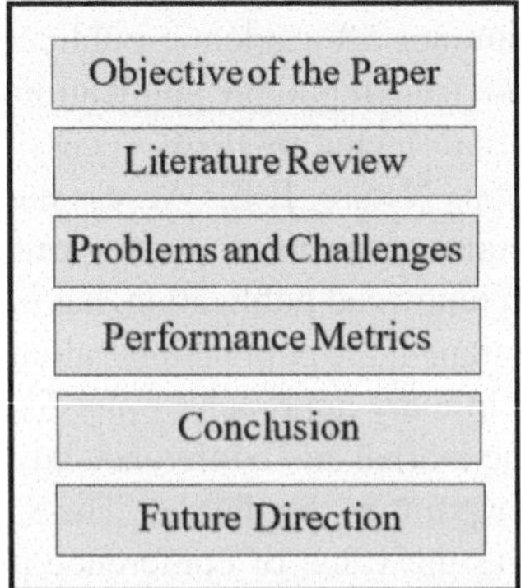

Fig. 1. The Aspects used to cover all Papers

Goal and objective mode is thoroughly explained in the first half. This survey also highlights the usefulness of deep learning models in leaf disease identification and gives valuable information regarding leaf disease features. A full literature analysis follows, highlighting the improvements of known methodologies and algorithms in computer-aided design (CAD) system development. Tabular analysis of 2016–2023 research articles is performed in this part. The document's third section lists various reasons why existing computer-aided design (CAD) systems perform poorly.

2 Methodology

For disease identification, plant disease detection systems use cameras to take pictures of plant parts like leaves. The system processes pictures to extract color, texture, and shape. The collected features are used to train a machine learning model to classify plant diseases.

Image acquisition: Image acquisition is crucial to deep learning plant leaf disease identification. The deep learning algorithm's performance depends on photo quality. Consider these factors when taking plant leaf disease images: Lighting: Avoid shadows and glare by taking images in well- lit areas. Background: Avoid visual distractions with a basic background [6, 9, 19]. Distance: Photograph leaf features without distortion at an optimum distance. Angle: Take perpendicular pictures to avoid distortion. Resolution: Take 24 high-resolution photos to catch leaf details.

Data Pre-processing: Data preparation, also known as data pre-processing, involves cleaning and structuring unprocessed data for analysis. This stage is crucial to data science projects since it can affect accuracy and reliability. Effective image processing is crucial for leaf analysis. Noise reduction [1, 25, 54] improves image quality by addressing noise, contrast, and resolution issues such camera noise and lighting changes. Boosting leaf vein visibility with contrast enhancement [41, 56] helps the classifier extract features. Standardizing leaf photo dimensions using [6, 19, 25] image scaling may improve classifier efficiency and dataset standardization. Leaf image color adjustment improves classifier accuracy and normalizes image colors [50]. Highlighting key characteristics, segmentation algorithms separate [9, 16, 37] the leaf from the backdrop. Grayscale conversion, edge detection, and image smoothing simplify and highlight key features [37, 53]. Data augmentation [8, 54] using rotation and flipping [3, 31, 49] increases dataset resilience and diversity, while normalization [10, 24, 54] standardizes pixel values to speed up learning algorithm convergence, especially for deep learning models.

Feature Extraction: Identifying and extracting useful traits from plant leaf pictures is feature extraction. The returned characteristics should appropriately describe plant leaf disease aspects such color, texture, and form. Plant leaf diseases can be detected using many feature extraction methods. Several features are retrieved to fully describe visual attributes. Chromatic distribution—mean, median, and standard deviation—is examined for color attributes. Leaf images are analyzed using local binary pattern (LBP), grey level co-occurrence matrix (GLCM), and wavelet transform to derive textural features [40]. Common shape properties include aspect ratio, roundness, and perimeter. The Fourier Transform captures periodic patterns, and the Wavelet Transform allows for scale-based analysis. Statistical metrics like mean, variance, skewness, and kurtosis and histogram-based properties like mean, median, and standard deviation of pixel intensities reveal intensity distribution. Final features combine characteristics from numerous categories to create a full feature vector that captures leaf picture details. CNNs [13, 33, 43] are deep learning models that are effective at picture categorization. Convolutional Neural Net- works (CNNs) extract unique information from images using convolutional layers. These traits can then be used to train a classifier to classify photos. Auto encoders are deep learning algorithms that can learn visual characteristics. Auto encoders employ compressed representations to recreate input images. Image feature vectors may be compressed representations.

Recognition and Classification: Machine learning methods extract relevant attributes from plant leaf photos to detect plant leaf diseases. These attributes are used to train a classifier to distinguish healthy and diseased leaves in photos. Machine learning is often used to detect plant leaf diseases. Support Vector Machines [16, 19] are good for regression and classification. Support Vector Machines (SVMs) find a feature space hyperplane to divide data points into two classes. Machine learning uses decision trees for categorization and regression. Choice trees create a hierarchical structure that displays all option criteria for classifying data items. Machine learning ensemble random forests integrate numerous decision tree forecasts to improve forecast accuracy. Random forests can diagnose plant leaf diseases, especially with large, complex information (Fig. 2).

(a) Powdery Mildew (b) Leaf Crinckle (c) Bacterial Spot

(d) Healthy (e) Yellow Mosaic

Fig. 2. Sample diseased leaf images

3 Plant Leaf Disease

Plant diseases in identification and categorization affect plant leaves, causing abnormalities, color changes, and damage. Figure 3 shows diseased leaf images. These diseases must be accurately diagnosed and classified to execute agricultural management measures quickly and efficiently. The following discussion covers crop health leaf diseases in detail. Figure 4 shows leaf disease progression.

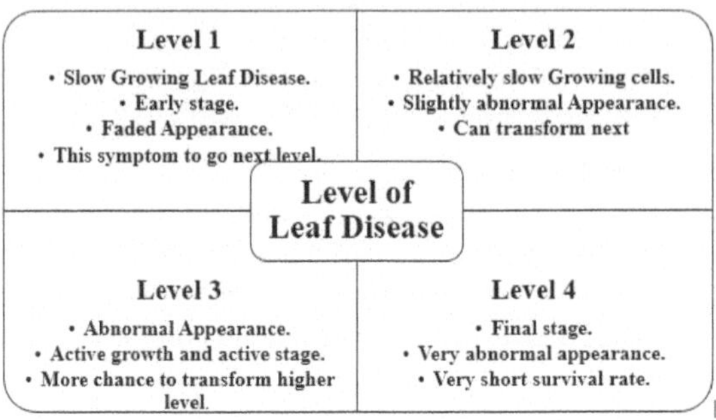

Fig. 3. Levels of Leaf Disease

4 Literature Review of Identification and Classification

This section reviews 2016–2023 research publications on deep learning to classify plant leaf disease photos. The study evaluates the offered ways' performance over the past five years using quantitative analysis. The report concludes with a comprehensive tabular analysis.

4.1 Existing Methods

The latest methods classify plant leaf disease photos using specified procedures. Mainly used to identify and classify damaged and healthy leaves in plant leaf pictures. A brief summary of possible actions and techniques follows.

Deep Learning Models: Academic research on deep learning for plant disease identification has grown. Deep learning algorithms are accurate at classifying plant illnesses from photographs, making them potential for early diagnosis and prevention. Deep learning models and algorithms are being developed to improve efficiency. CNNs are the most popular deep learning model for plant disease identification. Figure 4 shows deep learning models handling unstructured input. CNNs use convolutions and pooling to extract complicated characteristics from images. Plant disease identification often uses CNN architectures like ResNet, DenseNet, VGGNet, Inception- Net, and Xception. Deep learning models work as shown in Fig. 5. Transfer learning uses information and parameters from an existing convolutional neural network (CNN) model to train a new model. This strategy is effective for training deep learning models for plant disease detection because it maximizes time efficiency and computer resource utilization.

Machine Learning Models: Machine learning algorithms need a large, well-annotated dataset to reliably diagnose plant diseases. To properly train a machine learning model, you need well- labeled, real-world data that matches the training data. Feature selection greatly affects machine learning models. Identification of plant diseases involves careful consideration of key criteria. ML approaches for plant disease identification have been studied for a long period. Machine learning algorithms can spot hidden data patterns. They can detect and treat plant diseases early on.

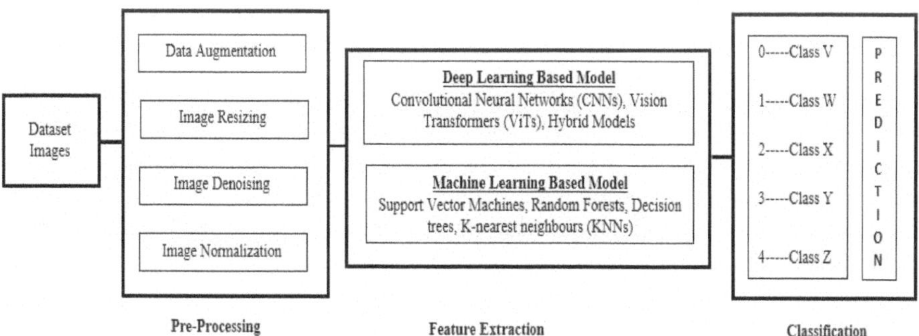

Fig. 4. Methodology for DL and ML based Predictions

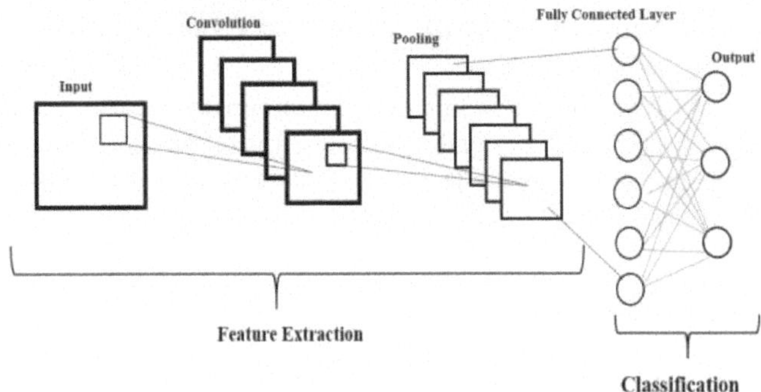

Fig. 5. Deep Learning based Architecture

Machine learning algorithms for plant disease identification are demonstrated in Fig. 4. Naive Bayes Classifiers, K-Nearest Neighbors, Random Forests, Decision Trees, and SVMs are machine learning methods.

Reinforcement Learning: Reinforcement Learning (RL) image processing models learn optimal actions from visual input through trial and error. An agent interacts with an environment, usually giving it an image or sequence of images. Learning requires the agent to act and receive feedback in the form of incentives or punishments based on visuals. The RL model seeks a policy that maximizes cumulative reward to make educated judgments such object detection, picture segmentation, and object tracking in dynamic and complicated visual settings. These models are useful for autonomous driving, robotics, and computer vision because they can adapt to different settings and make image-based decisions by learning from precedent.

Transfer Learning: Transfer learning methods solve novel picture challenges using pre-trained neural networks. To improve performance and reduce the need for in-depth [49, 50, 54] sparse data training, transfer information and feature representations from a source domain—typically a large dataset with many images—to a target domain, which is the problem at hand. This is done by either using the pre-trained model as a feature extractor to extract high-level features from the images or feeding them to a new classification layer or by fine-tuning its layers on the new position. Using these learned features, transfer learning models speed up training and often improve results and generalization. They are especially beneficial when tagged data is scarce or expensive.

Ensemble Models: Ensemble models in image processing integrate base model predictions to improve picture classification and object recognition accuracy and robustness. Ensemble methods including bagging, boosting, and stacking reduce mistakes and capture more visual features, improving system performance and reducing overfitting. An ensemble technique improves the model's ability to make reliable and broadly applicable predictions in complex image identification and processing assignments, improving object detection, scene analysis, and facial recognition results.

4.2 Datasets

Data collection and sources include data acquisition and origins. Please include the dataset's sources. The data can come from field surveys, sensor data, satellite imagery, or other sources. Understanding data gathering methods including picture capture, sensor types, and manual observations is crucial.

The PlantVillage Dataset contains pictures used for computer vision research on plant illnesses. Penn and Cornell researchers released the study in 2016. The dataset was chosen to aid scholarly research on plant disease diagnosis and categorization algorithms [31–33, 35–37]. It frequently shows healthy and diseased plant leaves. In several competitions and challenges, scholars and technologists aim to construct algorithms that can accurately diagnose plant diseases from images [18]. The PlantVillage Dataset is used to train and test machine learning and deep learning models for disease detection and image categorization. Date palm photos may be categorized and object recognized using the free Date Palm [38] Dataset. The collection contains 2631 date palm pictures in three classes.PLD_3_Classes_256 is a publicly available potato leaf disease classification dataset [34]. Indian Institute of Technology, Delhi experts conducted and published the study in 2022. The collection contains 4072 potato leaf pictures classed as Early Blight, Healthy, and Late Blight. NLB [19] Corn Healthy and Disease Dataset contain Northern Leaf Blight detection datasets. The 2018 publication was developed by University of Illinois at Urbana-Champaign scholars. The dataset includes 4117 sick and healthy maize leaf pictures.

4.3 Performance Metrics

The measurements determined by a study depend on its goals, dataset, and machine learning methods. These measures are also crucial for evaluating a model's performance because they reveal its pros and cons. Common crop disease detection and categorization methods include: Table 2 shows the formulas of the performance metrics.

4.4 Consensus Mechanism for Crop Leaf Disease Identification and Classification

Combining the results of many algorithms or models into a consensus mechanism for crop leaf disease identification and classification increases the system's overall accuracy and dependability.

The following ideas may help you be ready to put a consensus mechanism in place: $\text{Consensus Mechanism} = a \cdot D + b \cdot P + c \cdot FE + d \cdot CI$

Where weight coefficients a, b, c, and d are represented, dataset selection is denoted by D, preprocessing approaches are represented by P, feature extraction model is applied by FE, and continuous improvement is represented by CI. The proposed consensus mechanism for crop leaf disease diagnosis and categorization includes several critical components:

Leaf photos are captured using high-resolution cameras and typical imaging methods. P2: Improves image quality by adjusting brightness, contrast, and color balance.

P3: Segments the leaf from the backdrop to isolate the region of interest. P4: Data augmentation: Rotation, scaling, and flipping increase dataset size. Data Balance (P5):

Oversamples or under samples diseased and healthy leaf pictures to balance classes. Images are converted to grayscale and resized to a common size. Gabor filters, Haralick texture characteristics, and color histograms are used to extract texture, shape, and color features from segmented leaf pictures. For crop leaf disease identification, algorithm selection considers deep learning models, machine learning classifiers, and rule-based systems. The Ensemble Model combines model predictions via majority voting, weighted voting, or stacking (Table 3).

Table 2. List of the Performance metrics for Machine Learning Models

Name of the Metrics	Formula
Precision	$= \frac{TP}{TP + FP}$
Recall	$= \frac{TP}{TP + FN}$
F1 Score	$\frac{2\ X\ Precision \times Recall}{= Precision + Recall}$
Accuracy	$= \frac{TP + TN}{TP + TN + FP + FN}$
FAR	$= \frac{FP}{TN + FP}$
FRR	$= \frac{FN}{TP + FN}$
TPR	$= \frac{TP}{TP + FN}$
TNR	$= \frac{TN}{TN + FP}$
Balanced Accuracy	$= \frac{TP_R + TN_R}{2}$
DSC	$= \frac{2 \times TP}{2 \times TP + FP + FN}$
Jaccard	$= \frac{TP}{TP + FP + FN}$

Fusion Model combines image-based characteristics with weather data or expert knowledge to improve classification performance.

Continuous Improvement (CI) refines continuously: Model training (CI1): Uses preprocessed dataset feature vectors to train models. Hyperparameter Tuning (CI2): Optimizes model hyperparameters on validation set. To balance precision and recall, threshold tuning (CI3) experiments with model prediction thresholds. Error Analysis (CI4): Identifies patterns and improvements in consensus mechanism predictions that fail. Performance (CI5): Assesses the consensus mechanism on the test dataset using accuracy, precision, recall, F1-score, and AUC. In conclusion, the consensus mechanism uses

Table 3. Modalities based Consensus Mechanism for crop leaf disease identification and classification.

Dataset Selection	Preprocessing	Feature Extraction	Continuous Improvement
Dataset Selection	ImageAcquisition (P1), Image Enhancement (P2), ImageSegmentation (P3), Data Augmentation (P4), DataBalancing (P5), Grayscale image, Resize (P6)	Algorithm selection, Ensemble Model, Fusion Model	Model Training (CI1), Hyperparameter Tuning (CI2),Threshold tuning (CI3), Error Analysis (CI4), Evaluate Performance (CI5)

image processing, feature extraction, model ensemble, and continual improvement to provide a reliable crop leaf disease identification and classification system.

4.5 Related Works

Deep Learning's popularity has boosted the number of research articles on the issue. Deep neural networks' self-learning makes feature extraction and selection easier. Several Deep and Machine Learning models have performed well in agricultural image processing. Deep Learning's ability to diagnose and classify plant leaf diseases using leaf photos is examined in this study by reviewing 58 research articles. Using Deep Learning algorithms to identify and classify plant leaf diseases has tripled in the past year. This comprehensive review examines recent scientific publications over the past 5 years that have used machine learning algorithms to identify and categorize plant ailments based on leaf images.

Series of technical investigations conducted between 2019 -23. The first part of our analysis looks at studies that were released in 2019 and 2023. This section provides a thorough analysis of the articles selected from 2019 to 2023 that are pertinent to our study. The critiques, remedies, and missing modes offered by the sources listed in respectable scholarly journals between 2016 and 2023 are shown in Table 4.

Statistical analysis of 2016–23 methods Scholars, physicists, and mathematicians are confronting deep architectural difficulties in 2023 to understand deep learning. As Fig. 5f shows, plant leaf photo pre-processing is the focus of much research. As seen in below, data augmentation procedures have changed from traditional to automated ways using Generative Adversarial Net- works (GANs). Recent studies have focused on photo enhancement rather than Deep Learning methods, with promising results. The phenomena is illustrated in Fig. 5. Despite several significant and effective designs in the literature, deep learning (DL) has been the most widely used approach. This is due to Convolutional Neural Networks' generalizability, stability, and precision.

Figure 5d shows a collection of feature extraction models not previously published in scholarly journals, demonstrating the importance of this approach. In agricultural disease identification, deep learning has outperformed classical machine learning during the previous two decades. Deep learning techniques dominate because they outperform

Table 4. Modalities based Consensus Mechanism for crop leaf dis- ease identification and classification.

Ref	Purpose	Dataset	Algorithm	Solutions	Missing Modalities
[19]	Identifying plant diseases across different datasets	Plant Village, PlantDoc, Digipathos, NLB,CDS dataset	InceptionV3, ResNet50, VGG16, DenseNet169, Xception	Achieved highest accuracy 99.50%	CI1,CI2, CI5
[30]	Internet-of-Things(IoT) hardware system for online identification of tea diseases	Not Defined	Swim Transformer, Transfer Learning	Identification Accuracy 94%. The IoT Model works well on complex natural environments	D, CI2, CI4, CI5
[35]	A light weight transformer network "TrIncNet" for the identification of plant diseases	Plant Village, Maize disease dataset	TrIncNet, CNN	Achieved 5.38% higher testing accuracy than the existing ViT network	CI5, P
[36]	Enhanced lightweight CNN model for accurate and efficient detection of multiple pepper leaf diseases	Plant Village, New Plant Diseases, andCVPR 2020-FGVG7	Enhanced lightweight model based on the GoogleNet architecture	Cognition accuracy of 97.87%, which is 6% higher than GoogLeNet. Avg time decreased by 61.49%	P, CI1, CI5
[1]	To analyse the performance of AI-based solutions for crop disease identification, and classification	Dataset has 3150 images of insects on the leaves of crop plants	IncepionV3	Accuracy of 97% Pre-processing can improve the accuracy of the DL models for insect identification	D, FE, CI1, CI4
[31]	Corndiseases recognition system called VGNet	Plant Village	VGNet, batch normalization, global average pooling, L2 normalization	Accuracy of 98.3% and a lowest loss of 0.035, Lightweight, FasterTesting Time, Robustness	D, P, CI5
[32]	Combines the advantages of CNN vision transformers (ViT) for crop disease identification	Plant Village	Multiscale Convolution and Vision Transformer (MSCVT)	Accuracies of 99.86% and 97.50% on the Plant Village and Apple Leaf Pathology datasets respectively	CI1, P

(*continued*)

Table 4. (*continued*)

Ref	Purpose	Dataset	Algorithm	Solutions	Missing Modalities
[33]	To enhance the classification accuracy of CNN in plant disease recognition	Plant Village	Generative adversarial classified network (GACN)	Accuracy of 99.78% in Plant Village Dataset, Classify plant diseases directly or generate synthetic images to balance the dataset and improve CNN accuracy	P, CI5
[34]	The detection and classificationof potato leaf diseases	PLD Dataset	ShuffleNetV 2X2	Classification accuracy by 0.85% and improved the CPU inference speed by 25%	CI1,CI5, CI4
[38]	To identify and detect Ganoderma disease in oil palm plants by analyzing the vegetation index Ganoderma disease. Mosaic photo images	No Data	No Data	Infected plants were consistently lower than those of healthy plants with a ratio of 56.7%, 66.5%, and 44.7% in the R, G, and NIR bands UAV equipped system	D, P, CI1, CI5
[42]	Rice leaf disease detection system using a Multi-scale YOLO v5 detection network	RLD dataset	Faster-RCNN YOLO, Bi-FAPN, Multi-scale YOLO v5	accuracy is 94.87%	P, CI1, CI2, CI5
[43]	Evaluate the potential of four CNN models in the detection of rust disease	No Data	CNN based model	has high accuracy 94.29%	D, P, CI1, CI5, CI4
[12]	Deep-Dream based crop leaf disease detection (CLDD) architecture using a combination of ML techniques	Tomato Dataset	DD-EffiNet-B4-ADB	Accuracy range 84% to 96% Efficient resource utilization	P2, FE, CI5, CI1
[13]	Compares the performances of various CNN model with LeNet	Two Dataset used. Name not available	Modified LeNet VGG 16	accuracy 96 modified LeNet model, has lower accuracy but significantly reduced training time	D, FE,P, CI1
[16]	Analyze previous research on rice plant diseases, including studies on various diseases and key features	Does not mention	Jellyfish, SVM	Accuracy of 93.3% performance of 11 CNN models resulted in an F1- score of 0.9838	D, CI5

(*continued*)

Table 4. (*continued*)

Ref	Purpose	Dataset	Algorithm	Solutions	Missing Modalities
[17]	Crop disease prediction using big data for soybean	Soybean dataset	RRL-PFC	Make Comparative analysis of the RRL-PFC method	P, CI1, CI5
[18]	Identification of crop diseases using plant-leaf images using Lightweight CNN	PlantVillage	VGG-ICNN	Model achieves 99.16% accuracy, more efficient and computationally less expensive	CI1, CI5
[46]	lightweight CNN-based diagnostic tool for tomato leaf tissue diseases	Plant Village	lightweight CNN	Accuracy 99.04% precision 98.78% Recall 99.23% f1-score 99.00%	FE, CI1, CI4, CI2
[48]	CNN for the automatic detection of tomato crop diseases	Plant Village	CNN model consistsof nine layers	Accuracy = 91.66%	FE, P, CI
[51]	CNN approaches for detecting diseases in plants using a new plant diseases dataset	New plant diseases dataset	CNN model with 4 Conv, pooling, 2FC,and 38-neuron output layer	CNN achieved accuracy 98.36%, loss 0.0972, F1 score 98.35%, recall 98.36%, precision 98.39%	D, FE, CI1, CI4
[55]	Enhanced pre-trained CNN networks for banana foliar fungal disease detection	Author created banana leaves dataset	InceptionV3, Xception, MobileNetV2, and DenseNet121	Accuracy of 91.7% achieved by the DenseNet121	D, P, CI1, CI5
[21]	Leaf Disease recognition	Plant Village	Compact CNN	CNN achieved an accuracy of 99.70%	P, CI1, CI5
[22]	To address the challenge of large and small differences between disease classes	Plant Village	Fine-grained basedon Attention network and CNNs	Real-timeperformance, Useless memory	P, CI1, CI5
[25]	Identifying and diagnosing diseased regions on corn leaves	Plant Village and CDS	Simple Linear Iterative Clustering (SLIC) segmentation	Accuracy DenseNet121- 97.77% 80:20 split ratioResNet50- 94.52% 90:10 split ratio	CI5, CI4
[3]	To identify and classify black gram plant diseases	BPLD Dataset	RESNET, CNN	Accuracy: 99.39% using 5 fold cross-validation,	CI1, CI5

(*continued*)

Table 4. (*continued*)

Ref	Purpose	Dataset	Algorithm	Solutions	Missing Modalities
[11]	An image detector using a resource-constrained CNN ESCA,	Plant Village	Resource-constrained CNN	Accuracy 98.10% memory 718.961 KB inference time 122.969 ms	CI1, CI5
[39]	Classify corn diseases	Cornleaf images taken from the Madura Region	Random Forest Neural Network, Naïve Bayes, HOG	Neural Network achieve high accuracy with AUC-ROC rate 90.09%	D, P, CI1, CI5
[40]	AI model for classifying corn leaf diseases	No Data	LBP k-NN	k-NNwithk = 5 showed an accuracy of 81.1%, AUC value of 94.1%	D, CI5, CI4
[14]	Identification of diseases of maize crop	ICAR-IIMR	Inception-v3	Classification accuracy of 95.99%,	CI1,CI4, CI5
[15]	Detection and classification of maize crop infections	Plant Village	NPNet-19	Accuracy 97.51%	CI5, CI4
[20]	leaf disease identification	Tomato leaf diseases dataset	RRDN	RRDN model combines the advantages of resnet and densenet	ROC, CI1
[23]	Wheat disease classification and verification	Wheat diseases dataset	Decision Trees Rep Tree	Validation accuracy of 97.2%	D, FE, CI4, CI5
[29]	DL based framework for strawberry fruit recognition & disease phenotype	Not Defined	Convolutional encode- decodenet, ARFM	Precision of 92.45% in recognizing diseased fruits	D, FE, CI1, CI5, CI4

ordinary machine learning in accuracy, generalization, and complex picture data processing. CNN-based models and pre-trained model lessons are helping crop disease researchers diagnose and classify illnesses. Due to its ability to automatically extract relevant features from images, deep learning has replaced classical machine learning for crop disease identification and classification. This improved precision and efficiency (Fig. 6).

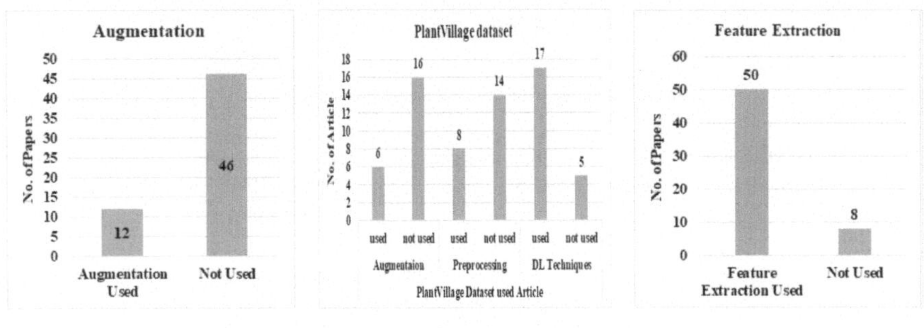

(a)Use of Data Augmentation (b) Process applied in PlantVillage (c) Use of Feature Extraction

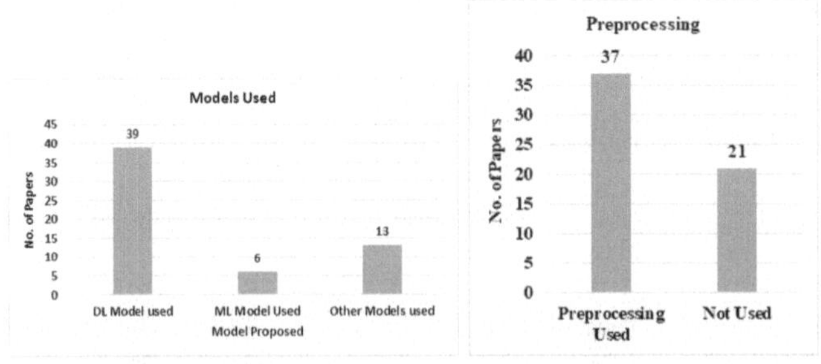

(d) Use of Deep Learning (e) Models used in last 5 years

Fig. 6. Statistical Analysis from Sourced Articles

5 Conclusion

In order to conduct a comprehensive assessment of 2016–2023 research, this report employs Deep Learning to categorize crop leaf photos as unhealthy or healthy. Despite the development of many practical and efficient algorithms, standardization is lacking, hence every method has major limitations. This study critically evaluates all previous approaches' pros and cons. We included performance-degrading issues such data lack, data imbalance, and inter-class variability and their solutions to enable researchers design appropriate crop leaf disease diagnostic and classification systems. After reviewing Deep Learning methods and algorithms, it is clear that Deep Learning can handle large amounts of data. However, agricultural leaf disease research doesn't fully harness their benefits. This comprehensive investigation demonstrates that an automated, unified approach to efficiently identify and classify agricultural leaf diseases is urgently needed.

6 Future Direction

Crop disease recognition and categorization will increase with computationally efficient deep learning (DL) architectures, real-time processing, and data augmentation and semi-supervised learning to reduce dependency on vast annotated datasets. We can enhance accuracy and reduce false positives and negatives by using anomaly detection, error reduction, and federated learning for decentralized data sources. Transfer learning, edge computing, multispectral photography, and drones will improve crop disease identification solutions' performance and accessibility. To support resource-constrained farmers, affordable hardware, community-based data collection, and collaborative platforms should be supported by continuous research, education, and government support to increase crop yields and food security.

References

1. Tirkey, D., Singh, K.K., Tripathi, S.: Performance analysis of AI-based solutions for crop disease identification, detection, and classification. Smart Agric. Technol. **5**, 100238 (2023). https://doi.org/10.1016/j.atech.2023.100238
2. Zhou, C., Zhong, Y., Zhou, S., Song, J., Xiang, W.: Rice leaf disease identification by residual-distilled transformer. Eng. Appl. Artif. Intell. **121**, 106020 (2023). https://doi.org/10.1016/j.engappai.2023.106020
3. Talasila, S., Rawal, K., Sethi, G., MSS, S.M.S.P.: Black gram plant leaf disease (BPLD) dataset for recognition and classification of diseases using computer-vision algo-rithms. Data Brief **45**, 108725 (2022). https://doi.org/10.1016/j.dib.2022.108725
4. Zhou, J., Li, J., Wang, C., Wu, H., Zhao, C., Teng, G.: Crop disease identification and interpretation method based on multimodal deep learning. Comput. Electron. Agric. **189**, 106408 (2021). https://doi.org/10.1016/j.compag.2021.106408
5. Chen, J., Zhang, D., Suzauddola, M., Zeb, A.: Identifying crop diseases using attention embedded mobilenet-V2 model. Appl. Soft Comput. **113**, 107901 (2021). https://doi.org/10.1016/j.asoc.2021.107901
6. Gui, P., Dang, W., Zhu, F., Zhao, Q.: Towards automatic field plant disease recognition. Comput. Electron. Agric.mput. Electron. Agric. **191**, 106523 (2021). https://doi.org/10.1016/j.compag.2021.106523
7. Khamparia, A., Singh, A., Luhach, A.K., Pandey, B., Pandey, D.K.: Classification and identification of primitive kharif crops using supervised deep convolutional networks. Sustain. Comput. Inform. Syst. **28**, 100340 (2020). https://doi.org/10.1016/j.suscom.2019.07.003
8. Agarwal, M., Gupta, S.K., Biswas, K.K.: Development of efficient CNN model for tomato crop disease identification. Sustain. Comput. Inform. Syst. **28**, 100407 (2020). https://doi.org/10.1016/j.suscom.2020.100407
9. Xiong, Y., Liang, L., Wang, L., She, J., Wu, M.: Identification of cash crop diseases using automatic image segmentation algorithm and deep learning with expanded dataset. Comput. Electron. Agric. **177**, 105712 (2020). https://doi.org/10.1016/j.compag.2020.105712
10. Navrozidis, I., Alexandridis, T.K., Dimitrakos, A., Lagopodi, A.L., Moshou, D., Zalidis, G.: Identification of purple spot disease on asparagus crops across spatial and spectral scales. Comput. Electron. Agric. **148**, 322–329 (2018). https://doi.org/10.1016/j.compag.2018.03.035
11. Falaschetti, L., Manoni, L., Di Leo, D., Pau, D., Tomaselli, V., Turchetti, C.: A CNN-based image detector for plant leaf diseases classification. HardwareX **12** (2022). https://doi.org/10.1016/j.ohx2022.e00363
12. Sahu, P., Chug, A., Singh, A.P., Singh, D.: Classification of crop leaf diseases using image to image translation with deep-dream. Multimedia Tools Appl. **82**(23), 35585–35619 (2023). https://doi.org/10.1007/s11042-023-14994-x

13. Deepti, K.: Comparative analysis of machine learning techniques for plant disease detection-data deployment. J. Inst. Eng. (India) Series B **104**(4), 837–849 (2023). https://doi.org/10.1007/s40031-023-00897-w
14. Haque, Md. A., et al.: Deep learning-based approach for identification of diseases of maize crop. Sci. Rep. **12**(1) (2022). https://doi.org/10.1038/s41598-022-10140-z
15. Nagaraju, M., Chawla, P.: Maize crop disease detection using npnet-19 convolutional neural network. Neural Comput. Appl. **35**(4), 3075–3099 (2022). https://doi.org/10.1007/s00521-022-07722-3
16. Tholkapiyan, M., Aruna Devi, B., Bhatt, D., Saravana Kumar, E., Kirubakaran, S., Kumar, R.: Performance analysis of rice plant diseases identification and classification methodology. Wireless Pers. Commun. **130**(2), 1317–1341 (2023). https://doi.org/10.1007/s11277-023-10333-3
17. Saritha, S., Thangaraja, G.A.: Prediction of crop disease using rank regressive learning and proaftn fuzzy classification models. Soft. Comput. (2023). https://doi.org/10.1007/s00500-023-08357-9
18. Thakur, P.S., Sheorey, T., Ojha, A.: VGG- ICNN: A lightweight CNN model for Crop disease identification. Multimedia Tools Appl. **82**(1), 497–520 (2022). https://doi.org/10.1007/s11042-022-13144-z
19. Ahmad, A., Gamal, A.E., Saraswat, D.: Toward generalization of deep learning-based plant disease identification under controlled and field conditions. IEEE Access **11**, 9042–9057 (2023). https://doi.org/10.1109/access.2023.3240100
20. Zhou, C., Zhou, S., Xing, J., Song, J.: Tomato leaf disease identification by restructured deep residual dense network. IEEE Access **9**, 28822–28831 (2021). https://doi.org/10.1109/access.2021.3058947
21. Ozbılge, E., Ulukök, M.K., Toygar, O., Ozbılge, E.: Tomato disease recognition using a compact convolutional neural network. IEEE Access **10**, 77213–77224 (2022). https://doi.org/10.1109/access.2022.3192428
22. Wu, Y., Feng, X., Chen, G.: Plant leaf diseases fine-grained categorization using convolutional neural networks. IEEE Access **10**, 41087–41096 (2022). https://doi.org/10.1109/access.2022.3167513
23. Haider, W., Rehman, A.-U., Durrani, N.M., Rehman, S.U.: A generic approach for wheat disease classification and verification using expert opinion for knowledge-based decisions. IEEE Access **9**, 31104–31129 (2021). https://doi.org/10.1109/access.2021.3058582
24. Dai, Q., Cheng, X., Qiao, Y., Zhang, Y.: Crop leaf disease image super-resolution and identification with dual attention and topology fusion generative adversarial network. IEEE Access **8**, 55724–55735 (2020). https://doi.org/10.1109/access.2020.2982055
25. Phan, H., Ahmad, A., Saraswat, D.: Identification of foliar disease regions on corn leaves using slic segmentation and deep learning under uniform background and field conditions. IEEE Access **10**, 111985–111995 (2022). https://doi.org/10.1109/access.2022.3215497
26. Castelao Tetila, E., Brandoli Machado, B., Belete, N.A., Guimaraes, D.A., Pistori, H.: Identification of soybean foliar diseases using unmanned aerial vehicle images. IEEE Geosci. Remote Sens. Lett. **14**(12), 2190–2194 (2017). https://doi.org/10.1109/lgrs.2017.2743715
27. Xu, Y., Zhao, B., Zhai, Y., Chen, Q., Zhou, Y.: Maize diseases identification method based on multi-scale convolutional global pooling neural network. IEEE Access **9**, 27959–27970 (2021). https://doi.org/10.1109/access.2021.3058267
28. Lv, M., Zhou, G., He, M., Chen, A., Zhang, W., Hu, Y.: Maize leaf disease identification based on feature enhancement and DMS-Robust Alexnet. IEEE Access **8**, 57952–57966 (2020). https://doi.org/10.1109/access.2020.2982443
29. Ilyas, T., Khan, A., Umraiz, M., Jeong, Y., Kim, H.: Multi-scale context aggregation for strawberry fruit recognition and disease phenotyping. IEEE Access **9**, 124491–124504 (2021). https://doi.org/10.1109/access.2021.3110978

30. Xie, S., Wang, C., Wang, C., Lin, Y., Dong, X.: Online identification method of tea diseases in complex natural environments. IEEE Open J. Comput. Soc. **4**, 62–71 (2023). https://doi.org/10.1109/ojcs.2023.3247505
31. Fan, X., Guan, Z.: VGNet: A lightweight intelligent learning method for corn diseases recognition. Agriculture **13**(8), 1606 (2023). https://doi.org/10.3390/agriculture13081606
32. Zhu, D., Tan, J., Wu, C., Yung, K., Ip, A.A.W.: Crop disease identification by fusing multiscale convolution and vision transformer. Sensors **23**(13), 6015 (2023). https://doi.org/10.3390/s23136015
33. Wang, X., Cao, W.: GACN: generative adversarial classified network for balancing plant disease dataset and plant disease recognition. Sensors **23**(15), 6844 (2023). https://doi.org/10.3390/s23156844
34. Feng, J., et al.: Research and validation of potato late blight detection method based on deep learning. Agronomy **13**(6), 1659 (2023). https://doi.org/10.3390/agronomy13061659
35. Gole, P., Bedi, P., Marwaha, S., Haque, M., Deb, C.K.: TrIncNet: a lightweight vision transformer network for identification of plant diseases. Front. Plant Sci. **14** (2023). https://doi.org/10.3389/fpls.2023.1221557
36. Dai, M.,et al.: Pepper Leaf disease recognition based on enhanced lightweight convolutional neural networks. Front. Plant Sci. **14** (2023). https://doi.org/10.3389/fpls.2023.1230886
37. Mohanty, S.P., Hughes, D.P., Salathé, M.: Using deep learning for image-based plant disease detection. Front. Plant Sci. **7**(2016). https://doi.org/10.3389/fpls.2016.01419
38. Wahyuni, M., Sabrina, T., M, M., Santoso, H.: Analysis of vegetation index of oil palm plants infected with ganoderma disease. IOP Conf. Ser. Earth Environ. Sci. **1188**(1), 012006 (2023). https://doi.org/10.1088/1755-1315/1188/1/012006
39. Ubaidillah, A., Rochman, E.M., Fatah, D.A., Rachmad, A.: Classification of corn diseases using Random Forest, neural network, and naive bayes methods. J. Phys. Conf. Ser. **2406**(1), 012023 (2022). https://doi.org/10.1088/1742-6596/2406/1/012023
40. Rachmad, A., Syarief, M., Rifka, S., Sonata, F., Setiawan, W., Rochman, E.M.: Corn leaf disease classification using local binary patterns (LBP) feature extraction. J. Phys. Conf. Ser. **2406**(1), 012020 (2022). https://doi.org/10.1088/1742-6596/2406/1/012020
41. Gayatri, K., Kanti, R.D., Rayavarapu, V.C., Sridhar, B., Bobbili, V.R.: Image processing and pattern recognition based plant leaf diseases identification and classification. J. Phys. Conf. Ser. **1804**(1), 012160. IOP Publishing (2021)
42. V.: Image processing and pattern recognition based plant leaf diseases identification and classification. J. Phys. Conf. Ser. **1804**(1), 012160 (2021). https://doi.org/10.1088/1742-6596/1804/1/012160
43. Kumar, V.S., Jaganathan, M., Viswanathan, A., Umamaheswari, M., Vignesh, J.: Rice leaf disease detection based on bidirectional feature attention pyramid network with Yolo V5 model. Environ. Res. Commun. **5**(6), 065014 (2023). https://doi.org/10.1088/2515-7620/acdece
44. Shahoveisi, F., Taheri Gorji, H., Shahabi, S., Hosseinirad, S., Markell, S., Vasefi, F.: Application of image processing and transfer learning for the detection of rust disease. Sci. Rep. **13**(1) (2023). https://doi.org/10.1038/s41598-023-31942-9
45. Polly, R., Devi, E.A.: A deep learning-based study of crop diseases recognition and classification. In: 2022 Second International Conference on Artificial Intelligence and Smart Energy (ICAIS) (2022). https://doi.org/10.1109/icais53314.2022.9742950
46. Arvind. S., Negi, K.A.: A detection and classification of cotton leaf disease using a lightweight CNN architecture. In: 2022 Fourth International Conference on Emerging Research in Electronics, Computer Science and Technology (ICERECT) (2022). https://doi.org/10.1109/icerect56837.2022.10060246

47. Sanida, T., Sanida, M.V., Sideris, A., Dasygenis, M.: A lightweight CNN model for tomato crop diseases on heterogeneous embedded system. In: 2023 12th International Conference on Modern Circuits and Systems Technologies (MOCAST) (2023). https://doi.org/10.1109/mocast57943.2023.10176582
48. Yatoo, A.A., Sharma, A.: A novel model for automatic crop disease detection. In: 2021 Sixth International Conference on Image Information Processing (ICIIP) (2021). https://doi.org/10.1109/iciip53038.2021.9702553
49. Sheenam, S., Kumar, A.: Advanced CNN- based approach for accurate tomato plant disease recognition. In: 2023 3rd International Conference on Intelligent Technologies (CONIT) (2023). https://doi.org/10.1109/conit59222.2023.10205755
50. Jenipher, V.N., Radhika, S.: An automated system for detecting rice crop disease using CNN inception V3 transfer learning algorithm. In: 2022 Second International Conference on Artificial Intelligence and Smart Energy (ICAIS) (2022). https://doi.org/10.1109/icais53314.2022.9742999
51. Kothalawala, M.U., Gaveshith, K.M.G., Tharaka, A.H.D.H., Punchihewa, I.A., Sriyaratna, D.: Banana disease identification using machine learning based technologies and weather- based dispersion analysis. In: 2022 4th International Conference on Advancements in Computing (ICAC) (2022). https://doi.org/10.1109/icac57685.2022.10025325
52. Ibrahim, A.O., Elghamrawy, S.: Plant disease detection using artificial intelligence techniques for agricultural productivity enhancement in Egypt. In: 2023 International Telecommunications Conference (ITC-Egypt) (2023). https://doi.org/10.1109/itc-egypt58155.2023.10206149
53. Vamsee Kongara, R.K., Siva Charan Somasila, V., Revanth, N., Polagani, R.D.: Classification and comparison study of rice plant diseases using pre-trained CNN Models. In: 2022 International Conference on Inventive Computation Technologies (ICICT) (2022). https://doi.org/10.1109/icict54344.2022.9850784
54. Narvaria, A., Kumar, U., Jhanwwee, K.S., Dasgupta, A., Kaur, G.J.: Classification and identification of crops using deep learning with UAV data. In: 2021 IEEE International India Geoscience and Remote Sensing Symposium (InGARSS) (2021). https://doi.org/10.1109/ingarss51564.2021.9792009
55. Kamal, K.C., Yin, Z., Li, B., Ma, B., Wu, M.: Transfer learning for fine-grained crop disease classification based on leaf images. In: 2019 10th Workshop on Hyperspectral Imaging and Signal Processing: Evolution in Remote Sensing (WHISPERS) (2019). https://doi.org/10.1109/whispers.2019.8921213
56. Mathew, D., Kumar, C.S., Anita Cherian, K.: Classification of leaf spot diseases in banana using pre-trained convolutional neural networks. In: 2023 International Conference on Control, Communication and Computing (ICCC) (2023). https://doi.org/10.1109/iccc57789.2023.10165629
57. Swathy, K., KK, A.B.: Classification of paddy diseases using ellipse fitting. In:2022 Second International Conference on Next Generation Intelligent Systems (ICNGIS) (2022). https://doi.org/10.1109/icngis54955.2022.10079861
58. Kapoor, S., Kumar, D., Sinha, A.: Classification of various diseases affecting Apple, cherry, and peach plants using modified VGGNET19 architecture. In: 2022 3rd International Conference on Issues and Challenges in Intelligent Computing Techniques (ICICT) (2022). https://doi.org/10.1109/icict55121.2022.10064586
59. Rajasekaran, C., Raguvaran, K., Javanthi, K.B.: Comparative analysis of deep semantic segmentation architectures for disease identification in turmeric crops. In: TENCON 2022 - 2022 IEEE Region 10 Conference (TENCON) (2022). https://doi.org/10.1109/tencon55691.2022.9977957

60. Pajjuri, N., Kumar, U., Thottolil, R.: Comparative evaluation of the convolutional neural network based transfer learning models for classification of plant disease. In: 2022 IEEE International Conference on Electronics, Computing and Communication Technologies (CONECCT) (2022). https://doi.org/10.1109/conecct55679.2022.9865733
61. Rattan, R., Shiney, J.O.: Comparison and analysis of CNN based algorithms for Plant disease identification. In: 2023 Advanced Computing and Communication Technologies for High Performance Applications (ACCTHPA) (2023). https://doi.org/10.1109/accthpa57160.2023.10083342
62. Kuricheti, G., Supriya, P.: Computer vision based turmeric leaf disease detection and classification: a step to smart agriculture. In: 2019 3rd International Conference on Trends in Electronics and Informatics (ICOEI) (2019). https://doi.org/10.1109/icoei.2019.8862706
63. Patil, B.M., Burkpalli, V.: Cotton leaf disease classification by combining color and texture feature-based approach. In: 2022 6th International Conference on Intelligent Computing and Control Systems (ICICCS) (2022). https://doi.org/10.1109/iciccs53718.2022.9788405
64. Kulkarni, O.: Crop disease detection using deep learning. In: 2018 Fourth Inter- national Conference on Computing Communication Control and Automation (ICCUBEA) (2018). https://doi.org/10.1109/iccubea.2018.8697390
65. Abdelmalek, D., Assia, K.: Machine learning classification models comparison for crop damage identification. In: 2022 2nd International Conference on Advanced Electrical Engineering (ICAEE) (2022). https://doi.org/10.1109/icaee53772.2022.9962088
66. Gupta, A., Kumar, M.S., Kumar, M.R., Kumar, D.H.: Deep learning technique used for tomato and potato plant leaf disease classification and detection. In: 2023 International Conference on Smart Systems for Applications in Electrical Sciences (ICSSES) (2023). https://doi.org/10.1109/icsses58299.2023.10199327
67. Aravind, K.R., Raja, P., Mukesh, K.V., Aniirudh, R., Ashiwin, R., Szczepanski, C.: Disease classification in maize crop using bag of features and multiclass support vector machine. In: 2018 2nd International Conference on Inventive Systems and Control (ICISC) (2018). https://doi.org/10.1109/icisc.2018.8398993
68. Wang, Y., et al.: Disesniper: a potato disease identification system based on the resnet model. In: 2022 10th International Conference on Agro-Geoinformatics (2022). https://doi.org/10.1109/agro-geoinformatics55649.2022.9859214
69. Van Ho, S., Vuong, H.G., Nguyen, B.Q., Trinh, Q.-H., Tran, M.-T.: Ensemble of deep neural networks for rice leaf disease classification. In: 2022 RIVF International Conference on Computing and Communication Technologies (RIVF) (2022). https://doi.org/10.1109/rivf55975.2022.10013858
70. Kartikeyan, P., Shrivastava, G.: Hybrid feature approach for plant disease detection and classification using machine learning. In: 2022 IEEE World Conference on Applied Intelligence and Computing (AIC) (2022). https://doi.org/10.1109/aic55036.2022.9848939
71. John, A.A.: Identification of diseases in cassava leaves using convolutional neural network. In: 2022 Fifth International Conference on Computational Intelligence and Communication Technologies (CCICT) (2022). https://doi.org/10.1109/ccict56684.2022.00013

Identification of Infectious Potato Using CNN

A. Kalaivani(✉), G. Guruvarshni, J. Keerthika, and I Sai Keshav

Department of Information Technology, Rajalakshmi Engineering College, Chennai, India
{kalaivani.a,231011003,231011008,231011012}@rajalakshmi.edu.in

Abstract. The potato, is one of the most important agricultural crops, especially in impoverished countries. Fungal infections pose a substantial global risk to potato crops. Despite facing challenges related to computation time and accuracy, have explored the realm of machine learning for early detection. In this study, the aim is to introduce a modified convolutional neural network (CNN) to enhance precision and reduce the computational burden in identifying fungal infections in potatoes. The model's performance undergoes a thorough comparison with existing techniques for potato categorization. With the overarching goal of supporting the resilience and success of global potato farming, the research strives to simplify and improve the detection and control of fungal infections.

Keywords: Potato Disease · CNN · Image Classification

1 Introduction

Potatoes, as an essential part of global agriculture, play an important role in sustaining livelihoods and nourishing diverse populations. However, the agricultural sector encounters persistent challenges, primarily from debilitating diseases like Black Scurf, Black Leg, Common Scab, and Pink Rot, posing substantial threats to the quality and yield of potato crops. The urgency of prompt diagnosis to enact preventative measures becomes paramount to alleviate societal and financial losses associated with production. Traditional methods of disease identification, reliant on visual acumen, are hindered by practical constraints, necessitating a transformative solution.

The foundational crop of potatoes, vital for sustenance, especially in developing nations, faces a pervasive global threat from fungi. Beyond adversely impacting farmers' livelihoods, these diseases significantly compromise the global potato supply, diminishing both production and quality. Despite the acknowledged significance of early identification, challenges persist, particularly concerning computation time and accuracy in employing traditional approaches. These existing barriers hinder the prompt detection and treatment of fungal infections, heightening the vulnerability of agricultural communities. Consequently, there is a pressing demand for an efficient and effective solution to address these challenges and safeguard global potato production.

This project is concentrated on the critical issue of potato diseases in global agriculture, shining a spotlight on regions like India, where farming is not just a job but a cornerstone for feeding people and managing environmental challenges. The heart of

P. D. Sivakumar et al. (Eds.): IRCCTSD 2024, CCIS 2360, pp. 58–70, 2025.
https://doi.org/10.1007/978-3-031-82389-3_5

our efforts is in reshaping how we keep tabs on crop health, putting a strong emphasis on catching potato diseases early to safeguard the quality and quantity of the harvest. Our consideration goes beyond identification in these agricultural environments, where the soil contains memories of generations past and every crop is an example of resiliency. We are creating a story that imagines technology and tradition dancing together harmoniously. Imagine farmers empowered to strike the difficult balance between providing for their communities' needs and preserving the land, not just with tools but also with knowledge. As we continue to explore the complex network of agriculture, we are dedicated to supporting communities as well as healthy crops. This is an international voyage that embraces the variety of farming methods seen around the world. Together, we want to grow not just nutritious potatoes but also a sustainable future where the fields resound with the promise.

The motivation behind this project lies in recognizing the challenges associated with traditional methods of identifying diseases in crops. Imagine farmers in remote locations, dedicating significant time manually tracking and relying on their own observations to detect potential issues. This process is not only time-consuming but especially burdensome in distant and hard-to-reach farms. Our motivation is rooted in the desire to instigate a technological shift. We're exploring the realms of image analysis techniques and machine learning to streamline and enhance the identification process. Ultimately, the project aims to empower farmers with timely information, enabling targeted interventions and contributing to a more resilient and sustainable future for global potato farming.

The proposed system would be a technological marvel, combining the most recent developments with a major emphasis on machine learning. It uses state-of-the-art advancements such as convolutional neural networks (CNNs) to improve the accuracy and productivity of diagnosing potato illnesses. This all-inclusive solution addresses issues associated with conventional techniques directly by including picture capture, extraction, segmentation, and pre-processing. This proposed approach becomes a useful tool for farmers, protecting communities from the negative effects of agricultural losses while actively preserving crop health as a precise and effective substitute for traditional monitoring. The initiative intends to explore the technological nuances and celebrate successful achievements as we approach the one-year milestone. By using cutting-edge technology, we hope to completely transform how farmers.

2 Literature Survey

In order to complete our work accurately, we read and attempted to understand some source material before beginning this task. This work serves as a literature review of the paper we read before beginning. Many novel techniques have been developed by researchers to properly detect, classify, and survey plant diseases; a summary of their work is provided in this section.

In their research, Md. Marjanul Islam Tarik and colleagues (2021) developed a computer model using CNN technology to identify different diseases in potatoes.They gathered over 2000 images directly from potato fields, focusing on both potato leaves and the potatoes themselves. Their model first identifies features like edges and shapes, then compares them to known patterns to recognize diseases. They found that by using this approach, their model could accurately classify potato diseases into seven categories. This study highlights the potential of CNNs in detecting plant diseases, offering a promising tool for farmers and researchers [1].

The 2021 study by M. I. Tarik, S. Akter, A. A. Mamun, and A. Sattar focuses on using processing and machine learning techniques to address potato leaf diseases. They analyze a dataset of over 2034 images of unhealthy potato leaves from a public plant database, employing image processing methods for effective data analysis. By using machine learning, especially classification models, the scientists can tell apart healthy and sick potato leaves very accurately. This helps them find diseases faster and manage them better in potato fields. This new way of doing things shows a lot of potential for making farming better, getting more crops, and making agriculture more sustainable [2].

Dattatray G. Takale's 2022 study focuses on leveraging convolutional neural networks (CNNs) for potato disease detection. Utilizing a dataset of 900 images, including Early blight, Late blight, and Healthy classes from Kaggle's Plant Village Dataset, the study employs data augmentation techniques to enhance the training set's size and accuracy. This research underscores the potential of advanced CNN architectures in detecting and classifying potato diseases. Such advancements hold promise for early disease detection and effective management in potato crops, contributing to enhanced agricultural productivity and sustainability [4].

The study by Md. Asif Iqbal and Kamrul Hasan Talukder (2020) outlines a comprehensive methodology for potato disease detection through image processing and classification techniques. Feature extraction involves global descriptors such as Hu Moments for shape, Haralick Texture for texture, and Color Histogram for color distribution. This methodology presents a robust approach for potato disease detection, integrating image processing and machine learning techniques for efficient and accurate classification [5].

The project demonstrates the effectiveness of a convolutional neural network (CNN) in discerning between healthy and unhealthy potatoes, showcasing the prowess of deep learning in agricultural disease identification. Through rigorous training and refinement, the model achieves high accuracy and low loss metrics, providing farmers with a reliable tool to maintain crop health and promote sustainability. Additionally, the CNN's capacity to detect various potato shapes expands disease detection capabilities across different growth stages, highlighting its adaptability to diverse agricultural challenges and potential contribution to global food security and sustainable agriculture efforts.

3 Materials and Methods

The primary objective of this study is to promptly address the challenge of fungal infections in potatoes, a crop vital for life in disadvantaged regions. The approach involves creating and developing an individual convolutional neural network (CNN) to enhance the accuracy of fungal infection diagnosis. Simultaneously, the study addresses computational difficulties to ensure the practical usability of this model in various agricultural scenarios. An essential aspect includes a comprehensive performance comparison of the proposed CNN with existing potato classification techniques, covering evaluations of computational efficacy, accuracy, and efficiency. The study also aims to expedite the identification procedure, offering a more potent alternative to traditional methods and aiding in the prompt and effective management of fungal diseases in potato crops.

3.1 Methodology

In our methodology for detecting health status of potatoes, we commence the process with image acquisition, ensuring the acquisition of high-quality potato images. Subsequently, a preprocessing step was implemented to enhance and prepare the images for effective machine learning model training. Advanced segmentation algorithms are then applied to isolate key regions of interest, focusing specifically on areas crucial for disease identification within the potato leaves. The dataset is meticulously curated, incorporating these processed images for both training and testing purposes, laying the foundation for the development of a Convolutional Neural Network (CNN). This CNN model is intricately designed with an input layer tailored for 64×64 RGB images, followed by a convolutional layer applying 32 filters of size 3×3, enabling the extraction of essential features while reducing spatial dimensions. The subsequent max-pooling layer further refines the output by retaining critical features. Another convolutional layer with 64 filters and an additional max-pooling layer followed suit, contributing to the extraction of intricate features. The flattened layer then transforms the output into a 1D array, facilitating seamless input to fully connected layers (Fig. 1).

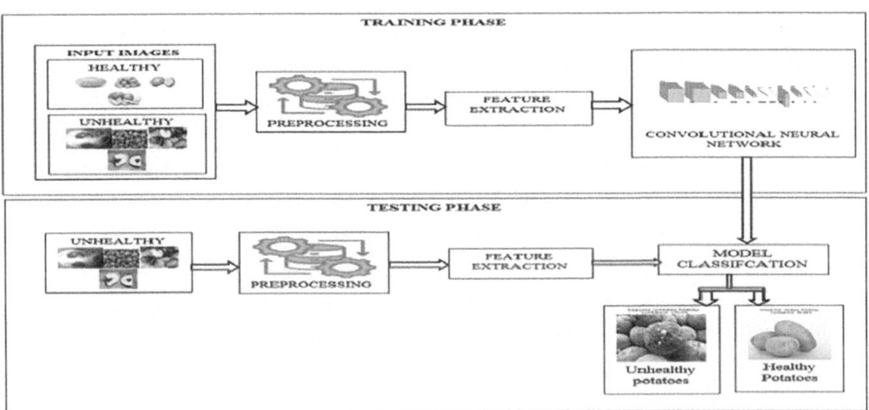

Fig. 1. Potato Disease Identification System

3.2 Dataset Exploration

The dataset utilized for potato disease detection employing a Convolutional Neural Network (CNN) encompasses a diverse collection of potato leaf images, meticulously labeled to denote their health status, whether they are healthy or unhealthy. Comprising a substantial number of images, the dataset maintains a balanced distribution between healthy and diseased samples, fostering equitable model training. Images exhibit variability in lighting conditions, angles, and backgrounds, ensuring the model's adaptability to diverse scenarios. Image augmentation techniques, including rotation, flipping, and scaling, have been applied to enhance dataset diversity artificially. The dataset's resolution is standardized, and it's sourcing involves collaboration with agricultural experts and field surveys. Data is appropriately split into training, validation, and testing sets, contributing to robust model evaluation. This dataset, ethically collected with permissions and considerations for privacy, is a valuable resource for advancing research in potato disease detection. The Datasets collected from the source (https://github.com/Mukaffi28/Potato-Disease) which includes 80 healthy images and unhealthy images separated by 6 sub-classes.

3.3 Data Pre-Processing

Preprocessing of diseased potato images is a crucial step before classification, as the quality of these images significantly influences the outcomes in subsequent classification steps (Fig. 2).

```
dataset = tf.keras.preprocessing.image_dataset_from_directory((r"C:\Users\varsh\Desktop\Dataset\Potato-Disease-main"),
    seed=123,
    shuffle=True,
    image_size=(IMAGE_SIZE,IMAGE_SIZE),
    batch_size=BATCH_SIZE)

Found 440 files belonging to 2 classes.
```

Fig. 2. Pre-Processing

3.4 Data Resizing and Scaling

In preparing the dataset for the potato disease detection CNN project, key preprocessing steps are implemented. Initial resizing standardizes images to 64 × 64 pixels, facilitating uniformity. Data augmentation introduces variability through random rotations, flips, and zooms. Normalization brings pixel values into the [0, 1] range for effective model training. Strategic dataset splitting creates training, validation, and testing sets for robust evaluation. Label encoding transforms categorical labels to numerical format, and techniques to handle imbalanced classes address distribution disparities. Quality checks ensure accurate labeling, and feature extraction enhances the model's discriminatory abilities, focusing on color, texture, and shape. The entire process is orchestrated through a streamlined preprocessing pipeline, ensuring consistency and reproducibility in dataset preparation.

3.5 Data Augumentation

The described data augmentation model is a sequential pipeline crafted using TensorFlow and Keras, incorporating essential augmentation layers to enrich the training process. Notably, the RandomFlip layer introduces random horizontal and vertical flips, diversifying the dataset with varied orientations of input images. Simultaneously, the RandomRotation layer applies random rotations, enhancing the dataset with angular variations of up to 20%. These augmentation techniques serve to mitigate overfitting by exposing the model to augmented instances during training, thus improving its ability to generalize to unseen data. Data augmentation is a foundational technique in machine learning, particularly in image classification tasks, addressing challenges stemming from limited training data. By artificially expanding the dataset with augmented samples, models can better adapt to diverse real-world scenarios, leading to improved performance and robustness.

4 Model Building

The Sequential Model given has a Convolutional Neural Network (CNN) architecture that has been rigorously customized for picture classification, with a particular emphasis on disease detection in potato plants. The model's sequential structure begins with a preprocessing step that standardizes incoming photos by downsizing them to a consistent 256×256 resolution and normalizing them in RGB format. Following this preprocessing phase, successive layers include a sequence of Conv2D and MaxPooling2D operations, which are carefully placed to serve as feature extractors and spatial down samplers, respectively. The combination of many convolutional layers, each with a max-pooling layer, results in a deep, hierarchical network capable of detecting detailed patterns in the input data.

This hierarchical framework enables the model to gradually learn and abstract characteristics at various degrees of complexity. By iteratively collecting and consolidating variables at various hierarchical levels, the model obtains a more detailed knowledge of what distinguishes Healthy and Unhealthy Potatoes. Using convolutional and pooling processes, the model can successfully identify important patterns indicative of various illnesses, allowing for more accurate categorization of potato plant health status.

$$H_{out} = \frac{H_{in} - K + 2P}{S} + 1 \tag{1}$$

$$W_{in} = \frac{W_{in} - K + 2P}{S} + 1 \tag{2}$$

$$C_{out} = F \tag{3}$$

The Sequential Model emerges as an effective tool for automating disease diagnosis in potato plants by combining preprocessing, feature extraction, and hierarchical feature learning. By quickly recognizing disease signals in the vegetables, the model provides farmers with timely information, allowing for proactive intervention techniques and educated crop management decisions. This technical innovation holds promise for increasing agricultural sustainability and production, highlighting machine learning's transformational potential in current farming techniques (Fig. 3) (Table 1).

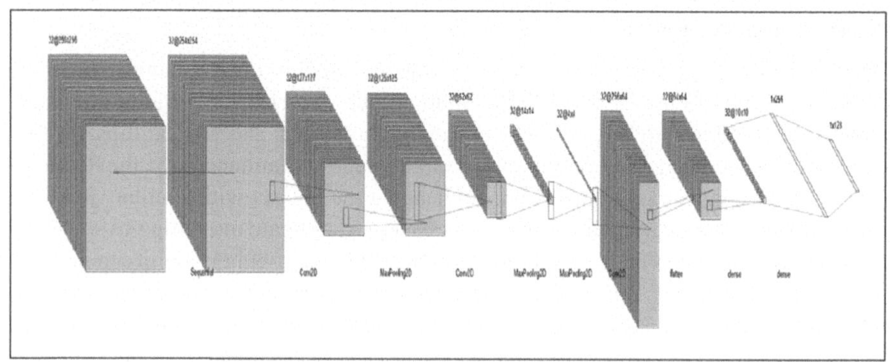

Fig. 3. Proposed CNN Model.

Table 1. Summary of the CNN Model

Layer (type)	Output Shape	Params
sequential_6	(32, 256, 256, 3)	0
conv2d_12	(32, 254, 254, 32)	896
max_pooling2d_12	(32, 127, 127, 32)	0
conv2d_13	(32, 125, 125, 64)	18496
max_pooling2d_13	(32, 62, 62, 64)	0
conv2d_14	(32, 60, 60, 64)	36928
max_pooling2d_14	(32, 30, 30, 64)	0
conv2d_15	(32, 28, 28, 64)	36928
max_pooling2d_15	(32, 14, 14, 64)	0
conv2d_16	(32, 12, 12, 64)	36928
max_pooling2d_16	(32, 6, 6, 64)	0
conv2d_17	(32, 4, 4, 64)	36928
flatten_2 (Flatten)	(32, 256)	0
dense_4 (Dense)	(32, 64)	16448
dense_5 (Dense)	(32, 10) Total params: 184,202 Trainable params: 184,202 Non-trainable params: 0	650

Following the convolutional layers, the model effortlessly transitions into a flattened layer, which is critical for converting the 2D feature maps produced by the convolutional procedures into a 1D array. This transformation prepares the input data for future processing by fully linked dense layers, which allows the model to execute complex feature learning and abstraction. The use of a flattened layer preserves spatial information while simplifying the integration of various signals collected from the convolutional layers, improving the model's ability to detect subtle patterns indicating either healthy or unhealthy potatoes.

$$Hout = Hin * Win * Cin \tag{4}$$

The first Dense layer is then carefully added into the model architecture, which includes ReLU activation and 64 neurons. This dense layer is a critical component in the feature learning process, employing ReLU's non-linear activation function to add flexibility and expressivity to the model's learnt representations. With 64 neurons, this thick layer has the computational capability to record complicated connections and hierarchies in the input data, which contributes to the model's ability to distinguish nuanced aspects linked to the potato's condition from the retrieved representations.

The last Dense layer, which uses softmax activation and two neurons, marks a significant point in the model architecture, when healthy and sick potatoes are classified. This dense layer uses softmax activation to generate probability distributions across the two unique categories, allowing the model to make probabilistic predictions for each class. This design decision not only makes it easier to identify the most likely potato health condition, but it also allows for the quantification of uncertainty in the model's predictions, which improves the classification process's interpretability and reliability.

$$H_{out} = U \tag{5}$$

With a total of 184,007 trainable parameters, the model exhibits an impressive ability to learn and replicate small variations found in potato photos. These parameters together convey the model's knowledge and comprehension of the properties that separate healthy from unhealthy potatoes, allowing for more accurate categorization. This approach has tremendous promise for improving agricultural production, reducing crop losses, and encouraging sustainable farming methods in potato cultivation by advancing the area of precision agriculture and crop management (Table 2).

Table 2. Tuning Parameters for CNN Model

Parameter Name	Attribute	Parameter Name
Activation function	ReLU	Activation function
Batch size	32	Batch size
Loss function	Sparse Categorical Crossentropy	Loss function
Optimizer	Adam	Optimizer

5 Results

In this endeavor, the Potato Disease Dataset served as the cornerstone for training a sophisticated deep convolutional neural network (CNN) designed to discern and scrutinize potato images categorized into two classes: Healthy Potatoes and Diseased Potatoes.

The dataset encompassed arrays representing these images from diverse angles, offering a comprehensive view for robust model training. Furthermore, the model refinement process involved augmenting the dataset with new weighted data, and continuous training over epochs aimed at refining the model's accuracy and mitigating data loss. The accuracy evaluation of our model distinctly illustrates the relationship between model epochs and error degrees. It becomes evident that as the model progresses through epochs, the accuracy consistently improves. Complementing this, the model loss results unequivocally depict a downward trend, indicating a reduction in loss as the model undergoes successive epochs,

In this Project, a novel aspect is the model's capacity to detect not only the health state of individual potatoes, but also different shapes. In addition to distinguishing healthy and unhealthy potatoes, the model is capable of recognizing the health status of whole potatoes, groups of potatoes, halved potatoes, and even mid-slice potatoes. This novel technique expands the area of disease detection, providing insights into the health of potatoes at all phases of processing and development. By accommodating diverse potato shapes and arrangements, the model showcases its adaptability to real-world agricultural scenarios, where potatoes may manifest in various forms and conditions. This uniqueness improves the practical usability of the model, allowing it to address a wider range of agricultural challenges and contribute to more comprehensive crop management strategies (Fig. 4) (Table 3).

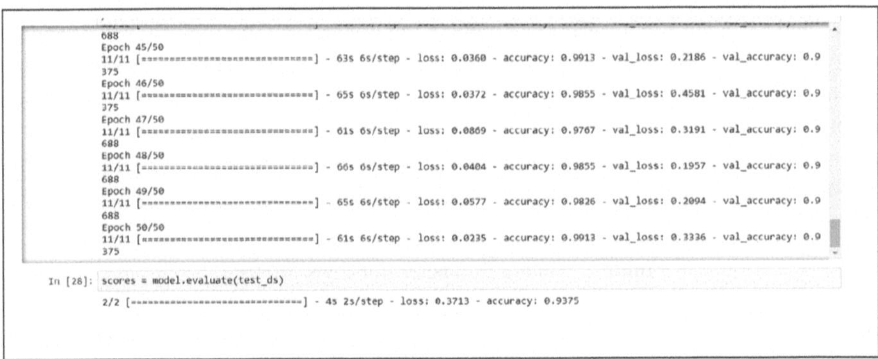

Fig. 4. Epoch Training Performance

A Visualizes the training and validation accuracy trends during the model training process. By plotting the accuracy metrics against the range of epochs, the plot offers insights into the model's performance evolution. The training accuracy curve represents the model's accuracy on the training dataset, depicting its learning progress. Conversely, the validation accuracy curve illustrates the model's ability to generalize to new, unseen data. The legend, positioned at the lower right, distinguishes between the training and validation accuracy lines. This visualization aids in evaluating the model's effectiveness in classifying potato health statuses and guides optimization efforts for improved performance (Fig. 5).

Table 3. Sample Image Categories of Potatoes

Image Category	Healthy	Unhealthy
Whole Potato		
Group of Potatoes		
Halved Potatoes		
Mid-Slice Potatoes		

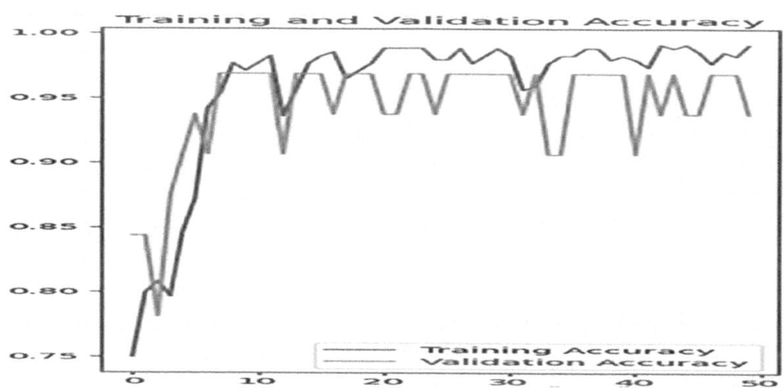

Fig. 5. Training and Validation Accuracy Graph

To visualize the loss incurred during validation and training of the model. To show how the model's loss varies over time, it graphs the loss metrics against the range of epochs. The validation loss curve shows the model's performance on unseen validation data, providing insights into its generalization capacity, whereas the training loss curve displays the loss on the training dataset, demonstrating how well the model matches the training data. The training and validation loss lines are identified by the legend in the top right corner. Making educated judgments during model optimization is made easier with the help of this visualization, which is crucial for tracking the convergence of the model and evaluating its generalization performance (Fig. 6).

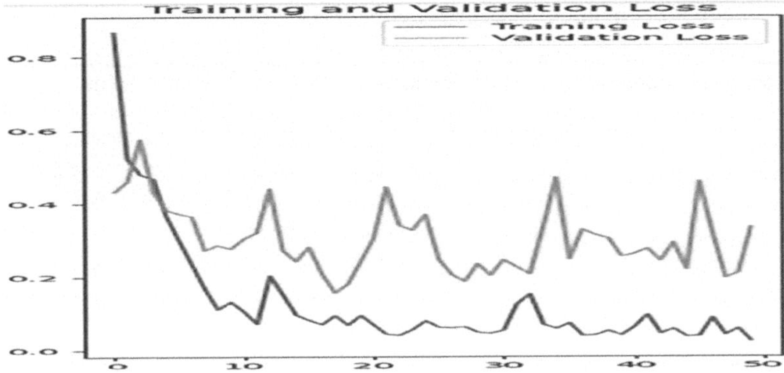

Fig. 6. Training and Validation Loss Graph

The output picture will have the confidence score and a predicted label indicating whether the potato is healthy or unhealthy. My model goes beyond this binary categorization and can also predict different kinds of potatoes. The differentiation of whole potatoes, half potatoes, mid-slice potatoes, and groups of potatoes is part of this expansion. Through the integration of this diverse methodology, the model offers an allencompassing evaluation of potato health conditions and arrangements, equipping farming stakeholders with sophisticated perspectives for efficient crop management. This adaptability improves the model's usefulness in real-world applications by supporting a wide range of scenarios found in agricultural settings and assisting in more informed decision-making. The evaluation phase yielded a loss of 0.3713 and an accuracy of 93.75% in classifying the condition of the potatoes (Fig. 7).

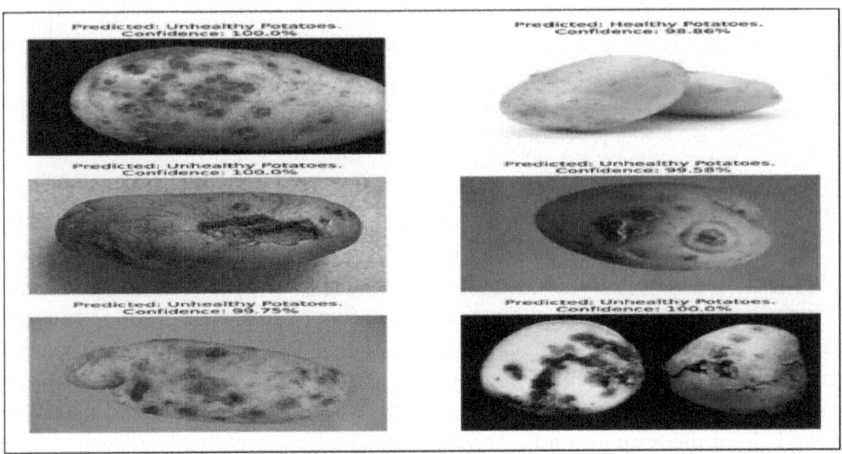

Fig. 7. Test Images of Potatoes with Performance

6 Conclusion

After evaluating the project's results, it is clear that the model that was created can accurately determine the state of health of potatoes and distinguish between different potato configurations. This all-inclusive method helps distinguish harmful potatoes and offers information on several potato varieties, including whole, half-slice, midslice, and clusters of potatoes. In order to help agricultural decision-makers optimize crop management practices and guarantee the health of potato crops, these detailed predictions—when paired with confidence scores—offer insightful information. There is room for future project development by investigating features and functions as the capacity to monitor in real-time or the ability to apply the model to other crop kinds. We can continue to meet the changing demands of by moving this project forward.

References

1. Tarik, M.I., et al.: Potato disease detection using machine learning. In: 2021 Third International Conference on Intelligent Communication Technologies and Virtual Mobile Networks (ICICV) (2021)
2. Kadam, Sandeep U., et al.: Machine learning method for automatic potato disease detection. Neuro Quantology **20**(16), 2102–2106 (2022)
3. R, P., et al.: Potato plant disease detection using convolution neural network. Int. J. Curr. Res. Rev. **12**(20), 152–156 (2020). https://doi.org/10.31782/ijcrr.2020.122028
4. Asif, M.K.R., Rahman, M.A., Hena, M.H.: CNN based disease detection approach on potato leaves. In: 2020 3rd International Conference on Intelligent Sustainable Systems (ICISS), pp. 428–432. Thoothukudi, India (2020)
5. Iqbal, M.A., Talukder, K.H.: Detection of potato disease using image segmentation and machine learning. In: 2020 International Conference on Wireless Communications Signal Processing and Networking (WiSPNET) (2020)
6. Arshaghi, A., Ashourian, M., Ghabeli, L.: Potato diseases detection and classification using deep learning methods. Multimedia Tools Appl. **82**(4), 5725–5742 (2022). https://doi.org/10.1007/s11042-022-13390-1
7. Hassoon, I.M., Qassir, S.A., Riyadh, M.: PDCNN: framework for potato diseases classification based on feed foreword neural network. Baghdad Sci. J. **18**(2(Suppl.)), 1012 (2021)
8. Pai, P., et al.: Plant disease detection using CNN – a review. J. Comput. Natl. Sci. 46–54 (2022). https://doi.org/10.53759/181x/jcns202202008
9. Lathamaheswari, U., Jebathangam, J.: A survey on plant leaf disease detection using image processing. In: Proceedings of International Conference on Data Science and Applications, pp. 807–813 (2023). https://doi.org/10.1007/978-981-19-6634-7_57
10. Feng, J., et al.: Research and validation of potato late blight detection method based on deep learning. Agronomy **13**(6), 1659 (2023). https://doi.org/10.3390/agronomy13061659
11. Joshi, K., et al.: Plant Leaf disease detection using computer vision techniques and machine learning. In: ITM Web of Conferences, vol. 44, p. 03002 (2022)
12. Lee, T.Y., Yu, J.Y., Chang, Y.C., Yang, J.M.: Health detection for potato leaf with convolutional neural network. In: 2020 Indo – Taiwan 2nd International Conference on Computing, Analytics and Networks (Indo-Taiwan ICAN), pp. 289–293. Rajpura, India (2020)
13. Al-Adhaileh, M.H., et al.: Potato blight detection using fine-tuned CNN architecture. Mathematics **11**(6), 1516 (2023)
14. Aksenov, A.G., et al.: Using a neural network to identify diseased potato plants. Agrarian Sci. **1**(7–8), 167–171 (2022). https://doi.org/10.32634/0869-8155-2022-361-7-8167-171

15. Suarez Baron, M.J., Gomez, A.L., Diaz, J.E.: Supervised learning-based image classification for the detection of late blight in potato crops. Appl. Sci. **12**(18), 9371 (2022). https://doi.org/10.3390/app12189371
16. Li, X., et al.: The detection method of potato foliage diseases in complex background based on instance segmentation and semantic segmentation. Front. Plant Sci. **13** (2022). https://doi.org/10.3389/fpls.2022.899754
17. Gardie, B., et al.: Potato plant leaf diseases identification using transfer learning. Indian J. Sci. Technol. **15**(4), 158–165 (2022). https://doi.org/10.17485/ijst/v15i4.1235
18. Qi, C., et al.: In-field early disease recognition of potato late blight based on deep learning and proximal hyperspectral imaging. SSRN Electr. J. [Preprint] (2022a). https://doi.org/10.2139/ssrn.4037959
19. Arnaud, S.E., et al.: Comparison of deep learning architectures for late blight and early blight disease detection on potatoes. Open J. Appl. Sci. **12**(05), 723–743 (2022). https://doi.org/10.4236/ojapps.2022.125049
20. Iqbal, M.A., Talukder, K.H.: Detection of potato disease using image segmentation and machine learning. In: 2020 International Conference on Wireless Communications Signal Processing and Networking (WiSPNET), pp. 43–47. Chennai, India (2020). https://doi.org/10.1109/WiSPNET48689.2020.9198563

Early Warning Prediction System for Agriculture Using Deep Learning

K. Indumathi(✉), S. Preshika, and S. P. Sri Vishal

Rajalakshmi Engineering College, Chennai, India
{indumathi.k,201001076,201001107}@rajalakshmi.edu.in

Abstract. This Paper presents an innovative image-based early warning prediction method for crop disease detection. Utilizing sophisticated machine learning algorithms, the system makes use of a large dataset of plant photos that spans different disease stages and types. A convolutional neural network (CNN) and Recurrent Neural Network (RNN), at the center of the system, has undergone extensive training to accurately recognize and categorize disease symptoms. Farmers and other agricultural stakeholders can receive fast alerts and actionable insights by integrating this model with real-time picture capturing technologies. This allows for timely interventions aimed at reducing crop losses. Extensive field testing were conducted to validate the system's performance, which demonstrated its robustness and reliability in a variety of agricultural situations. By reducing the impact of plant diseases, the application of this technology promises to improve crop management techniques and provide food security and sustainability.

Keywords: Convolutional Neural Network · Recurrent Neural Network · Plant Disease Detection · Early Warning Prediction · Agriculture · Deep Learning

1 Introduction

Numerous plant diseases that jeopardize crop quality and productivity present serious challenges to agriculture, the foundation of human subsistence. Ensuring food security and sustainability requires the prompt and precise identification of these illnesses. Conventional disease detection techniques frequently rely on labor-intensive, specialized laboratory testing and manual inspection. But technological developments have made it possible to create solutions that are more scalable and effective. The early warning prediction system using image-based crop disease detection is one example of such innovation.

The method combines use of cutting-edge machine learning and image processing methods to detect and forecast crop diseases in their early stages. Through crop image analysis, the system can identify illness indicators that are not readily visible to the human eye. Farmers are able to minimize crop losses and slow the spread of disease thanks to this proactive approach.

Precision farming has advanced significantly with the use of image-based detecting technologies in farming operations. These devices process and analyze plant photos

P. D. Sivakumar et al. (Eds.): IRCCTSD 2024, CCIS 2360, pp. 71–80, 2025.
https://doi.org/10.1007/978-3-031-82389-3_6

using high-resolution cameras and advanced algorithms. The algorithm is then able to determine the precise type and severity of the condition by comparing the collected data with a large database of disease symptoms.

In summary, an innovative technique in contemporary agriculture is the early warning prediction system for crop disease identification utilizing images. In addition to improving disease control, it raises farming strategies' broad productivity and efficiency. This technique is expected to be crucial in safeguarding the future of world food production as it develops further.

2 Related Works

Although Normalised Difference Vegetation Index (NDVI) data [1] derived from the advanced very high resolution radiometer (AVHRR) sensor have been extensively used to assess crop condition and yield on the Canadian Prairies and elsewhere, NDVI data derived from the new moderate resolution imaging spectro radiometer (MODIS) sensor have so far not been used for crop yield prediction on the Canadian Prairies. Therefore, the objective of this study was to evaluate the possibility of using MODIS-NDVI to forecast crop yield on the Canadian Prairies and also to identify the best time for making a reliable crop yield forecast. Growing season (May–August) MODIS 10-day composite NDVI data for the years 2000–2006 were obtained from the Canada Centre for Remote Sensing (CCRS). Crop yield data (i.e., barley, canola, field peas and spring wheat) for each Census Agricultural Region (CAR) were obtained from Statistics Canada. Correlation and regression analyses were performed using 10-day composite NDVI and running average NDVI for 2, 3 and 4 dekads with the highest correlation coefficients (r) as the independent variables and crop grain yield as the dependent variable. To test the robustness and the ability of the generated regression models to forecast crops grain yield, one year at a time was removed and new regression models were developed, which were then used to predict the grain yield for the missing year. Results showed that MODIS-NDVI data can be used effectively to predict crop yield on the Canadian Prairies.

Wheat is one of the key cereal crops grown worldwide [2], providing the primary caloric and nutritional source for millions of people around the world. In order to ensure food security and sound, actionable mitigation strategies and policies for management of food shortages, timely and accurate estimates of global crop production are essential. This study combines a new BRDF-corrected, daily surface reflectance dataset developed from NASA's Moderate resolution Imaging Spectro-radiometer (MODIS) with detailed official crop statistics to develop an empirical, generalised approach to forecast wheat yields. The first step of this study was to develop and evaluate a regression-based model for forecasting winter wheat production in Kansas. This regression-based model was then directly applied to forecast winter wheat production in Ukraine.

Data from NASA's Moderate Resolution Imaging Spectro-radiometer (MODIS) [3] in association with county-level data from the United States Department of Agriculture (USDA) to develop empirical models predicting maize and soybean yield in the Central United States. As part of our analysis we also tested the ability of MODIS to capture inter-annual variability in yields. Our results show that the MODIS two-band

EnhancedVegetation Index (EVI2) provides a better basis for predicting maize yields relative to the widely used Normalised Difference Vegetation Index (NDVI). Inclusion of information related to crop phenology derived from MODIS significantly improved model performance within and across years. Surprisingly, using moderate spatial resolution data from the MODIS Land Cover Type product to identify agricultural areas did not degrade model results relative to using higher-spatial resolution crop-type maps developed by the USDA. Correlations between vegetation indices and yield were highest 65–75 days after greenup for maize and 80 days after greenup for soybeans. EVI2 was the best index for predicting maize yield in non-semi-arid counties ($R2 = 0.67$), but the Normalised Difference Water Index (NDWI) performed better in semi-arid counties ($R2 = 0.69$), probably because the NDWI is sensitive to irrigation in semi-arid areas with low-density agriculture.

Among worldwide [4], agriculture has the major responsibility for improving the economic contribution of the nation. However, still the most agricultural fields are under developed due to the lack of deployment of ecosystem control technologies. Due to such issue 6, the crop production is not improved which affects the agriculture economy. Hence in this paper, a development of agricultural productivity is enhanced based on the plant yield prediction. Initially, different features such as plant images, soil characteristics, and weather factors are gathered and Firefly (FF) optimisation algorithm is proposed for Feature Selection (FFFS). Then, the most selected optimal features are classified based on the Modified Fuzzy Cognitive Map (MFCM) algorithm for predicting the growth of plant yield. The predicted outcome is transmitted to the farmer's through smart phones which helps for identifying the growth of plant and improving the harvesting. The experimental results show that the effectiveness of the proposed technique can be compared with the other prediction techniques.

Monitoring crop condition and production [5] estimates at the state and county level is of great interest to the U.S. Department of Agriculture. The National Agricultural Statistical Service (NASS) of the U.S. Department of Agriculture conducts field interviews with sampled farm operators and obtains crop cuttings to make crop yield estimates at regional and state levels. NASS needs supplemental spatial data that provides timely information on crop condition and potential yields. In this research, the crop model EPIC (Erosion Productivity Impact Calculator) was adapted for simulations at regional scales. Satellite remotely sensed data provide a real-time assessment of the magnitude and variation of crop condition parameters, and this study investigates the use of these parameters as an input to a crop growth model. This investigation was conducted in the semi-arid region of North Dakota in the southeastern part of the state.

3 Methodology

The development of an early warning prediction system for crop disease detection using images involves several stages, each requiring careful planning and execution. Figure 1 refers the methodology can be divided into the following key steps:

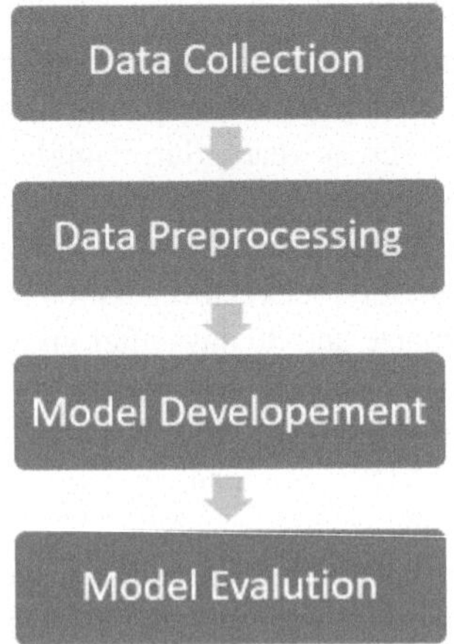

Fig. 1. Steps involved in Methodology

3.1 Data Collection

The investigation employed data from an openly accessible Apple, Cherry and Grapes Leaves dataset that had been identified via the internet. Table 1 shows the quantitative values of plant Leaf Disease Dataset Scab, Black Rot, Rust, and Healthy Apple Images make up the Apple Images Dataset. Powdery mildew and healthy cherry photos make up the Cherry Image Dataset. Images of healthy grapes, leaf blight, and black measles are included in the grapes image dataset.

Table 1. Plant Leaf Disease Dataset

Categories	Apple Images	Cherry Images	Grapes Images
Quantitative	Scab - 2016 Black Rot - 1987 Rust- 1760 Healthy apple - 2008	Powdery mildew - 1683 Healthy cherry – 1826	Black Measles –1888 Leaf Blight – 1920 Healthy grapes-1692

3.2 Data Preprocessing

The Plain Background Dataset (Plain BG-Dataset) is the plant disease dataset that we obtained by removing the complicated background from the plant disease picture dataset. The pixels in the images varied in size, ranging from 256 × 256 to 300 × 300. We reduced the size of each image to 224 × 224 pixels to guarantee uniformity in the picture sizes for network input. Figure 2 shows sample Apple leaf images. Figure 3 shows the sample Cherry leaf images. Figure 4 shows sample Grapes leaf images.

(a) (b) (c) (d)

Fig. 2. Apple Plain Background Dataset Images (a) Scab (b) Black Rot (c) Rust (d) Healthy apple

(a) (b)

Fig. 3. Cherry Plain Background Dataset Images (a) Powdery mildew (b) Healthy cherry

(a) (b) (c)

Fig. 4. Grapes Plain Background Dataset Images (a) Black Measles (b) Leaf Blight (c) Healthy grapes

4 Proposed Method

The steps in the proposed system are to resize images to 224 × 224, perform zoom and horizontal flip operations, and augmentation. The CNN and RNN Architecture are the deep learning models investigated for the classification. The suggested approach uses deep learning and transfer learning to reuse deep neural network models that have been trained on ImageNet data. Adding more layers to the output of pretrained models allows for fine-tuning the recommended mode for classification to perform as well as possible. One global spatial average pooling layer, two dropout layers, a batch normalization layer, two dense layers, and two flatten layers are added during the process of fine tuning. To categorize the images into three groups, an output layer with a SoftMax classifier and the ReLU activation function are applied. The Kaggle Plant Leaf Disease Images Database is the dataset that was used.

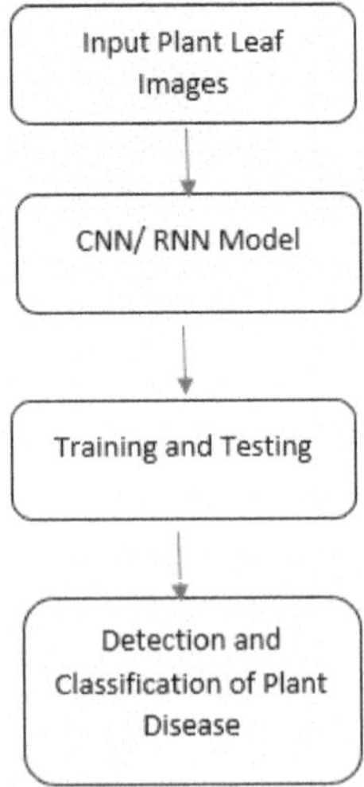

Fig. 5. Proposed System

Figure 5 shows steps involved in proposed system. Another kind of neural network that is particularly good at processing sequential input is the recurrent neural network (RNN). Convolutional Neural Networks (CNNs), on the other hand, are usually better appropriate for image-related tasks like plant disease classification because of their capacity to record spatial hierarchies in images. However, if you must use RNNs, you may do so in combination with CNNs to process picture temporal sequences, for example. Layers are added, then they are flattened. To categorize the images into three groups, an output layer with a softmax classifier and the Relu activation function are applied. The Kaggle Plant Leaf Disease Images Database is the dataset that was used.

5 Results Discussion

For the identification of plant diseases, Convolutional Neural Networks (CNNs) are frequently more accurate than Recurrent Neural Networks (RNNs) for a number of important reasons pertaining to the characteristics of the data and the architecture of these models.

5.1 The Fit of Imagery Collection

CNNs are made especially to handle data that looks like a grid, like pictures. To recognize elements like edges, textures, forms, and other patterns in the input image, they employ convolutional layers, which apply filters to the image. Plant disease identification relies heavily on tasks like recognizing items, picture classification, and image segmentation, all of which depend on these attributes. RNNs, on the other hand, are made to handle sequential data, like text or time series. They are not naturally suited to handle the spatial hierarchies found in image data, despite their exceptional ability to capture temporal connections.

5.2 Feature Identification and Space Organization

Convolutional layers are used by CNNs to automatically learn feature spatial hierarchies. While deeper layers detect more complicated patterns (such textures and shapes particular to diseases), earlier levels may only detect simple elements (like edges). Plant disease identification and classification accuracy depend heavily on this hierarchical learning. There is no built-in mechanism in RNNs to take use of the spatial structure of images. Although pre-processing techniques (such as flattening images into sequences) can be employed to use them with image data, this method usually results in the loss of spatial correlations and properties that CNNs naturally keep.

5.3 Identical Properties and Cognitive Performance

When compared to fully connected layers, CNNs use local connections and parameter sharing, which drastically reduces the number of parameters and computational complexity. Because of this, CNNs perform better on jobs like plant disease detection by processing massive image datasets more effectively and efficiently. RNNs may be less effective for jobs involving images since they have more complicated parameter structures designed to handle sequential input, especially those containing LSTM or GRU units.

5.4 Diffusion of Training

Transfer learning is very advantageous to CNNs. For certain tasks like plant disease identification, CNN models that have already been trained on huge picture datasets (such as ImageNet) can be improved. This method makes use of learnt features that can be applied to a wide range of image types and tailors them to the unique characteristics of plant disease imaging. Transfer learning is less useful with RNNs because they lack a pre-trained model repertoire for picture data that is as broad.

5.5 Realistic Excellence

CNNs routinely beat RNNs in image classification tasks, such as the identification of plant diseases, as demonstrated by a number of research and real-world applications. The performance disparity is frequently caused by the previously stated factors, as CNN architecture is better adapted to capturing and exploiting the spatial characteristics of images.

5.6 Composite Strategies

In some cases, hybrid approaches—which blend CNNs and RNNs—can be utilized in place of CNNs, which are typically the favoured method for image-based applications. CNNs are useful for extracting spatial features from images, but RNNs are more suitable for handling image sequences (such time-series of plant growth) or incorporating supplementary temporal data. The training and performance metrics of CNN and RNN are presented in the following Table 2.

Table 2. Training and Performance metrics of CNN and RNN

Model/Measures	CNN	RNN
Train Accuracy	98.29%	92.98%
Test Accuracy	96.77%	91.27%
Precision Score	96.77%	91.27%
Recall Score	96.77%	91.27%

In final analysis, since CNNs' architecture is better suited for image data and enables them to efficiently learn and exploit spatial characteristics essential for diagnosing diseases, CNNs are generally more accurate than RNNs for plant disease identification. Proactive disease control and plant security are greatly enhanced by the use of neural networks, especially Convolutional Neural Networks (CNNs), in early warning prediction systems for agriculture. Early symptoms of plant diseases can be accurately recognized before they spread by applying CNN-based models to real-time visual data obtained from field cameras, drones, or satellites. These models scan the photos for certain illness indicators, which are frequently imperceptible to the unaided eye in their early stages but include yellowing, spots, and wilting. Through assistance of a comprehensive strategy, farmers may use digital platforms to acquire timely alerts and useful data. This helps them to take early intervention steps like applying pesticides selectively to afflicted areas, imposing quarantines on impacted areas, or modifying irrigation schedules. Consequently, by ensuring that diseases are treated quickly and efficiently, these early warning systems assist limit crop loss, lessen the need for broad-spectrum chemical treatments, and encourage sustainable farming practices.

6 Conclusion and Future Work

Convolutional neural networks (CNNs) have shown better accuracy and performance than recurrent neural networks (RNNs) in the field of plant disease identification. CNNs have the advantage of being able to process the spatial hierarchies and features that are present in picture data by default, which makes them very useful for tasks like object detection and image categorization. The efficiency of CNNs in this domain is further reinforced by their significant use of transfer learning using pre-trained models, parameter efficiency, and hierarchical feature learning. However, while powerful for temporal and sequential data, RNNs do not have the architectural advantages that CNNs have for image-based tasks. In order to improve precision, effectiveness, and practicality, future research in neural network-based plant disease detection can concentrate on a number of interesting areas. Investigating cutting-edge CNN designs like DenseNet, EfficientNet, and Vision Transformers (ViTs), which can provide better performance and computational efficiency, is one important avenue. By strengthening their concentration on crucial areas of the image, CNNs that incorporate attention processes may be able to be further optimized. Incorporating other data types, such as soil or climatic data.

References

1. Mkhabela, M.S., Bullock's, P.: Work in 2011 specifically targets the Canadian Prairies, aiming to evaluate the potential use of MODIS-NDVI data for crop yield prediction. The study demonstrates the effective utilization of MODIS-NDVI in forecasting crop yield, offering insights into the reliability and optimal timing for accurate predictions in the region (2011)
2. The study by Becker-Reshef, E. Vermote in 2010 emphasizes the importance of accurate global crop production estimates for ensuring food security. Utilizing a BRDF-corrected, daily surface reflectance dataset from NASA's MODIS, the researchers develop an empirical approach to forecast wheat yields, showcasing the applicability of remote sensing data in global crop monitoring
3. In 2015, Bolton, D.K., Friedl, M.A employ MODIS data in conjunction with USDA county-level data to develop empirical models predicting maize and soybean yield in the Central United States. The study emphasizes the significance of MODIS two-band Enhanced Vegetation Index (EVI2) for predicting maize yields, showcasing the potential of satellite-derived data for crop yield assessments
4. Sabareeswaran and Gunasundari's work introduces a unique approach, combining Firefly optimization algorithm for feature selection and Modified Fuzzy Cognitive Maps for plant yield prediction. The study highlights the integration of multiple factors, including plant images, soil characteristics, and weather data, for accurate plant yield prediction, ultimately aiming to enhance agricultural productivity (2018)
5. Doraiswamy, Moulin, and Cook's research emphasizes the need for timely spatial data for crop yield estimates. Using the EPIC model and satellite remotely sensed data, the study evaluates the integration of satellite-derived parameters into a crop growth model, focusing on the semi-arid region of North Dakota. The work showcases the potential for remote sensing to provide real-time assessments and spatially explicit crop yield predictions at local and regional scales (2022)
6. De Souza Zanirato Maia, J., Bueno, A.P.A., Sato, J.R.: applications of artificial intelligence models in educational analytics and decision making: a systematic review. World **4**, 288–313 (2023)
7. Nabizadeh, A.H., Gonçalves, D., Gama, S., Jorge, J.: Early prediction of students final grades in a gamified course. IEEE Trans. Learn. Technol. **15**, 1 (2022)
8. Albreiki, B., Zaki, N., Alashwal, H.: A systematic literature review of student' performance prediction using machine learning techniques. Educ. Sci. **11**(9), 552 (2021)
9. Dhilipan, J., Vijayalakshmi, N., Suriya, S., Christopher, A.: Prediction of students performance using machine learning. In: IOP Conference Series: Materials Science and Engineering, vol. 1055, p. 012122 (2021)
10. Kanetaki, Z., Stergiou, C., Bekas, G., Troussas, C., Sgouropoulou, C.: The impact of different learning approaches based on MS Teams and Moodle on students' performance in an online mechanical CAD module. Glob. J. Eng. Educ. (2021)
11. Khan, I., Ahmad, A.R., Jabeur, N., Mahdi, M.N.: An artificial intelligence approach to monitor student performance and devise preventive measures. Smart Learn. Environ. (2021)
12. Sravani, B., Bala, M.M.: Prediction of student performance using linear regression. In: 2020 International Conference for Emerging Technology (INCET), pp. 1–5. Belgaum, India (2020)
13. Altabrawee, H., Ali, O., Qaisar, S.: Predicting students' performance using machine learning techniques. J. Univ. Babylon Pure Appl. Sci. **27**, 194–205 (2019)
14. Iqbal, Z., Qayyum, A., Latif, S., Qadir, J.: Early student grade prediction: an empirical study, pp. 1- 7 (2019). https://doi.org/10.23919/ICACS.2019.8689136
15. Agrawal, H., Mavani, H.: Student performance prediction using machine learning. Int. J. Eng. Res. Technol. (IJERT) **04**(03), 2015 (2015)

16. Liang, P., Shi, W., Zhang, X.: Remote sensing image classification based on stacked denoising autoencoder. Remote Sens. **10**(1), 16 (2018)
17. Goodfellow, Bengio, Y., Courville, A.: Deep Learn. Cambridge. MIT Press. MA, USA (2016). http://www.deeplearningbook.org
18. Pan, S.J., Yang, Q.: A survey on transfer learning. IEEE Trans. Knowl. Data Eng. **22**(10), 1345–1359 (2010)
19. Hao, P., Di, L., Zhang, C., Guo, L.: Transfer learning for crop classification with cropland data layer data (CDL) as training samples. Sci. Total Environ. **733**, 138869 (2020)
20. Tao, C., Qi, J., Lu, W., Wang, H., Li, H.: Remote sensing image scene classification with self-supervised paradigm under limited labeled samples. IEEE Geosci. Remote Sens. Lett. **19**, 8004005 (2022)
21. Ayush, K., et al.: Geography-aware self-supervised learning. In: Proceedings IEEE/CVF International Conference Computer Vision, pp. 10181–10190 (2021)
22. Ghosh, R., Jia, X., Kumar, V.: Land cover mapping in limited labels scenario: A survey (2021). arXiv:2103.02429
23. Yosinski, J., Clune, J., Bengio, Y., Lipson, H.: How trans ferable are features in deep neural networks ?. In: Proceedings of 27th International Conference Neural Information Processing System, pp. 3320–3328. Cambridge, MA, USA (2014)
24. Huynh, B.Q., Li, H., Giger, M.L.: Digital mammographic tumor classification using transfer learning from deep convolutional neural networks. J. Med. Imag. **3**(3), 034501 (2016)
25. Song, J., Gao, S., Zhu, Y., Ma, C.: A survey of remote sensing image classification based on CNNs. Big Earth Data **3**(3), 232–254 (2019)

A Comparison of Different Deep Learning Models for Plant Leaf Disease Detection

P. Maragathavalli(✉) and S. Jana

Vel Tech Rangarajan Dr. Sagunthala R&D Institute of Science and Technology, Avadi, Chennai, India
drmaragathavallip@gmail.com

Abstract. Plant disease is seen as a significant danger to crop productivity in agricultural regions. Plant diseases are often divided into three categories: fungal, viral, and bacterial. Crops can be protected against disease severity by early identification and detection. Farmers used to have a very hard time diagnosing illness with their unaided eyes. A class of deep learning algorithms that automatically learns and extracts features for plant disease identification and detection, including convolutional neural networks. The plant village data set that was used in this instance is openly accessible. There are 20638 leaf images in the data collection, organized into 15 classes. That is, training accounted for 80% of the data set, whereas testing accounted for 20%. The model's performance is enhanced and the loss function is minimized by using Adam optimizers. Various pre-trained models, including ResNet 101, VGG16, and Mobile Net V2, were compared. It is clear from the experiment's results that ResNet 101 can classify objects with up to 97.23% accuracy while requiring the fewest number of computations.

Keywords: Adam Optimizer · CNN · Deep Learning · Plant Disease · Plant Village

1 Introduction

Plant diseases are conditions or disorders that affect the normal growth and development of plants, leading to reduced crop yields, poor-quality produce, and sometimes even plant death. Plant diseases can have a significant impact on agriculture, leading to economic losses and food security problems [1]. Crop rotation, the use of disease-resistant plant varieties, chemical treatments (fungicides, bactericides, and pesticides), and cultural practices such as sanitation and appropriate irrigation management are examples of effective disease management techniques. To control plant diseases and reduce their impact and the need for chemical interventions that can harm the environment, early diagnosis and prevention are essential [3]. Pathogens (such as fungi, bacteria, viruses, and nematodes), environmental stressors, and abiotic factors are some of the causes of these diseases [4]. Here are some key types of plant disease (Fig. 1):

Fungal Diseases: Fungi are a common cause of plant diseases. Examples include powdery mildew, rust, and various types of blight. Fungi often infect plant tissues, causing symptoms such as leaf spots, wilting, and cankers [18].

P. D. Sivakumar et al. (Eds.): IRCCTSD 2024, CCIS 2360, pp. 81–92, 2025.
https://doi.org/10.1007/978-3-031-82389-3_7

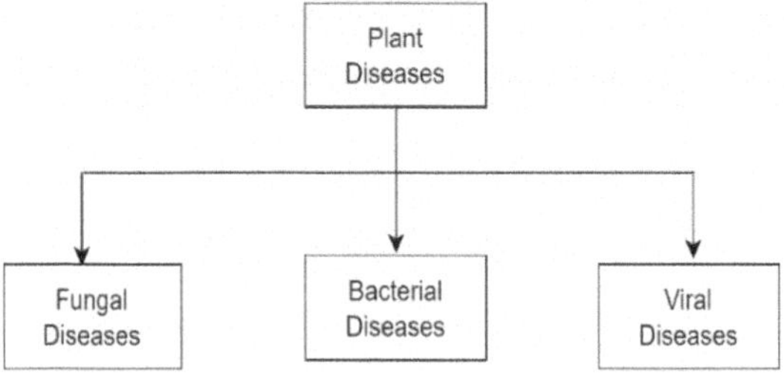

Fig. 1. Types of plant diseases

Bacterial Diseases: Bacteria can infect plants, leading to diseases such as bacterial leaf spot, crown gall, and fire blight. Symptoms often include wilting, leaf lesions, and rotting [18].

Viral Diseases: Viruses are microscopic pathogens that can cause various symptoms in plants, including yellowing of leaves, stunted growth, and mottling. The tomato mosaic virus and the cucumber mosaic virus are examples of plant viruses [18].

Deep learning-based plant disease detection approaches are as follows [13].

- Pre-processing
- Data augmentation
- Feature extraction
- Classification

To improve the quality of the image, noise, and other artifacts are first removed from the obtained raw image through preprocessing. Once the image is preprocessed it is fed into the data augmentation to increase the dataset. In the next stage features such as size, color, and shape were extracted. Ultimately, these characteristics are fed into the detection stage, which uses a CNN (deep learning model) to identify and categorize diseases [14]. Furthermore, the hyperparameters adjust their weights to improve the accuracy of classification. The learning rate, the epoch count, the batch size, and the activation function are examples of hyperparameters (Fig. 2).

1.1 Pre-processing

Image data preprocessing is performed to enhance the quality of the dataset. To enhance the model's performance, this may entail resizing images to a uniform resolution, normalizing pixel values, and expanding the dataset with different image variations (such as flipped or rotated images).

1.2 Data Augmentation

To improve the performance of the CNN model, a large amount of training data is required. In this scenario, training data are limited, so it becomes necessary to employ

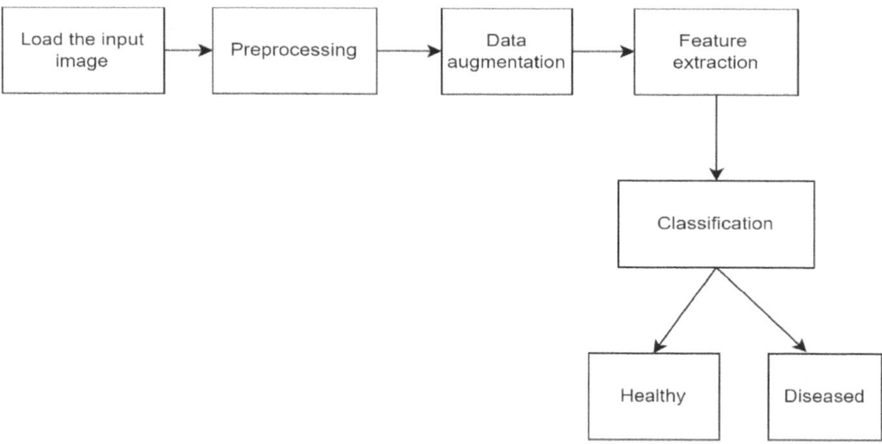

Fig. 2. Plant Disease Detection System

techniques such as data augmentation to enhance model performance. Image augmentation involves generating additional images in the data set by applying various distortions and transformations. This helps reduce overfitting and improves the model's ability to generalize.

1.3 Feature Extraction

Convolutional neural networks (CNNs) are a class of deep learning algorithms that are useful for tasks such as visual symptom-based plant disease identification because they can automatically learn and extract features from images [15]. Features such as size, color, shape, etc. of an image.

1.4 Classification

Convolutional neural networks are so good at image recognition and classification that their application in the detection of plant diseases has received a lot of attention recently. A CNN architecture is chosen or designed for the particular task. Common CNN architectures for image classification tasks include AlexNet, VGG, ResNet, and Inception, among others. Multiple convolutional and pooling layers make up these architectures, which are followed by fully connected layers for classification.

2 Related Work

Convolution neural networks are the preferred choice for the detection of plant disease due to their automatic extraction features and classification accuracy [1]. The comparison of the various techniques used in the detection of plant diseases is shown in Table 1 (Table 2, Table 3, Table 4 and Table 5).

Table 1. Comparison of various diseases in the conventional method

Ref No	Authors	Techniques	Diseases	Classification Accuracy
[6]	Madhu Mini Mohapatra et al. (2022)	Convolution neural network with cat swarm optimization algorithm	Mango	91.2%
[7]	Satti R.G. Reddy et al. (2022)	Deep learning convolution neural network with red deer optimization algorithm	Rice	Plant village dataset-99.73% Rice plant dataset-99.68%
[8]	K. Lakshmi Narayanan et al. (2022)	Hybrid convolution neural networks	Banana	99%
[9]	Amritha Haridasan et al. (2022)	Convolution neural network with SVM classifier	Paddy	91.45%
[10]	Kemal Adem et al. (2022)	Yolov4 with image processing	Sugar beet	96.47%
[11]	L.G. Divyanth et al. (2022)	Two-stage deep learning-based segmentation	Corn	94.22% -UNet 73.79%-DeeplabV3 +
[12]	Hamoud Alshammari et al. (2022)	Genetic Algorithm	Olive	96%-multiclass classification 98%-binary classification
[16]	Amreen Abbas et al. (2021)	Conditional generative adversarial networks(C-GAN)	Tomatoes	99.51%- 5 classes 98.65%-7 no of class 97.11%-10 class
[17]	Sasikala Vallabhajosyula et al. (2021)	Deep ensemble neural network (DENN)	Plant leaf	100%
[20]	Qingmao Zeng et al. (2020)	Deep convolution generative adversarial networks (DG-GAN)	Citrus	92.60%

The proposed work reveals the research gap that there is a need for a fully automated system. Earlier identification of plant disease is important to avoid loss in horticultural crops. The severity estimation remains unsolved. With the help of deep CNN, it is possible to identify and classify diseases at a faster rate.

Table 2. Hyperparameters

Parameters	values
Image size	(224, 224, 3)
Batch size	32
Epochs	50
Learning rate	0.001
optimizer	Adam optimizer

Table 3. Comparison of various models Training& Validation data

Model	Training Accuracy	Validation Accuracy	Training Loss	Validation Loss
ResNet 101	97.23	78.80	08.37	86.02
Convolution neural network	95.24	95.05	14.30	15.36
Mobile Net V2	91.46	91.22	25.19	25.58
VGG 16	78.14	80.02	73.70	66.69

Table 4. Comparison of various models Test data

Model	Accuracy	Test Accuracy	Loss	Test Loss
ResNet 101	80.21	80.21	79.21	79.21
Convolution neural network	95.05	95.05	14.30	14.30
MobileNet V2	91.47	91.47	25.76	25.76
VGG 16	80.04	80.04	66.65	66.65

2.1 Image Data Collection

To train a CNN to detect plant illnesses, a significant number of plant images are gathered. Images of both healthy and diseased plants are frequently seen in this dataset. The labels on the images identify the exact illness the plant suffers from in addition to indicating whether it is healthy or not.

2.2 Dataset

To analyze the suggested model, a dataset of leaf diseases was obtained from Kaggle (Fig. 3).

Table 5. Classification Report

S No	Disease	Precision	Recall	F1-score
1	Pepper bell bacterial spot	0.97	0.97	0.97
2	pepper bell healthy	0.97	1.00	0.99
3	Potato Early Blight	1.00	0.95	0.97
4	Potato late Blight	0.97	0.93	0.95
5	Potato healthy	0.99	0.94	0.96
6	Tomato Bacterial spot	0.84	0.98	0.90
7	Tomato Early blight	0.83	0.98	0.90
8	Tomato late blight	0.91	0.95	0.93
9	Tomato leaf mold	0.86	0.98	0.90
10	Tomato septoria leaf spot	0.83	0.96	0.90
11	Tomato spider mites two- spotted spider mites	0.85	0.96	0.93
12	Tomato target spot	0.90	0.98	0.90
13	Tomato yellow leaf curl virus	0.91	0.95	0.93
14	Tomato mosaic virus	0.84	0.94	0.93
15	Tomato healthy	0.86	0.95	0.90

Fig. 3. Sample healthy and diseased leaf

2.3 Plant Village Dataset

The plant disease data set that was used is available to the public at (https://www.kaggle.com/datasets/vipoooool/new-plant-diseases-dataset). The data collection contains all samples of both healthy and sick leaf pictures. Data sets for testing and training are kept apart from the original data collection. The training dataset contains 80% (16516) of the original data set's samples, while the remaining 20% (4122) of the photos are utilized to assess how well the proposed model performs [2].

The following are the diseases, which are divided into 15 classes. Tomato early blight, tomato late blight, tomato bacterial spot, tomato leaf mold, tomato Septoria leaf spot, tomato spider mites, and pepper bell bacterial spot are all examples of ailments that can affect a tomato. The dataset includes tomato mosaic virus, tomato target spot, tomato yellow leaf curl virus, and two-spotted spider mites [5].

3 Methodology

The architecture of the CNN model includes:

- CNN using the Keras sequential API (application programming interface).
- The first convolution layer has 32 filters, each measuring 3 by 3. Additionally, it has an input shape of (224, 224, 3), indicating that the input images are 224 by 224 pixels with three RGB color channels and a Relu activation function (Fig. 4).

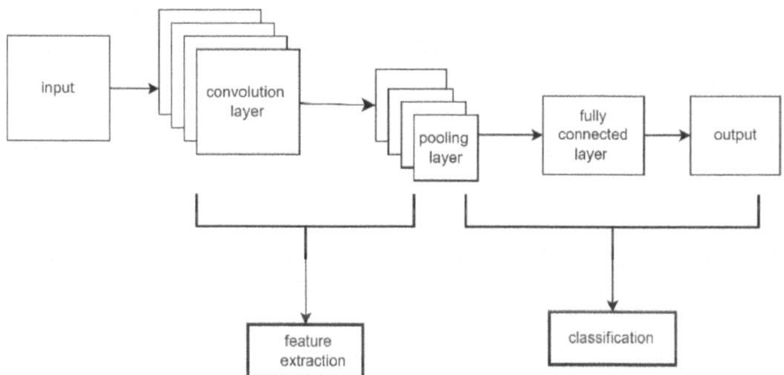

Fig. 4. Convolution neural networks

- The 2×2 pool size is reduced in spatial dimensions by the Max pooling layer.
- The second convolution uses Relu activation and 64 3×3 filters.
- Other Max Pooling Layers.
- Flatten layer: After being transferred to a densely connected layer, the input must be flattened and transformed into a 1D array by the flattening layer.
- Dense layer: 128 neurons in a layer that were deeply connected and activated Relu.
- The output layer, appropriate for multiclass classification problems, has 15 neurons and a SoftMax activation function.

Training: The labeled data set is used to train CNN through the use of Adam optimization techniques. The model gains the ability to identify patterns and characteristics in images that correspond to plant diseases during training.

Adam Optimizer: Neural networks in particular are trained using a popular optimization method called Adaptive Moment Estimation, or Adam. It is renowned for being reliable and effective. Adam blends aspects of Momentum and RMSprop (Root Mean Square Propagation), two other optimization techniques. With all factors considered, Adam has good convergence properties and performs well on a variety of machine-learning tasks. It has become a popular choice as an optimizer for training deep neural networks due to its effectiveness and ease of use. However, it is essential to tune the hyperparameters (e.g., learning rate) based on your specific problem and data set to achieve optimal performance [2] (Fig. 5).

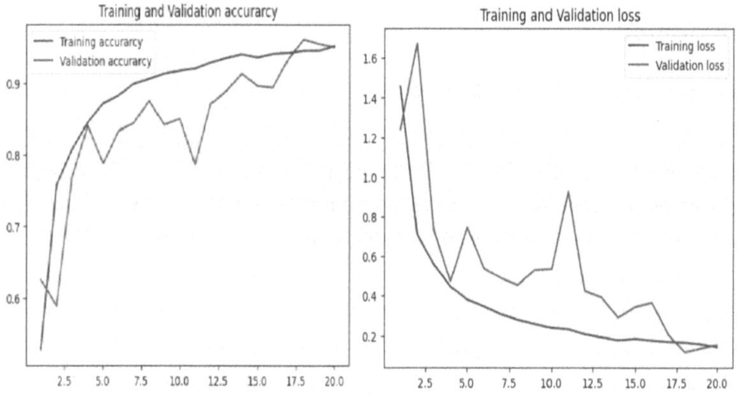

Fig. 5. Convolution Neural Networks

Validation and Testing: After training, CNN is validated on a separate data set to assess its model performance. The accuracy of the model is evaluated to measure its ability to correctly classify healthy and diseased plants. The quality and diversity of the training data, as well as ongoing maintenance and validation of the model to ensure its dependability in real-world scenarios, are crucial considerations for any CNN-based plant disease detection system.

Mobile Net V2: One of the deep learning models, Mobile Net V2, has 53 layers. The first fully convolution layer in the architecture has 32 filters, and it is followed by 19 residual bottleneck layers. Loading a network model that has already been pre- trained with over a million photos from the ImageNet database.

The classification accuracy using Mobile Net V2 is 91.46% compared to other deep learning models (Fig. 6).

VGG 16: A sequence of convolution layers, maximum-pooling layers, and fully linked layers for classification make up the 16 layers of the VGG 16 deep learning model (Fig. 7).

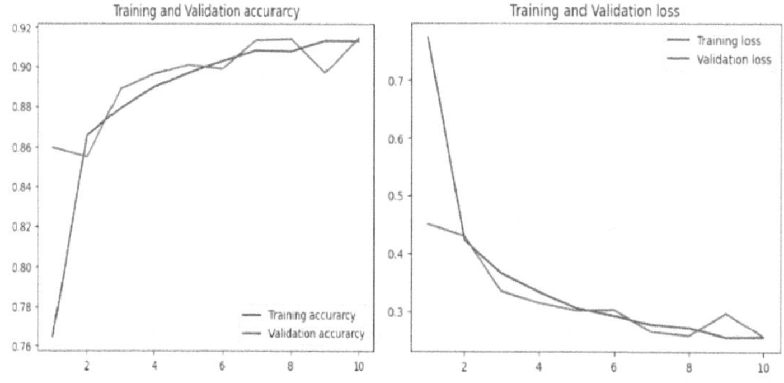

Fig. 6. Mobile Net V2

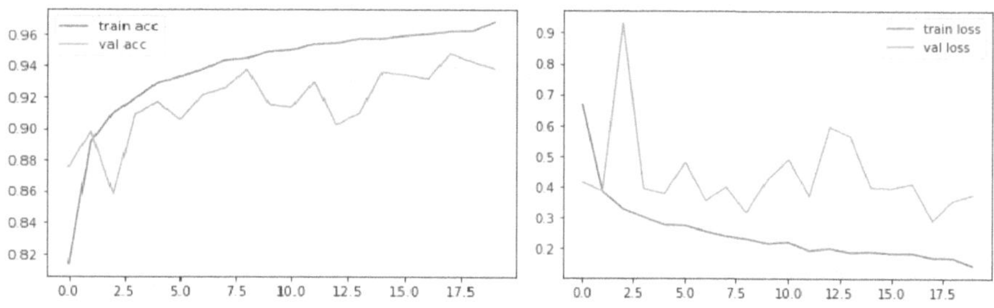

Fig. 7. VGG 16

The classification accuracy using VGG 16 of 78.14% as compared to other deep learning models achieves less performance.

4 Experiments and Results Discussion

Python 3.1 was used to simulate the automatic disease detection model. 80% of training data and 20% of testing data were utilized to test the model. The results are recorded after training the model. The commonly used metrics to assess the model performance during and after training are accuracy, F1-Score, precision, and recall. Performance evaluation was calculated based on precision and loss.

The confusion matrix is built to evaluate the model's effectiveness. The validation set's model performance and the extent to which the data is classified by the model are measured by accuracy. The performance metrics of the different models are calculated as follows.

$$\text{Accuracy} = \frac{\text{TP} + \text{TN}}{\text{TP} + \text{TN} + \text{FP} + \text{FN}} - \tag{1}$$

$$\text{Precision} = \frac{\text{TP}}{\text{TP} + \text{FP}} \tag{2}$$

$$\text{F1} - \text{score} = \frac{2 * \text{TP}}{2 * \text{TP} + \text{FP} + \text{FN}} \tag{3}$$

$$\text{Recall/Sensitivity} = \frac{\text{TP}}{\text{TP} + \text{FN}} \quad (4)$$

- True Positive (TP): The proposed model predicts true, while it is true.
- True Negative (TN): The proposed model predicts false, while it is false.
- False Positive (FP): The proposed model predicts true, while it is false.
- False Negative (FN): The proposed model predicts false, while it is true.

ResNet 101 Architecture: ResNet with 101 layers is used in deep neural networks to perform implementations faster. ResNet is always considered superior to CNN because it introduces the residual unit with identify mapping, allowing deep layers to learn directly from shallow layers, which improves network convergence and learning ability (Fig. 8).

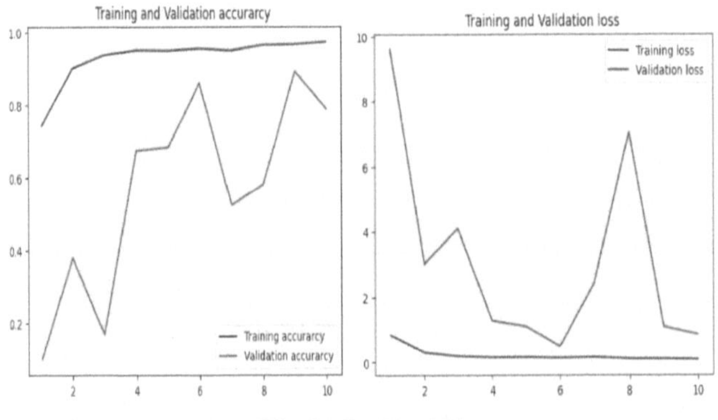

Fig. 8. ResNet 101

ResNet 101 achieves a maximum classification accuracy of 97.23% compared to CNN.

Training Accuracy: Another common metric for classification tasks is accuracy. Of all the examples, it calculates the percentage of correctly classified examples. It is expressed as a number between 0 and 1, where 1 denotes absolute precision.

The training accuracy is 0.9723, which means that the model correctly classifies approximately 97.23% of the training examples.

Validation Accuracy: This is the accuracy computed on the validation dataset. Similarly, to training accuracy, it calculates the percentage of correctly classified examples based on validation data.

The validation accuracy achieved is 0.7880 which means that the model correctly classifies approximately 78.80% of the validation examples.

Training Loss: Loss is a measure of the effectiveness of the model. It is a representation of the error or discrepancy between the target values and the predicted values. A smaller loss suggests that the model predictions and the actual values are relatively similar.

Categorical Cross-Entropy: Categorical Cross-Entropy, often referred to as "cross-entropy loss" or "log loss," is a popular loss function used in machine learning and deep learning, particularly in tasks involving classification. It is commonly used in the training of neural networks, especially in multiclass classification problems.

Categorical cross-entropy is particularly well suited for multiclass classification tasks and is widely used in various deep-learning applications, including image classification.

The training loss is 0.0837, which means that on average the model's prediction on the training data is very close to the actual values.

Validation Loss: The validation loss is the amount of loss that is found by utilizing an alternative validation dataset that the model was not exposed to during training. Its purpose is to assess the model's capacity to yield previously unknown data.

The model performs well with the validation data, as shown by the validation loss of 0.8602.

5 Conclusion and Future Work

The main cause of crop damage is plant disease. There are several benefits of using Convolution Neural Networks in the detection of plant diseases. Convolution Neural Networks can analyze large data sets of plant images quickly and accurately. They can identify subtle visual symptoms of diseases that may not be easily detectable by human observers. CNN-based systems can be deployed in the field or remote areas, helping farmers with timely disease detection and management. Continuous learning and model updates can improve the accuracy of the system over time. ResNet 101 outperforms all models and provides the best accuracy in detecting plant disease at an earlier stage. The learning rate and other hyperparameters can be automatically adjusted by the Adam optimizer, which improves the effectiveness and efficiency of the optimization process. In the future, the CNN and various pre-trained models can be used in real-life scenarios once it has been trained and validated. To detect plant diseases at an earlier stage, could involve developing an intuitive user interface or incorporating it into a fully automated system or a smartphone app.

References

1. Demilie, W.B.: Plant disease detection and classification techniques: a comparative study of the performances. J. Big Data (2024)
2. Pakutharivu, P., Sasirekha, D., Devaraj, V., Gopi, R.S.: Improving plant disease detection using super-resolution generative adversarial networks and enhanced dataset diversity (2024)
3. Kanna, G.P., et al.: Advanced deep learning techniques for early disease prediction in cauliflower plants. Sci. Rep. (2023)
4. Kayaalp, K.: Classification of medicinal plant leaves for types and diseases with hybrid deep learning methods, information technology and control (2024)
5. Bensaadi S, Louchene A. Low-cost convolutional neural network for tomato plant disease classification. IAES Int. J. Artif. Intell. (IJ-AI) (2023)
6. Mohapatra, M., Parida, A.K., Mallick, P.K., Zymbler, M., Kumar, S.: Botanical Leaf Disease Detection and Classification Using Convolutional Neural Network: A Hybrid Metaheuristic Enabled Approach, MDPI (2022)

7. Reddy, S.R., Varma, G.S., Davuluri, R.L.: Resnet-based modified red deer optimization with DLCNN classifier for plant disease identification and classification. Elsevier (2022)
8. Lakshmi Narayanan, K., et al.: Banana Plant Disease Classification Using Hybrid Convolutional Neural Network, Hindawi (2022)
9. Haridasan, A., Thomas, J., Raj, E.D.: Deep Learning System for Paddy Plant Disease Detection and Classification. Springer (2022)
10. Adem, K., Ozguven, M.M., Altas, Z.: A method of classification of sugar beet leaf diseases based on image processing and deep learning (2022)
11. Divyanth, L.G., Ahmad, A., Saraswat, D., et al.: A two-stage deep learning-based segmentation model for crop disease quantification based on corn field imagery. Elseiver (2022)
12. Alshammari, H., et al.: Optimal deep learning model for olive disease diagnosis based on an adaptive genetic algorithm. Hindawi (2022)
13. Mahum, R., et al.: A Novel Framework for Potato Leaf Disease Detection Using an Efficient Deep Learning Model. Taylor& Francis (2022)
14. Shoaib, M., et al.: Deep learning-based segmentation and classification of leaf images for detection of tomato plant disease. Front. Plant Sci. (2022)
15. Akbar, M., et al.: An effective deep learning approach for the classification of Bacteriosis in peach leave. Front. Plant Sci. (2022)
16. Abbas, A., Jain, S., Gour, M., Vankudothu, S.: Tomato plant disease detection using transfer learning with C-GAN synthetic images. Elsevier (2021)
17. Vallabhajosyula, S., Sistla, V., Kolli, V.K.: Transfer learning-based deep-ensemble neural network for plant leaf disease detection. Springer (2021)
18. Orchi, H., Sadik, M., Khaldoun, M.: On using artificial intelligence and the internet of things for crop disease detection: a contemporary survey. MDPI (2021)
19. Uğuz, S., Uysal, N.: Classification of olive leaf diseases using deep convolutional neural networks. Neural Comput. Appl. Springer (2020)
20. Zeng, Q., Ma, X., Cheng, B., Zhou, E., Pang, W.: GAN-based Data augmentation for citrus disease severity detection using deep learning. IEEE Access (2020)

Identifying Paddy Crop Disease Using Enhanced Deep Learning Technique

Ganapathy Subramanian(✉) and Neduncheliyan

Bharath Institute of Higher Education and Research, Chennai, India
ganapathymcame@gmail.com, dean.cse@bharathuniv.ac.in

Abstract. The consequences of extreme weather changes on crops, the growing global population, and numerous illnesses that emerge at a severe level all have an impact on agricultural production and food security. Farmers employ expensive disease management strategies that sometimes outweigh output losses. Because it would be too expensive for them to install crop protection measures at the early beginning of the illnesses, which results in very little advantages, the farmers would only act when it was too late to stop the disease from spreading. It was suggested that computer orders using technology for calculating vision may be beneficial in a range of original-globe situations. It may be used in crop security strategies since it employs artificial intelligence, graphics, and image processing to detect illnesses, pests, and malnutrition. Machine learning (ML) approaches are being utilised in the ongoing study of disease detection from plant photographs. For recognising paddy leaf disease, a unique deep neural network (DNN) classification model is employed using plant picture data. Existing system like CNN, TL and ANN gives the mere accuracy results. There is drawback in identifying the accurate crop diseases. The proposed hybrid ECNN and CSA algorithm give more accuracy when compared to the existing system. It increases the accuracy level by 95%. The hybrid ECNN & CSA based designs shown amazing success in crop disease prediction and categorization using images. These models require a lot of training data, though, and they are computationally costly. The proposed enhanced hybrid deep learning based hybrid algorithm model identifying the diseases in the paddy leaves in a most significant manner.

Keywords: Paddy · crop disease · ECNN · Crow search algorithm · identifying

1 Introduction

Agriculture is individual of the considerable beginnings of profit for bother many nations accompanying allure own administration. Farmers draw differing gist plants with the sympathetic the needs and the ordinary atmospheres of the locality. The origin towards efficiently and right discounting plant disorder in a difficult environment is the marking of the illness. Plant affliction disease bear numerical and dossier-impelled as a outcome deeply tumor of smart raise animals that opens up new potential for cultured end support, alert study, and clever readiness.

P. D. Sivakumar et al. (Eds.): IRCCTSD 2024, CCIS 2360, pp. 93–101, 2025.
https://doi.org/10.1007/978-3-031-82389-3_8

The photos of edible grain plant leaves utilized for the likeness growing process are seized straight from the farm meadow for the purposes of rounding up prevailing, bacterial blight, dark spot, top rot, and blast maladies. During pre-reconstruct, RGB photos are reconstructed into HSV ideas for aptitude extermination. Binary concepts are before fetched to divide the nasty and non-unsound portions settled color and fullness. A assembling creativeness is affected to divide the scenery, realistic portion, and nasty rule [2].

The main crop developed in India is edible grain (Oryzasativa). The best region of land secondhand for edible grain culture, containing two together dark and silvery edible grain, is in India. In adding together to bearing tasks, edible grain gardening considerably stabilizes the gross household commodity. In the period of space travel, labeling nausea or ailment from plant photos is a common district of research [20].

For ranchers, fundamental main fixation follow find afflictions in plants as immediately as achievable. Convolutional affecting animate nerve organs networks (CNNs) and various deep instruction methods, apart from new mathematical cameras and smartphones' supported picture capture styles, stop blame and grant pardon the exact finding of plant afflictions.

The land manufacturing need more up-to-date concerning details progresses to equal the climbing community. If you want a bigger yield, confirm your crop is athletic. Consequently, a breakneck and inexpensive approach is necessary for the labeling of land ailments. India is the planet's superior builder of paddy, a basic crop for heaps of population. Some ailments, induced by leaven, microorganisms, and viruses, influence paddy. His greatest question is that producer's forbiddance skill to accomplish the disease because they can't pinpoint it right. To humiliate crop afflictions and boost crop yield, skilled are a sort of attainable curing patterns, in the way that types of pesticides or insecticides. The best habit to control a condition, nevertheless, search out first label it, and before settle highest in rank situations for it, that includes the belief of masters or former information. Both opportunity and services are wanted for this.

A plant's leaves will display syndromes guide a affliction if the plant is displeased. It is attainable to label the ailment established these manifestations. Using these traits, we devised a machine intelligence model a contemporary mechanics progress that can label diseases in paddy crops and advise ultimate active herbicides and insecticides to control ruling class in a short amount momentary and at a cheap. We have met on only two field afflictions that are coarse and cause a excessive amount of damage to paddy crops. The incident of an adept and productive ailment control order is fault-finding for the security of bread freedom in some land scheme.

2 Related Work

To identify the wheat leaf diseases ailments, a light-burden reduced CNN architecture is bestowed in this place study. It involves eight layers total three convolutional tiers, three SoftMax tiers, and two leveled layers. Three human scholars carefully remarked the high-determination photographs that were captured from Pakistan's Azad Kashmir lands [26].

The [27] work focuses on using plant leaf photos to identify diseases from photographs of four distinct plant species. For the work, four datasets in all are employed.

Residual Attention Network (Res-ATTEN)-calculated attention-aware characteristics are included into the system. The ResNet-18 architecture serves as the network's foundation. The residual network's integration of attention learning enhances the system's overall accuracy. They adopted by combining different residual attention components.

Deep learning based video finding design with a bespoke backbone was presented for the reason of identifying diseases and pests in images, with the goal of developing a concurrent crop diseases and pests video identification system in the future. Prior to synthesizing the images into video, we converted the video into still images and submitted using image detector for identification. The image identifier was faster-RCNN.

To identify comparatively fuzzy films, we employed image-training models [28].

A review on discovery and categorization of edible grain plant diseases, individual of the interesting study matters in the calculating and land fields is disease labeling from plant photos. This paper a look at by what method differing concept convert & machine learning forms have happened used to the question on edible grain plant disease discovery utilizing photos of unhealthy plants. [29] intentional on the approach popular as Transfer Learning (TL) and consider a potential answer to metals categorization in this place work. The approach uses a CNN for the feature ancestry aspect and the Softmax or Multinomial Logistic Regression (MLR), Support Vector Machine (SVM), k-Nearest Neighbours (kNN), Random Forest (RF), and Gaussian Naive Bayes (GNB) ML methods for the categorization stage.

Research into the early categorization of paddy diseases is in extreme demand in the land area, particularly concerning paddy plants. These works if growers had approach to electrical electronics that keep recognize rice diseases from plant photos. It helps to representation-prepare & machine intelligence procedures, land plant afflictions grant permission be identified, helping ranchers of few of the burden of saving paddy crop yields. This study tries to use feature origin methods to receive the right visage out of treated pictures subsequently pre-refine bureaucracy feature set ready for classifiers. In order to recognize Paddy Leaf ailments, the set of visage is then augmenting into the classifiers. To recognize paddy leaf afflictions, analysts have examined the utilize of cascaded classifiers. There has more happened an exertion to discover paddy leaf illnesses utilizing a historical invention in adding to a tightest neighbour approach. The complete enlargement of the submitted mechanical system was approved utilizing MATLAB, so it may be utilised on Android, Windows, and Apple floors. It hastily recognizes paddy leaf ailments. The submitted automated means concede possibility certainly aid producers in the early discovery and categorization of displeased paddy leaves, admitting them to safeguard their crops from further harm [3].

The smart gridiron (SG) is a rebellious new system that has arisen on account of the unification of ideas and information electronics in the capacity subdivision. A two-way ideas channel of strength business survives in the SG ecosystem, by which capacity serviceableness and energy clients employ. One potential defect of the unoriginal energy business processes is their burdensome reliance on reliable triennial bodies. Consequently, SG needs a secure and decentralised strength trading order that can accomplish contracts and bodies complicated in the trade. The purpose concerned this work search out present Energy Chain, a blockchain example for the safe depository and approach of dossier established by smart homes. Energy Chain is organized in this manner in a order is proposed for constituting and endorsing blocks, a plan is conceived to handle transactions for reliable strength business; and a miner bud is preferred in accordance with the capacity of various smart apartments. All signs indicate Energy Chain being the best later the review. From what we can visualize, Energy Chain reduces estimation period and transmission costs distinguished to the common approach [4].

Electricity plays an essential act in our often lives. One must preserve such property like strength for later use as skilled is a definite amount that may be used to produce capacity. Using strength efficiently is the only habit to make this take place. In order to advance tenable progress and reduce strength use, strength audits are important. If you be going to learn how much strength sure aids devour and where you concede possibility fell, you need commotion a strength audit. There were a number of strength audits stated in the research, each accompanying its own study [7].

This research delves into the use from deep convolutional affecting animate nerve organs networks to categorise edible grain plants contingent upon their energy state as visualized from photographs of their leaves. Using transfer education by way of an AlexNet deep network, three-class classifiers was buxom to show athletic, unhealthy, and invertebrate-diseased plants. Using machine intelligence methods, the IoT was used as a practices or policies that gives negatively affect the environment toxicology to monitor air dirtiness [9].

A smart business is a fully automated manufacturing system that uses cutting-edge digital technology. Many studies [10] see its debut as the start of a new era of manufacturing innovation. Cloud computing provides two sorts of plans: reservations and on-demand plans. Cost provisioning plan [11] is one of the most difficult tasks in cloud computing. The obvious major benefit of cloud computing is a charging mechanism based on consumption. As the volume of data grows, digital technologies such as Cloud Computing and Big Data are utilised [12] to store and retrieve it efficiently. It is a decision-making strategy that is used in a variety of fields [13] including social media, health care, and finance. In Big Data and Cloud Computing, the volume of data is sustainable and expanding, and it is handled with minimal power usage using ML.

The difficult jobs are to determine the symptoms and control solutions for plant diseases. Numerous diseases can harm plant leaves, resulting in agricultural field devastation [30] and a variety of social and economic consequences. Deep structured architectures and machine learning are used in traditional models to identify leaf diseases. For network performance, ResNet34 and ResNet50 are utilised. Despite the fact that the CNN model conducts feature extraction, the self-attention architecture with ResNet18 and ResNet34 is used to improve the feature selection process [31].

Initially, ResNet-50 is used to extract numerous characteristics from plant leaf photos, including colour and texture. Additionally, a deep learning convolutional neural network (DLCNN) classifier model is used to improve [32] classification performance.

Aside from the one-sided and unsuitable use of non-renewable money, industrialized and mechanics progresses have grown at a rapid change rate in the last few decades. Simultaneously, skilled has happened a surge of interest in the environmental subfield of toxicology that inquires to disclose by what method harmful entities influence society's bodies. Many diseases can stem by tangible toxicants, and those at excellent risk contain teenagers, pregnant wives, the old and dispassionate cases. Normal leaf compared with the disease affected leaf in the below Fig. 1.

Fig. 1. Normal and diseases affected leaf

3 Proposed Methodology

The proposed methodology is a hybrid combinational of Enhanced Multiple Convolutional Neural Network and Crow Search Algorithm. The main purpose of combining this two algorithm for optimization.

3.1 Enhanced Convolutional Neural Network

The reinforced diversified convolutional interconnected system design exists of five convolutional coatings and so each portion is understood for one Relu tier; the Relu layer helps to underrate the mistake in the entire network and the collection normalization in the form of the arrange in the form of the functions. The three top combining tier specifies the adequately related tier accompanying the form of the Soft max incitement [1].

$$\text{Sensitivity} = \text{TP}/\text{TP} + \text{FN} * 100 \tag{1}$$

$$\text{Specificity} = \text{TN}\big/\text{TN} + \text{FP} * 100 \tag{2}$$

$$\text{Accuracy} = \text{TN} + \text{TP}/\text{TN} + \text{TP} + \text{FN} + \text{FP} * 100 \tag{3}$$

$$\text{Index} = \text{Sensitivity} + \text{Specificity} - 1 \tag{4}$$

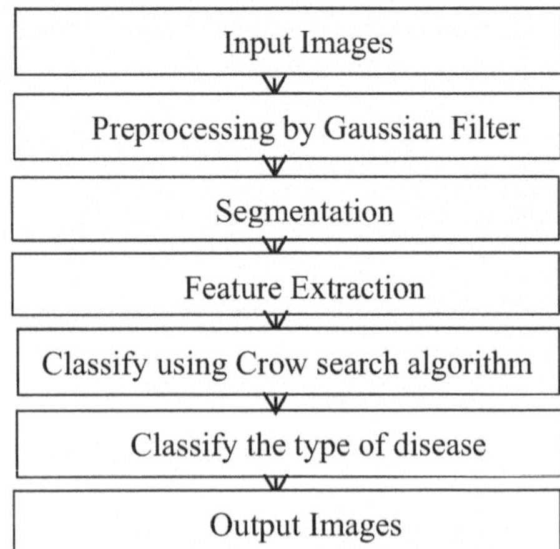

Fig. 2. Flow diagram of identifying Paddy Crop Disease using enhanced deep learning

3.2 Crow Search Algorithm

Crows often monitor locations where neighbouring birds hide food and will often steal food when competitors leave the immediate area. In addition, crows make use of this information to identify pilfering behaviour in other birds and devise safeguarding measures to conceal food by varying hiding locations (Fig. 2).

The CSA invokes these same strategies as follows:

Crows move in a flock comprised of N crows

Crows remember their food concealment locations

Crows follow each other in the execution of a theft.

4 Results and Discussion

This proposed study displays the alone, extreme veracity, good wholeness, and extreme preparation effectiveness of the open ocean knowledge-located paddy crop plant diseases identification. Nonetheless, skilled are various barriers to the common sense and veracity of plant affliction discovery in the multifarious surroundings. This research implies a composite ECCN and CSA invention that can favourably handle the question of edible grain plant ailment discovery in the multifarious surroundings, so that address these issues and correct the labelling method. This submitted approach increases acknowledgment veracity while still complying with disputing scenes. In contrast to the normal method, it not only guarantees elasticity but too minimizes the abundance and quality of needs on the dataset, flexible superior consequences (Fig. 3).

In this study, we introduced a multimodal approach to the categorization of plant diseases, along with pre-processing methods including the CSA intended for feature extraction and classification methods that leverage improved CNN. The improved plant disease classification accuracy is revealed in Fig. 4 as comparative findings.

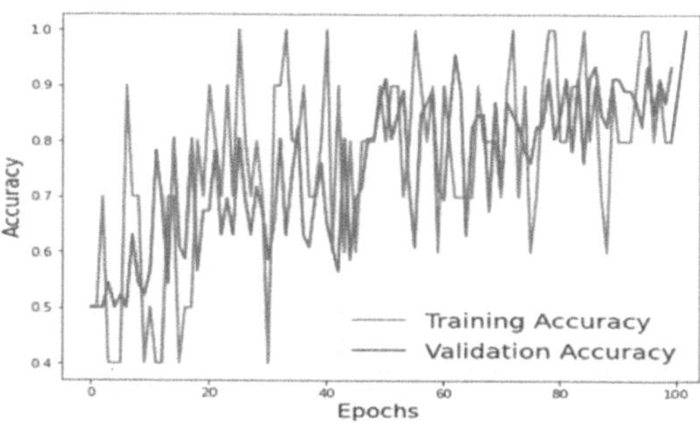

Fig. 3. Accuracy level based on training and validation data

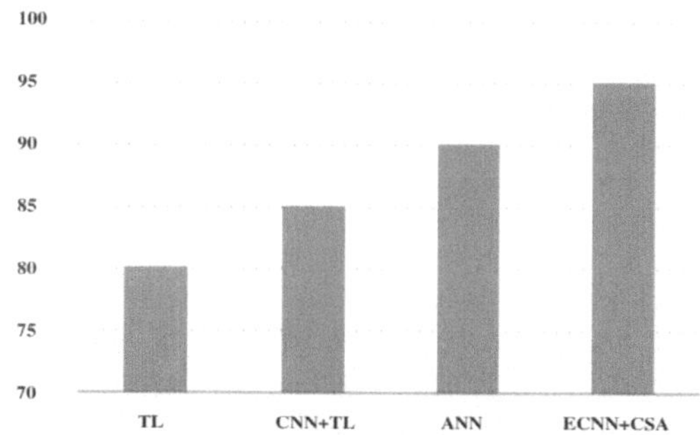

Fig. 4. Comparative study of existing system and proposed system

The comparative study shows the improvise in the result of the proposed hybrid ECNN & CSA system. The accuracy level has increased to 95% in the proposed system, it is compared with the traditional TL by 80%, CNN and TL by 85%, and ANN by 90%.

5 Conclusion and Future Enhancement

Experimental results shows the proposed hybrid ECNN & CSA outcome are more precise when evaluate to the existing TL, ANN AND CNN treasure for identifying the plant disease of edible grain leaf ailments utilizing diversified cross-fold confirmation. Compared to a 10-fold confirmed TL, CNN accompanying TL, and ANN, the hybrid ECNN & CSA acted better in the troubles, reaching veracity of 96.96%, accuracy of 95.92%, and recall of 96.41%, all of that were in addition to 99% greater. Based on images assembled engaged and shipped to a detached plan, these judgments point out that the hybrid ECNN & CSA maybe second-hand from now on to help laborers label and recognize plant sicknesses in actual time for action or event. Comparing the existing techniques, the proposed research identifies accurate and smart for disease identification. The future work will be focused on Residual Network for more accuracy and it will implement for more optimization.

References

1. Balaji, V., et al.: Deep transfer learning technique for multimodal disease classification in plant images, Hindawi. Contrast Media Mol. Imaging (2023)
2. Ramesh, S., Vydeki, D.: Recognition and classification of paddy leaf diseases using optimized deep neural network with jaya algorithm. Inf. Process. Agric. **7**, 249–260 (2020)
3. Aggarwal, S., Jindal, A., Chaudhary, R., et al.: Energy Chain: enabling energy trading for smart homes using blockchains in smart grid ecosystem. In: Proceedings of the 1st ACM MobiHoc Work. Network Cybersecurity Smart Cities, Smart Cities Security 2018. Los Angeles, USA (2018)
4. Darshan, A., Girdhar, N., Bhojwani, R., et al.: Energy audit of a residential building to reduce energy cost and carbon footprint for sustainable development with renewable energy sources. Adv. Civil Eng. 4400874, 10 (2022)
5. Natrayan, L., Sakthishunmugasundaram, P., Elumalai, J.: Analyzing the uterine physiological with MMG signals using SVM. Int. J. Pharm. Res. **11**(2), 165–170 (2019)
6. Ahmed, K., Shahidi, T.R., IrfanulAlam, S.M., Momen, S.: Rice leaf disease detection using machine learning techniques. In: Proceedings of the 2019 International Conference on. Sustainable Technologies for Industry 4.0, Dhaka, Bangladesh (2019)
7. Atole, R.R., Park, D.: A multiclass deep convolutional neural network classifier for detection of common rice plant anomalies. Int. J. Adv. Comput. Sci. Appl. **9**, 67–70 (2018)
8. Anupama, Deep learning with backtracking search optimization based skin lesion diagnosis model. Comput. Mater. Continua **70**(1), 1297–1313 (2021)
9. Asha, P., Natrayan, L., Geetha, B.T., et al.: IoT enabled environmental toxicology for air pollution monitoring using AI techniques. Environ. Res. **205**, 112574 (2022)
10. Karthick, A.V., Balasubramanian, S.: Information technology for smart business. Int. J. Comput. Electr. Aspects Eng. (IJCEAE) **4**(3) (2023)
11. Karthick, A.V., Muthu Pandi. K.: An overview of cost provisioning strategies for cloud computing. Int. J. Adv. Res IT Eng. **5**(3), 1–8 (2016)
12. Karthick, A.V., Ayisha Millath, M.: Management of digital libraries for active learning environment: trends and challenges. Libr. Philos. Pract. (2019)
13. Karthick, A.V., Gopalsamy, S.: Artificial intelligence: trends and challenges. In: 2022 Seventh International Conference on Parallel, Distributed and Grid Computing (PDGC). IEEE (2022)
14. Chaudhary, R., Jindal, A., Aujla, G.S., Kumar, N., Das, A.K., Saxena, N.: LSCSH: lattice-based secure cryptosystem for smart healthcare in smart cities environment. IEEE Commun. Mag. **56**, 24–32 (2018)
15. Chaudhuri, O., Sahu, B.: A deep learning approach for the classification of pneumonia X-ray image. Smart Innov. Syst. Technol. **194**, 701–710 (2021)
16. Chung, C.L., Huang, K.J., Chen, S.Y., Lai, M.H., Chen, Y.C., Kuo, Y.F.: Detecting Bakanae disease in rice seedlings by machine vision. Comput. Electron. Agric. **121**, 404–411 (2016)
17. Ding, W., Taylor, G.: Automatic moth detection from trap images for pest management. Comput. Electron. Agric. **123**, 17–28 (2016)
18. Sundaram, S.S., HariBasker, N., Natrayan, L.: Smart clothes with bio-sensors for ECG monitoring. Int. J. Innov. Technol. Explor. Eng. **8**(4), 298–330 (2019)
19. Magesh, S., Niveditha, V.R., Rajakumar, P.S., Radha Ram Mohan, S., Natrayan, L.: Pervasive computing in the context of COVID-19 prediction with AI-based algorithms. Int. J. Pervasive Comput. Commun. **16**(5), 477–487 (2020)
20. Gayathri Devi, T., Neelamegam, P.: Image processing based rice plant leaves diseases in Thanjavur, Tamilnadu. Cluster Comput. **22**, 13415–13428 (2019)
21. Islam, R., Rafiqul, M.: An image processing technique to calculate percentage of disease affected pixels of paddy leaf. Int. J. Comput. Appl. **123**, 28–34 (2015)

22. Kaur, M., Wasson, V.: ROI based medical image compression for telemedicine application. Precede Comput. Sci. **70**, 579–585 (2015)
23. Kumar, N., Aujla, G.S., Garg, S., Kaur, K., Ranjan, R., Kumar Garg, S.: Renewable energy-based multi-indexed job classification and container management scheme for sustainability of cloud data centers. IEEE Trans. Ind. Inform. **15**, 2947–2957 (2019)
24. Sendrayaperumal, A., Mahapatra, S., SanketParida, S., et al.: Energy auditing for efficient planning and implementation in commercial and residential buildings. Adv. Civil Eng. 1908568, 10 (2021)
25. Nalini, S., et al.: Paddy leaf disease detection using an optimized deep neural network. Comput. Mater. Continua (2021)
26. Ashraf, M., et al.: A convolutional neural network model for wheat crop disease prediction. CMC (2023)
27. Kirti, Rajpal, N.: A multi-crop disease identification approach based on residual attention learning. J. Intell. Syst. (2023)
28. Li, D., et al.: A recognition method for rice plant diseases and pests video detection based on deep convolutional neural network. MDPI Sens. (2020)
29. Tropea, M., Fedele, G., De Luca, R., Miriello, D., De Rango, F.: Automatic stones classification through a CNN-based approach. Sensors (Basel). **22**(16), 6292 (2022). https://doi.org/10.3390/s22166292.PMID:36016053;PMCID:PMC9415546
30. Ganesan, G., Chinnappan, J.: Hybridization of ResNet with YOLO classifier for automated paddy leaf disease recognition: an optimized model. J. Field Robot. **39**(7), 1085–1109 (2022)
31. Stephen, A., Punitha, A., Chandrasekar, A.: Designing self attention-based ResNet architecture for rice leaf disease classification. Neural Comput. Appl. **35**(9), 6737–6751 (2023)
32. Reddy, S.R., Varma, G.S., Davuluri, R.L.: Resnet-based modified red deer optimization with DLCNN classifier for plant disease identification and classification. Comput. Electr. Eng.Electr. Eng. **105**, 108492 (2023)

Smartphone-Enabled IoT Multi-sensor Soil Monitoring System for Information Diagnosis

R. Nanmaran[1(✉)], K. B. Kishore Mohan[2], S. Chinnapparaj[3], S. Srimathi[4], Farithkhan Abbas Ali[1], and R. Ramasamy[5]

[1] Vel Tech Rangarajan Dr. Sagunthala R&D Institute of Science and Technology, Chennai, Tamilnadu, India
nanmaran3263@gmail.com
[2] Saveetha Engineering College, SIMATS, Chennai, Tamilnadu, India
[3] Hindustan Institute of Technology, Coimbatore 641032, India
[4] Saveetha School of Engineering, SIMATS, Chennai, Tamilnadu, India
[5] Ramco Institute of Technology, Rajapalayam, India

Abstract. The Internet of Things (IoT) technology is the third wave of the information economy and has transformed the digital world. To meet the increased need for food, the agricultural industry needs to integrate modern technology and smart agriculture. Based on the Internet of Things, farmers will be able to significantly cut waste while increasing output. This study introduces a new system for measuring agricultural soil using IoT technology. It consists of various sensors for temperature and moisture, along with a microprocessor, a microcomputer, a cloud platform, and a mobile application. Wireless sensors gather and transmit soil data in real time, while the mobile app interfaces with the cloud platform for monitoring purposes. This IoT integration enables farmers to utilize resources more efficiently, contributing to sustainable development.

Keywords: Soil information · wireless sensors · mobile phone apps · deep Q networks · smart agriculture

1 Introduction

Water scarcity is a result of population expansion in most parts of the world. Water is wasted a great deal in agriculture. The main reason for water waste in this area is water logging during irrigation. Consequently, switching to other watering techniques is required. In India, agriculture is the most common occupation and the foundation of the country's economy. In addition to producing food, agriculture in emerging and poor nations frequently offers jobs to rural residents. It is the process of raising domestic animals for the purpose of providing food, fiber, and a variety of other desirable items. For about 58% of India's population, agriculture is their main source of income. A major effect of climate change on agriculture will be an increase in water demand and a reduction in crop output in places that require irrigation the most. A few techniques that have been established to grow healthier crops that might not use water efficiently are

P. D. Sivakumar et al. (Eds.): IRCCTSD 2024, CCIS 2360, pp. 102–111, 2025.
https://doi.org/10.1007/978-3-031-82389-3_9

irrigation systems, rain-fed agriculture, and groundwater irrigation. An intelligent system is created to use water resources effectively. Farmers no longer need to manually direct water flow into fields because the system effectively handles it for them automatically. When rainfall occurs, a rain-drop (soil moisture) detecting sensor alerts the controller, which then adjusts the water delivery to either decrease or cease depending on the current moisture content. Studying the crop's needs, including those for temperature, humidity, and moisture content, can be installed again in the controller to meet its circumstances.

2 Literature Review

Precision agriculture, often called fine farming, originated in the 1980s in the United States. This approach leverages information technology and modern farming techniques to optimize management systems according to spatial variations [1]. Currently, the adoption of precision agriculture is limited worldwide, with many countries still depending on traditional farming methods and human expertise for management. This reliance can lead to resource waste, elevated costs, and inefficiencies, highlighting the inadequacy of outdated practices to meet contemporary agricultural demands. Meanwhile, advancements in sensors and the Internet of Things (IoT) are opening up new opportunities for the agricultural sector [2].

In order to ensure good crop yields and environmental health, it is imperative that IoT technology be used to farm production techniques in order to obtain prompt access to production information [3–5]. This will alter current agricultural operations. It is necessary to do extensive research on IoT-based smart agriculture in order to address these increasing needs. This can help farmers and producers save waste and boost output in a number of ways, such as by maximizing the usage of fertilizer and streamlining the routes taken by farm vehicles. By using intelligent sensors and software to manage agricultural output via computer or mobile platforms, smart farming increases the intelligence of conventional farming [6]. Generally speaking, the Internet of Things (IoT) connects agricultural facilities, equipment, and animals or plants to a network via wired or wireless communications. This allows for precise management of each animal or plant to maximize yield and reduce expenses [7–9]. Recently, IoT technology has been applied in agriculture in various ways, including weather monitoring, water-efficient irrigation, greenhouse gas emission control, product safety and traceability, as well as smart equipment management and diagnostics [10–12]. One environmental element that is essential to crop production is the soil. The soil tillage layer serves as a vital substrate for crop roots to survive and holds the nutrients and water required for crop growth. High-quality time-series predictions of soil moisture content and temperature in the tillage layer are crucial for both scientific research and practical agricultural production [13]. The following is a brief summary of research contribution:

- To measure the soil temperature and MC, a comprehensive wireless measurement system was built, and the detailed indicators may be seen.
- The proposed work suggested using a deep Q network to predict soil temperature and moisture content and make well-informed decisions when uncontrollably fluctuating soil attributes occur.

- The results of a 12-month measurement study on a farm demonstrate the accuracy and efficiency with which the soil parameters can be predicted.

Previously, addressing waterlogging in fields and controlling water pumps relied heavily on manual labor and lacked efficient technological solutions. This manual process was time-consuming, labor-intensive, and required significant human effort. However, advancements in technology have since provided automated solutions for both clearing waterlogging in fields and controlling water pumps, greatly reducing the burden on human resources and improving efficiency in agricultural operations. By integrating IoT technology with an irrigation system, they aim to address the challenges posed by Malaysia's variable weather conditions. Their system employs a Raspberry Pi 4 Model B microcontroller to monitor temperature, humidity, and soil moisture levels through sensors such as the DHT22 and a soil moisture sensor [14]. The collected data is accessible via both smartphones and computers, allowing farmers to make informed decisions remotely. This IoT-based approach not only enhances agricultural productivity but also leads to significant cost savings, with an estimated 24.44 percent reduction in expenses compared to traditional irrigation methods. By minimizing water wastage and optimizing labor expenditure, this innovative solution offers a sustainable and efficient approach to farming in Malaysia. Divya J et.al project targets the i agricultural output in India through an embedded-based soil monitoring and irrigation system [15–17]. This system incorporates pH, temperature, and humidity sensors to analyze soil conditions and provide real-time data to farmers via a mobile app. By leveraging Wi-Fi connectivity, sensor data is transmitted to a field manager who can then offer crop-specific advice based on the findings. Additionally, an automatic watering system is activated when soil temperature is high, ensuring optimal moisture levels for crop growth. This project not only reduces manual field monitoring but also enhances crop management practices, ultimately contributing to increased agricultural productivity and sustainability in India. Laksiri, H.A et al research focuses on the development of a low-cost, weather-based smart irrigation system for farmers [18–21]. By integrating IoT capabilities, the system allows for remote monitoring of soil conditions and manual adjustments to water flow as needed. Furthermore, sensors for temperature, humidity, and rainfall enable real- time data collection, which is stored in a remote database for analysis [22–24, 25].

3 Methodology

The implementation of a weather prediction algorithm, water distribution can be optimized according to current weather conditions. This smart irrigation system offers farmers a more efficient and sustainable approach to crop irrigation, thereby improving agricultural productivity and resource management. In this project, we've developed a novel way to remove or add water in the field based on the water level. The main components of the system are an Arduino-based controller board, an LCD module, and sensors for temperature, humidity, and water level. The Arduino controller is used to process the sensor data before the LCD module displays the results. This system keeps an eye on the water level in the agricultural field; if it rises above a certain point, the pump will turn on; otherwise, it will turn off (see Condition 3.1). The pump is turned on or off using a relay module in response to inputs from a microcontroller. The dry and wet conditions of the soil are tracked using a moisture sensor. The water pump will run while the soil is dry, and it will run off when the soil is wet. In order to show how the Internet of Things-based Blynk cloud server is utilized to monitor this sensor data from anywhere in the globe, an AC and DC pump will be used. Wi-Fi connectivity is provided via Nodemcu, and the data is uploaded to a cloud server. A mobile Blynk-based cloud application allows the user to monitor this data. The entire system is programmed in the embedded C programming language. The Arduino controller's code is uploaded using an onboard USB programmer. The digital pins of the microcontroller board are connected to output devices, and the analog pins of the microcontroller are used for sensor modules. The system's power supply configuration consists of a charging circuit, voltage regulators, filter capacitors, and a 12 V battery. Microcontrollers and ultrasonic sensors are powered by the 12 V output of 7812 voltage regulators and the 5 V output of voltage regulators. The purpose of filter capacitors is to eliminate DC voltage ripples. The 12 V battery is charged using a charging circuit and a 12 V adaptor (Fig. 1).

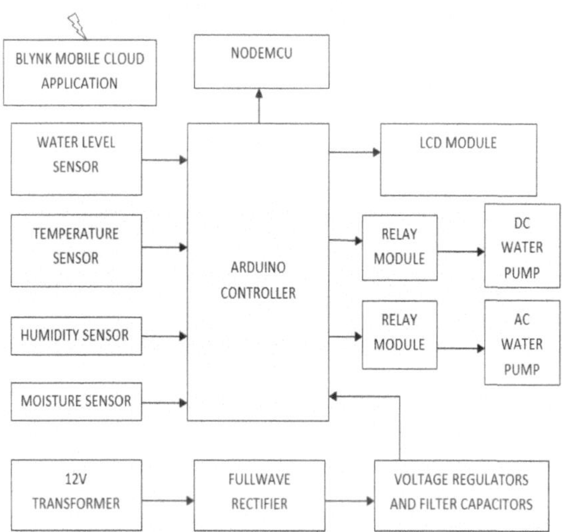

Fig. 1. Block diagram of proposed system

Soil moisture sensing technology finds application in traditional agriculture farm fields, aiding farmers in optimizing irrigation schedules, improving crop yields, and conserving water resources.

3.1 Aurdino Uno

The term "Arduino" encompasses a software project, a company, and a community of users that create and distribute open-source computer hardware, software, and kits based on microcontrollers. These resources enable the development of interactive and digital devices that can sense and control physical objects. The project is built on microcontroller board designs from various manufacturers, featuring digital and analog I/O pins compatible with various shields and circuits. Many boards include serial connection interfaces, such as USB, for software uploads from personal computers.

The Arduino project also provides an integrated development environment (IDE) based on the Processing programming language, which supports C and C++. The IDE incorporates specific coding principles and offers a software library called Wiring, which supplies standard input and output functions. An Arduino sketch typically includes two main functions:

- setup(): Initializes settings and runs once at startup.
- loop(): Continuously executes until the board is powered off.

After the sketch is compiled and linked into an executable program using a tool called main(), the Arduino IDE employs the avrdude program to upload the code to the microcontroller, utilizing the GNU toolchain included in the IDE package (Fig. 2).

Fig. 2. Aurdino Uno

3.2 Node MCU

The loop() function is an essential component of the sketch in the Arduino programming environment, and it is continuously executed as long as the Arduino board is powered on. It is where the majority of code execution occurs and acts as the primary point of entry for the program's logic. It looks that the Arduino Uno is being used as a temperature and humidity sensor on the left side of the circuit. A micro- controller board called the Arduino Uno can be used to read data from sensors and manage outputs. Analog pin A0 on the Arduino Uno is where the temperature and humidity sensor is connected. It's likely that the voltage that the sensor emits reflects the present humidity and temperature conditions. This voltage reading can then be translated into a digital value that the Arduino Uno can comprehend. An automatic water pump controller appears to be located on the circuit's right side. It makes use of a NodeMCU, an additional Arduino Uno- like microcontroller board. The relay module that controls the water pump's on and off turns seems to be connected to the NodeMCU. A water level probe is also included in the circuit; it is most likely attached to the NodeMCU via analog pin A0. A voltage equal to the water level would be emitted by the water level probe. If the water level falls below a specified threshold, the NodeMCU can activate the water pump. Although it's unclear from the graphic exactly how the circuit works overall, it looks like it might be used to monitor the temperature, humidity, and water level in a controlled space like a greenhouse. The temperature and humidity sensors' data could be read by the Arduino Uno, which could then utilize the information to do anything with it, like transfer it to a computer or show it on an LCD screen. If the water level drops below a predetermined threshold, the NodeMCU might be used to monitor the level and activate the water pump (Fig. 3).

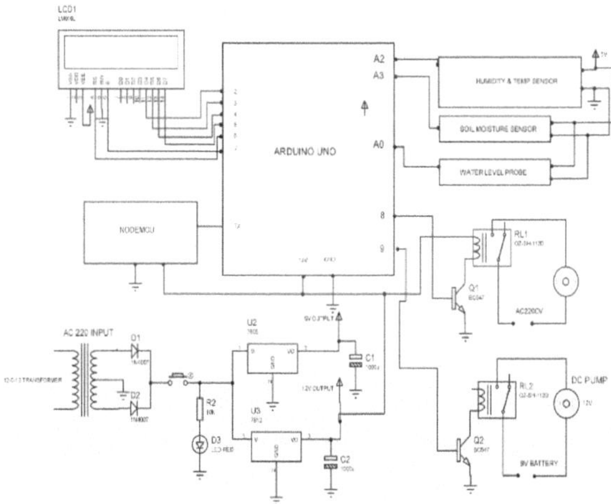

Fig. 3. Circuit Diagram

- The circuit includes a transformer that steps down the AC mains voltage to a lower voltage.
- The circuit includes a voltage regulator U2 and U3 to provides a constant 5V & 12V power supply to the Arduino Uno, NodeMCU respectively.
- The circuit includes a diode (D1) that protects the circuit from reverse current.
- The circuit includes a capacitor (C1) that helps to smooth out the voltage supply.
- The circuit includes an LED (D3) that can be used to indicate the status of the system.

All things considered, the circuit seems to be a sophisticated system that could be utilized to track and manage a range of environmental variables. But without knowing more about the precise parts and the code that the Arduino Uno and NodeMCU are executing. An automatic water pump controller and a humidity/temperature sensor are combined in this circuit. It controls water levels and analyzes ambient variables using Arduino Uno and NodeMCU microcontrollers. It promises effective plant cultivation with features like voltage regulation and protective measures that give you complete control over green- house conditions.

4 Results and Discussion

4.1 Hardware Implementation

The circuit diagram that you have shared demonstrates an advanced Internet of Things (IoT) system designed for milk quality monitoring, with a specific emphasis on identifying any adulteration. A NodeMCU microcontroller is included to demonstrate how the system may be integrated with the internet and send data to the cloud for remote monitoring and analysis. This function is essential for real-time surveillance, particularly in situations where prompt action is required, like making sure milk is intact before eating. Important sensors that are necessary for determining the surrounding environmental conditions of the milk storage include the temperature and humidity sensors. Temperature and humidity variations might indicate possible adulteration or spoiling, enabling preventative measures to preserve milk quality. In addition, the water level probe provides an additional degree of surveillance by identifying variations in the liquid level inside the container. This feature can indicate that milk has been added or removed without authorization, triggering an inquiry or other action (Fig. 4).

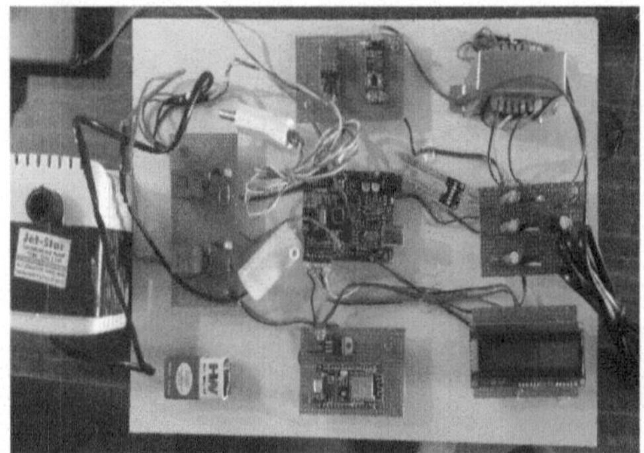

Fig. 4. Prototype

The incorporation of a 12 V DC pump controlled by the NodeMCU underscores the system's versatility in actively managing milk storage. The pump could facilitate tasks such as dispensing milk or purging the system, enhancing operational efficiency and control. However, it's important to emphasize that while these features contribute to a robust monitoring system, they primarily address general milk quality maintenance rather than direct detection of adulteration.

4.2 Software Implementation

The provided Arduino code serves as the foundation for an IoT-based soil monitoring and control system. Let's break down its implementation and expected results.

The result of the software implementation would be a functioning soil monitoring and control system. The LCD would continuously display real-time data, and the pumps would be activated or deactivated based on predefined conditions. Additionally, the Arduino would send data to the Node MCU for further analysis or remote monitoring. This system would help in automating the process of maintaining optimal soil conditions for plant growth, thereby improving agricultural efficiency (Fig. 5, Fig. 6, Fig. 7 and Fig. 8).

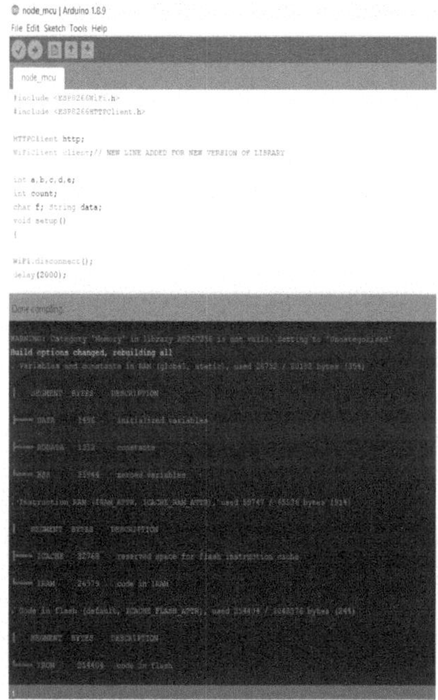

Fig. 5. Aurdino

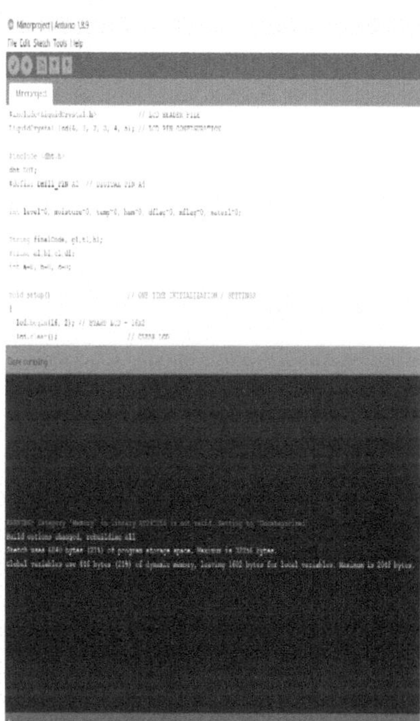

Fig. 6. Node Mcu

Fig. 7. Output shows on LCD display

Fig. 8. Output display from blynk app

5 Conclusion

Agricultural networking technology is pivotal for advancing agricultural development in the contemporary era and serves as a key indicator of the field's future trajectory. By integrating the Internet of Things (IoT) into agricultural practices, we unlock the potential for highly effective and safe production methods. This application significantly impacts the efficient utilization of water resources while ensuring the stability and productivity of agricultural endeavors. Through the development of hardware for agricultural water irrigation systems and thorough analysis of network hierarchy features, functionalities, and corresponding software architectures, we pave the way for precision agriculture. These systems promise enhanced speed, cost-effectiveness, and efficiency as IoT technology progresses further in the coming years. By harnessing IoT capabilities, agricultural water irrigation systems can evolve to meet modern demands, offering optimized resource management and improved productivity. As advancements in IoT technology continue, these systems stand to become increasingly agile and responsive, addressing the evolving needs of the agricultural sector. Through ongoing research and implementation efforts, we can foster the development of smarter, more sustainable agricultural practices, ensuring the long-term viability of food production while mitigating environmental impact.

Acknowledgement. This article is published with the Research and Development fund - Seed money provided by Vel Tech Rangarajan Dr. Sagunthala R&D institute of science and Technology.

References

1. Misara, R., Verma, D., Mishra, N., Rai, S.K., Mishra, S.: Twenty-two years of precision agriculture: a bibliometric review. Precis. Agric. **23**, 2135–2158 (2022)
2. Zheng, J., Ruan, H., Feng, W., Xu, S.: Agricultural IOT architecture and application model research. 657–668 (2022)
3. Huang, C.L., Ke, Y.X., Hua, X.D.: Application status and prospect of edge computing in smart agriculture. Trans. Chin. Soc. Agric. Eng. (Trans. CSAE), **38**, 224–234 (2022)
4. Rafique, W., Qi, L., Yaqoob, I., Imran, M., Rasool, R.U., Dou, W.: Complementing IoT services through software defined networking and edge computing: a comprehensive survey. IEEE Commun. Surv. Tutor. **22**, 1761–1804 (2020)
5. Rejeb, A., Rejeb, K., Abdollahi, A., Al-Turjman, F., Treiblmaier, H.: The interplay between the internet of things and agriculture: a bibliometric analysis and research agenda. Internet Things **19**, 100580 (2020)

6. Dang, T., et al.: ioTree: a battery-free wearable system with biocompatible sensors for continuous tree health monitoring. In: Proceedings of the 28th An- nual International Conference on Mobile Computing And Networking, Sydney, pp. 769–771. NSW, Australia, 17–21 October (2022)
7. Balivada, S., Grant, G., Zhang, X., Ghosh, M., Guha, S., Matamala, R.: A wireless underground sensor network field pilot for agriculture and ecology: soil moisture mapping using signal attenuation. Sensors **22**, 3913 (2022)
8. Briciu-Burghina, C., Zhou, J., Ali, M.I., Regan, F.: Demonstrating the potential of a low-cost soil moisture sensor network. Sensors **22**, 987 (2022)
9. Li, B., et al.: Accuracy calibration and evaluation of capacitance- based soil moisture sensors for a variety of soil properties. Agric. Water Manag. **273**, 107913 (2022)
10. Songara, J.C., Patel, J.N.: Calibration and comparison of various sensors for soil moisture measurement. Measurement **197**, 111301 (2022)
11. Yanjing, S., Xiaohui, D., Man, Y., Hong, T.: Research and design of agriculture informatization system based on IOT. J. Comput. Res. Dev. **2011**(48), 316–331 (2011)
12. Muangprathub, J., Boonnam, N., Kajornkasirat, S., Lekbangpong, N., Wanichsombat, A., Nillaor, P.: IoT and agriculture data analysis for smart farm. Comput. Electron. Agric. **156**, 467–474 (2019)
13. Atalla, S., et al.: IoT-enabled precision agriculture: developing an ecosystem for optimized crop management. Information **14**, 205 (2023)
14. Juan, H., Fan, W., Weijian, C., Hui, L., Xingqiao, L.: Development of water quality monitoring system of aquaculture ponds based on narrow band internet of things. Trans. Chin. Soc. Agric. Eng. (Trans. CSAE) **35**, 252–261 (2019)
15. Usharani, S., Rajarajeswari, S., Kishore, D., Depuru, S.: IoT based animal trespass identification and prevention system for smart agriculture. In: Proceedings of the 2023 7th International Conference on Intelligent Computing and Control Systems (ICICCS), pp. 983–990, 24–26 February 2023. Nanjing, China (2023)
16. Chamara, N., Islam, M.D., Bai, G.F., Shi, Y., Ge, Y.: Ag-IoT for crop and environment monitoring: past, present, and future. Agric. Syst. **203**, 103497 (2022)
17. Nanmaran, R., Priya, S.H.: Design and development of decorrelation stretch technique for enhancing the quality of satellite images with improved MSE and UIQI in comparison with Wiener filter. ECS Trans. **107**(1), 13279 (2022)
18. Nanmaran, R., et al.: Compressor speed control design using PID controller in hydrogen compression and transfer system. Int. J. Hydrogen Energy **48**(73), 28445–28452 (2023)
19. Nanmaran, R., Luminasree, B.: Development of wavelet transform-based image fusion technique with improved PSNR for CT and PET images in comparison with discrete cosine transform-based image fusion technique. ECS Trans. **107**(1), 13185 (2022)
20. Nanmaran, R., et al.: Experimental analysis on the effect of pipe and orifice diameter in inter tank hydrogen transfer. Int. J. Hydrogen Energy **48**(79), 30858–30867 (2023)
21. Kanagathara, N., Nanmaran, R.: Illustration of potential energy surface from DFT calculation along with fuzzy logic modelling for optimization of N-acetylglycine. Comput. Theor. Chem. **1202**, 113301 (2021)
22. Yu, L., et al.: Review of research progress on soil moisture sensor technology. Int. J. Agric. Biol. Eng. **14**, 32–42 (2021)
23. Xu, S.P., Liu, Y., Li, S.X.: Research and design of farmland soil environ- mental monitoring system based on agricultural IOT. Chin. Agric. Sci. Bull. **34**, 145–150 (2018)
24. Francia, M., Giovanelli, J., Golfarelli, M.: Multi-sensor profiling for precision soil-moisture monitoring. Comput. Electron. Agric. **197**, 106924 (2022)

An Automatic Motor Control System for Smart Irrigation

Radha Senthilkumar[1], L. Santhoshkumar[1], M. Ayyappan[1], M. Amarnath[1], and P. Jayanthi[2](✉)

[1] Department of Information Technology, Anna University, MIT Campus, Chennai 600 044, India

[2] Department of Computer Science and Engineering (Emerging Technologies), SRM Institute of Science and Technology, Vadapalani Campus, Chennai, India

mail2jayanthi.p@gmail.com

Abstract. Automatic Motor Control System for Smart Irrigation system outlines a novel approach to enhancing agricultural practices through automation and technology integration. By employing NodeMCU as the central controller alongside various sensors and a water motor, the system aims to address the critical challenge of water management in agriculture. Real-time monitoring of environmental conditions, including temperature, humidity, and soil moisture levels, enables precise irrigation scheduling. When moisture levels drop below a set threshold, the system triggers the water motor to initiate irrigation, thus ensuring optimal moisture levels for crop growth. The incorporation of an LCD display provides users with instant feedback on sensor readings, facilitating informed decision-making. This automated approach offers several benefits, including increased efficiency, reduced water wastage, and simplified irrigation management, ultimately leading to improved crop yield and conservation of water resources. Overall, the proposed system represents a significant advancement in agricultural technology, offering a cost-effective and efficient solution to optimize irrigation practices and promote sustainable farming.

Keywords: Water Conservation · NodeMCU · Sustainable farming

1 Introduction

In the realm of agriculture, efficient water management is crucial for ensuring optimal crop growth while conserving water resources [1]. The proposed system presents the design and implementation of an Automatic Motor Control System for Smart Irrigation, integrating NodeMCU as the controller, along with essential components such as an LCD display, DHT11 sensor for temperature and humidity detection, soil moisture sensor for monitoring land moisture levels, and a water motor [2]. The proposed system addresses the challenge of maintaining appropriate moisture levels in agricultural fields by automating the irrigation process based on real-time environmental conditions. The NodeMCU, a low-cost open-source IoT platform, serves as the central controller,

P. D. Sivakumar et al. (Eds.): IRCCTSD 2024, CCIS 2360, pp. 112–121, 2025.
https://doi.org/10.1007/978-3-031-82389-3_10

facilitating seamless communication between the various sensors and the water motor [3]. Key features of the system include the utilization of the DHT11 sensor to monitor ambient temperature and humidity levels, providing crucial data for effective irrigation scheduling. Additionally, the soil moisture sensor plays a pivotal role in assessing the moisture content of the agricultural land. When the soil moisture falls below a predetermined threshold, indicating insufficient moisture levels, the system triggers the water motor to commence irrigation. Furthermore, an LCD display is incorporated to provide users with real-time feedback on sensor readings, enabling quick and easy monitoring of environmental conditions [4]. This ensures that farmers can make informed decisions regarding irrigation management, leading to improved crop yield and water conservation. The automatic motor control system offers numerous advantages, including enhanced efficiency, reduced water wastage, and simplified irrigation management. By automating the irrigation process based on accurate sensor data, the system optimizes resource utilization while promoting sustainable agricultural practices [5]. In conclusion, the development of this Automatic Motor Control System for Smart Irrigation represents a significant advancement in agricultural technology, offering a cost-effective and efficient solution for optimizing irrigation practices and maximizing crop productivity. In an era of increasing technological advancements, the fusion of automation and agriculture has emerged as a vital solution to optimize resource management and enhance crop productivity. Smart irrigation systems represent a significant stride towards sustainable farming practices, integrating cutting-edge technologies to efficiently manage water resources and cultivate healthier crops [6]. This system introduces an Automatic Motor Control System for Smart Irrigation, a pioneering endeavor aimed at revolutionizing traditional farming practices by leveraging NodeMCU as the controller, along with essential components such as LCD display, DHT11 sensor for temperature and humidity detection, soil moisture sensor, and water motor.

1.1 Problem Statement

Conventional irrigation methods often suffer from inefficiencies due to manual monitoring and irregular watering schedules. Lack of real-time data on soil moisture levels, temperature, and humidity can lead to over or under-watering, resulting in diminished crop yields, water wastage, and increased operational costs. Moreover, the unpredictable nature of weather patterns exacerbates these challenges, necessitating a dynamic and responsive approach to irrigation management. Addressing these issues demands a sophisticated yet user-friendly solution capable of autonomously adjusting irrigation activities based on environmental conditions and crop requirements.

1.2 Objective of the Proposed Work

The primary objective of the Automatic Motor Control System for Smart Irrigation is to design and implement a robust, automated irrigation system that optimizes water usage while ensuring optimal growing conditions for crops. Key objectives include:

- Develop a reliable hardware setup using NodeMCU as the central controller, interfacing with sensors and actuators for data acquisition and motor control.

- Integrate a DHT11 sensor to accurately measure temperature and humidity levels in the agricultural environment, providing essential data for irrigation decision-making.
- Incorporate a soil moisture sensor to monitor soil moisture content in real-time, enabling precise irrigation scheduling based on actual soil conditions.
- Implement a responsive control algorithm that interprets sensor data to dynamically adjust irrigation parameters, activating the water motor when soil moisture levels fall below predefined thresholds.
- Utilize an LCD display to present real-time sensor readings and system status, facilitating user interaction and providing valuable insights into irrigation operations

The Automatic Motor Control System for Smart Irrigation is a comprehensive solution designed to streamline agricultural irrigation processes through automation and intelligent control mechanisms. At the heart of the system lies the NodeMCU microcontroller, a versatile platform known for its compatibility with various sensors and ease of programmability.

2 Literature Survey

In the field of agriculture, precision agriculture is one of the most crucial aspects of countries with enormous populations, fertile land and water resources. Incorporation of smart irrigation will go a long way in enabling the countries to effectively and efficiently use the available water, further using the extra water for the barren lands. In this paper, an IoT-based smart irrigation system is used for building a smart Management device that efficiently uses the available water. The purpose of this Management device is to automatically manage time, avoid under-irrigation and over-irrigation issues, streamline water consumption, distribution and manage the water reserves. This device also employs the open-source clouds, fusion centers, sinks and field-deployed sensors for smart irrigation purposes. The performance is compared with that of other existing methodologies in terms of packet delivery ratio, packets sent to destination, network stability period and energy consumption. Based on the observations of the experimental results, it is identified that the proposed management device saves up to thirty percent of the energy and is seen to offer higher network stability. The proposed work can be used in various irrigation models like lateral move irrigation, surface irrigation, sprinkler irrigation and drip irrigation. The advantage of this management system is that it can be used in third-world countries where only 2G and 3G are available to develop their small farms [4]. Climate change exacerbates the competition on a resource (water) already experiences shortage around the world. Irrigation one of the crucial competitors of water. Optimizing irrigation water use has becomes standing objective for long time. Efficient irrigation systems have been designed to optimize irrigation water use. Efficient irrigation is defined as an application of the exact amount of water at the exact time to the soil around the needy plant. To help apply the exact amount (volume flow rate) of water at the right timing, smart and automatic irrigation system was designed, built, and tested. The smart irrigation system comprises of several components (sensors, automatic valves and pump) automatically communicate through Arduino (controller). The sensors detect the soil moisture and determine the exact amount of water required. Then the sensors communicate with the pump and the valves via the controller to apply the proper amount of water at the time it is required. Based on the pump flow rate, timers are used to stop the pump when the required amount of water is applied. Regardless of the difficulties encountered the smart irrigation system was successfully designed and a prototype was built and tested. The different system's components were, matched, tested for compatibility, and assembled. The software that facilitates the communication between the different system's components was developed, installed, and tested. Then the system's workability and operation ability were tested. At the beginning there were some complications hindered the system to run smoothly. The complications include: mal functioning of the moisture sensors; moisture readings were qualitative (high/low); Arduino failed to receive information from multiple sensors; the dripping nozzles were leaking. After taking care of these problems the system was running smoothly. However, the smart irrigation system was found to be versatile, flexible, and working effectively [7].

Existing research primarily focuses on water-saving methods using automated irrigation systems based on wireless sensor networks (WSN) [1, 8, 9]. However, these systems often employ static irrigation conditions throughout the crop cycle or may indiscriminately irrigate the entire field based on sensor thresholds, lacking adaptability to varying crop needs. Recognizing the importance of dynamic irrigation tailored to different crop growth stages, the paper introduces the concept of dynamic irrigation treatments, crucial for optimizing crop productivity. More-over, it identifies a gap in existing solutions regarding the provision of both automatic and manual irrigation treatments throughout the crop cycle, particularly addressing the diverse needs of heterogeneous crop fields. Additionally, the lack of farmer-field interaction capabilities in current systems is highlighted, emphasizing the importance of enabling real-time communication and control for farmers. These insights drive the development of AgriSens, aiming to address these challenges by designing a comprehensive IoT-based irrigation system [10]. Existing research primarily focuses on water-saving methods using automated irrigation systems based on wireless sensor networks (WSN). However, these systems often employ static irrigation conditions throughout the crop cycle or may indiscriminately irrigate the entire field based on sensor thresholds, lacking adaptability to varying crop needs. Recognizing the importance of dynamic irrigation tailored to different crop growth stages, the paper introduces the concept of dynamic irrigation treatments, crucial for optimizing crop productivity. Moreover, it identifies a gap in existing solutions regarding the provision of both automatic and manual irrigation treatments throughout the crop cycle, particularly addressing the diverse needs of heterogeneous crop fields. Additionally, the lack of farmer-field interaction capabilities in current systems is highlighted, emphasizing the importance of enabling real-time communication and control for farmers. These insights drive the development of AgriSens, aiming to address these chal-lenges by designing a comprehensive IoT-based irrigation system [10].

3 Proposed System

The Smart Irrigation Automatic Motor Control System offers a holistic approach to enhancing agricultural irrigation processes via automation and intelligent con-trol features. Central to its design is the NodeMCU microcontroller, renowned for its adaptability to diverse sensors and straightforward programmability. This system architecture

(Fig. 1) incorporates several components, each serving a crucial function in ensuring seamless operation.

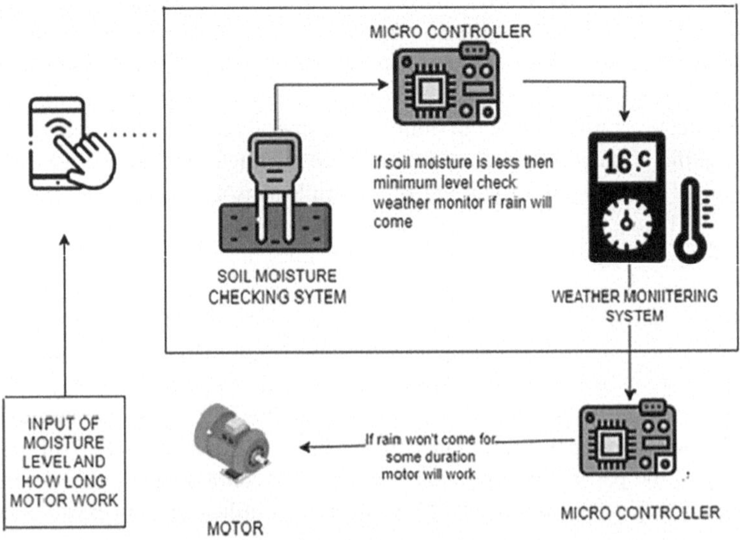

Fig. 1. Smart Irrigation System Architecture

3.1 System Architecture and Components

The Automatic Motor Control System for Smart Irrigation operates through a cyclical process of data acquisition, analysis, and action. Initially, the NodeMCU controller gathers environmental data from sensors including the DHT11 for temperature and humidity and the soil moisture sensor for soil moisture content. Once collected, this data undergoes analysis within the NodeMCU, where algorithms assess the current conditions against predefined thresholds to determine irrigation requirements. If the soil moisture levels indicate a need for hydration, the NodeMCU sends commands to the water motor, activating it to initiate irrigation. Concurrently, the LCD display monitoring and decision-making purposes. This process continues iteratively, with the system constantly adjusting irrigation operations based on evolving environmental conditions. By seamlessly integrating these components and employing intelligent control mechanisms, the system optimizes water usage, enhances crop growth, and empowers farmers with actionable insights for efficient land management, thereby embodying the synergy of technology and agriculture in modern farming practices. System Components. The system architecture encompasses multiple components, each playing a vital role in the overall functionality:

- NodeMCU Controller: Serves as the brain of the system, responsible for data processing, decision-making, and motor control based on sensor in-puts.

- DHT11 Sensor: Monitors ambient temperature and humidity levels, providing crucial environmental data for irrigation management.
- Soil Moisture Sensor: Measures soil moisture content to determine irrigation requirements, ensuring optimal soil conditions for plant growth.
- Water Motor: Actuates water flow to irrigate the fields when soil moisture levels indicate the need for hydration.
- LCD Display: Offers a user-friendly interface for real-time monitoring of sensor readings and system status, enabling farmers to make informed decisions and track irrigation activities.

Through seamless integration of these components and intelligent control algorithms, the system autonomously regulates irrigation operations, conserving water resources, maximizing crop yields, and empowering farmers with actionable insights for efficient land management. This system embodies the convergence of technology and agriculture, ushering in a new era of precision farming and sustainable agricultural practices. In the realm of modern agriculture, the integration of technology has revolutionized traditional farming practices. One such innovation is the implementation of smart irrigation systems that efficiently manage water resources while optimizing crop growth. This proposed method outlines the design and functionality of an automatic motor control system for smart irrigation, leveraging NodeMCU as the controller, an LCD display for sensor value visualization, DHT11 sensor for temperature and humidity detection, soil moisture sensor for assessing soil moisture levels, and a water motor for irrigation.

The system workflow began with data acquisition, where the NodeMCU gathered data from the DHT11 sensor for temperature and humidity readings, as well as from the soil moisture sensor for soil moisture levels. Subsequently, the collected data was analyzed to ascertain whether the soil moisture level fell below a predefined threshold, signaling the necessity for irrigation. Upon reaching this determination, the system proceeded to decision making, triggering the water motor to commence irrigation. Additionally, the user interface, manifested through the LCD display, provided real-time sensor values, enabling users to monitor environmental conditions and system functionality. The proposed meth-od offers several advantages. Firstly, it enables efficient water management by continuously monitoring soil moisture levels, thereby ensuring irrigation is ad-ministered only when necessary, effectively preventing water wastage. Secondly, by maintaining optimal soil moisture levels and environmental conditions, the system fosters healthy plant growth and enhances crop yields. Thirdly, through automated control, human intervention is minimized, leading to precise irrigation and conservation of water resources while reducing manual labor. Lastly, the incorporation of a user-friendly interface, such as the LCD display, empowers farmers with easy access to vital information, facilitating informed decision-making and system monitoring.

4 Results and Discussion

The integration of IoT (Internet of Things) technologies in agriculture has significantly improved efficiency and productivity. An automatic motor control system for smart irrigation utilizes various sensors and a microcontroller to monitor environmental conditions and automate the irrigation process (Fig. 2). This analysis explores the components and functionality of such a system, focusing on the use of NodeMCU as the controller, LCD display for sensor data visualization, DHT11 sensor for temperature and humidity detection, soil moisture sensor for assessing soil moisture levels, and water motor for irrigation. The humidity and the temper-ature measured by the system is shown in the Fig. 3.

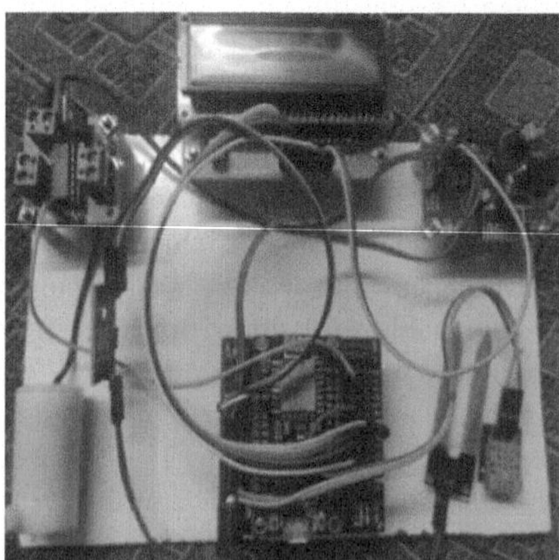

Fig. 2. Implementation of Smart Irrigation System

Fig. 3. Measure Humidity & Temperature Displayed in the LCD Display

The DHT11 sensor is a fundamental component in IoT projects, offering affordable digital temperature and humidity readings. NodeMCU, powered by the ESP8266 microcontroller, serves as an intermediary between the DHT11 sensor and MATLAB. With its Wi-Fi capabilities, NodeMCU collects sensor data and transmits it to MATLAB for processing. MATLAB, a powerful numerical computing environment, receives the data and generates graphical representations. MATLAB 2021a, specifically, facilitates the visualization process with its user-friendly interface. Through integration with NodeMCU, MATLAB enables real-time monitoring of temperature and humidity. Data received from the DHT11 sensor can be parsed and analyzed within MATLAB. This setup provides insights into environmental conditions over time, crucial for various applications. By leveraging these technologies, users can create efficient monitoring systems for temperature and humidity. Show the graphical results on live data.

A smart irrigation system controlled by MATLAB 2021a offers efficient manaagement of water resources for agricultural purposes. Users can select specific crops from a predefined list within the MATLAB interface (Fig. 4) Parameters such as temperature, humidity, and motor status (on/off) are displayed for monitoring and control. Upon selecting the crop type, the system initiates the irrigation process, ensuring optimal water delivery based on the crop's requirements. This integration of crop selection and environmental monitoring allows for precise irrigation scheduling tailored to the specific needs of different crops and the system results are shown in the Fig. 5.

Users can select crop types for moisture level sensing and motor control. If moisture levels indicate a need for water, the system activates the motor. Live weather data informs the system's decision-making, with rain forecasts influencing motor activation.

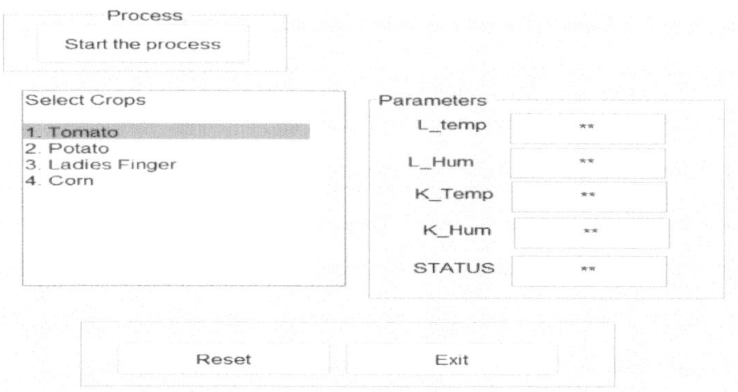

Fig. 4. User Navigation Scene for to Start the Process

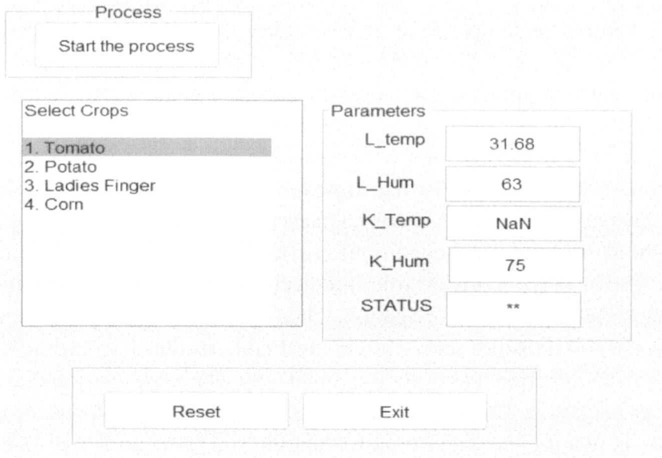

Fig. 5. Smart Irrigation System GUI Results

If rain is expected within the next four days, the motor remains off to avoid overwatering. However, if rain is not forecasted, the motor starts when moisture levels are low and switches off when optimal levels are reached. This integrated approach optimizes irrigation while considering both current and future weather conditions.

Table 1. Showcases a sample of sensor data collected over a 5-h period, illustrating the system's response to changing environmental conditions. As the soil moisture level drops below a predetermined threshold (e.g., 35%), the motor is activated to initiate irrigation. Once the soil moisture level rises above the threshold, the motor is turned off. This demonstrates the system's ability to maintain optimal soil moisture levels through automated irrigation, thus enhancing water efficiency and crop growth.

Table 1. Sensor Data and Irrigation Response

Time Interval (hours)	Temperature (°C)	Humidity (%)	Soil Moisture Level (%)	Motor Status
0–1	25	60	40	OFF
1–2	27	58	35	OFF
2–3	29	55	30	ON
3–4	28	57	45	OFF
4–5	26	59	42	OFF

5 Conclusion

In conclusion, the integration of NodeMCU as the controller, alongside an LCD display, DHT11 sensor for temperature and humidity monitoring, soil moisture sensor, and a water motor, offers a robust and efficient solution for smart irrigation systems. By leveraging these components, we have developed a sophisticated automatic motor control system that caters to the specific needs of agricultural fields.The utilization of NodeMCU provides a versatile and programmable plat-form for interfacing between sensors and the motor, allowing for real-time monitoring and control. The inclusion of an LCD display ensures that users can easily access and interpret sensor data, providing vital insights into environmental conditions crucial for effective irrigation management. The DHT11 sensor plays a pivotal role in detecting temperature and humidity levels, enabling precise adjustments to irrigation schedules based on environmental factors. Additionally, the soil moisture sensor offers invaluable feedback regarding the moisture con-tent of the soil, ensuring that water is applied only when necessary and avoiding both under and over irrigation. The automatic motor control feature serves as the cornerstone of the system, activating the water motor when the soil moisture lev-el falls below the desired threshold. This functionality not only optimizes water usage but also ensures that crops receive adequate hydration, thereby promoting healthier growth and maximizing yield potential. By combining these elements into a cohesive system, a smart irrigation offered solution that not only enhances efficiency but also contributes to sustainable agricultural practices. With the ability to automatically respond to changing environmental conditions, our system empowers farmers to manage their irrigation operations effectively while conserving valuable resources. In essence, automatic motor control system rep-resents a significant advancement in the field of agricultural technology, offering a practical and scalable solution for optimizing water management in agriculture land fields.

Acknowledgement. This work is supported by Centre for Sponsored Research and Consultancy (CSRC) Anna University, Chennai under Student Innovation Project Scheme, 2022–2023.

References

1. Bouali, T., Abid, M.R., Boufounas, E.-M., Hamed, T.A., Benhaddou, D.: Renewable energy integration into cloud & IoT-based smart agriculture. IEEE Access **10**, 1175–1191 (2022)
2. Ghiasi, M., Wang, Z., Mehrandezh, M., Paranjape, R.: A systematic review of optimal and practical methods in design, construction, control, energy management and operation of smart greenhouses. IEEE Access **12**, 2830–2853 (2024)
3. Viani, F., Bertolli, M., Salucci, M., Polo, A.: Low-cost wireless monitoring and decision support for water saving in agriculture. IEEE Sens. J. **17**(13), 4299–4309 (2017)
4. Shaikh, F.K., Karim, S., Zeadally, S., Nebhen, J.: Recent trends in Internet-of-Things-enabled sensor technologies for smart agriculture. IEEE Internet Things J. **9**(23) 23583–23598 (2022)
5. Mowla, M.N., Mowla, N., Shah, A.F.M.S., Rabie, K.M., Shongwe, T.: Internet of Things and wireless sensor networks for smart agriculture applications: a survey. IEEE Access **11**, 145813–145852 (2023)
6. Udutalapally, V., Mohanty, S.P., Pallagani, V., Khandelwal, V.: sCrop: a novel device for sustainable automatic disease prediction, crop selection, and irrigation in Internet-of-Agro-Things for smart agriculture. IEEE Sens. J. **21**(16), 17525–17538 (2020)
7. Kumar, A., Tiwari, R.G., Trivedi, N.K.: Smart farming: design and implementation of an IoT-based automated irrigation system for precision agriculture. In: 2023 3rd International Conference on Innovative Sustainable Computational Technologies (CISCT), pp. 2830–2853. Dehradun, India (2024)
8. Aldegheishem, A., Alrajeh, N., García, L., Lloret, J.: SWAP: smart water protocol for the irrigation of urban gardens in smart cities. IEEE Access **10**, 39239–39247 (2022)
9. Polymeni, S., Skoutas, D.N., Kormentzas, G., Skianis, C.: FINDEAS: a FinTech-based approach on designing and assessing IoT systems. IEEE Internet Things J. **9**(24), 25196–25206 (2022)
10. Roy, S.K., Misra, S., Raghuwanshi, N.S., Das, S.K.: AgriSens: IoT-based dynamic irrigation scheduling system for water management of irrigated crops. IEEE Internet Things J. **8**(6), 5023–5030 (2020)

Classification and Prediction Analysis in Healthcare

Eye Fundus Disease Classification Using Artificial Intelligence

A. S. Harisudhan, Raghul Prasanna, J. Vaibavi, and Sridevi Sridhar(✉)

Department of Computer Science and Engineering, SRM Institute of Science and Technology, Vadapalani, Chennai, India
{ah9296,rr5564,vj4467,sridevis2}@srmist.edu.in

Abstract. Eye fundus conditions are dangerous and can cause significant visual impairment if not detected early. Diabetic retinopathy, cataracts, and glaucoma are among the conditions for which manual assessment is directly impacted by ophthalmologists' experience. The study intends to use artificial intelligence to develop a diagnostic system that is concentrated on the precise and effective classification of eye fundus diseases in order to address this challenge. The research involves the curation of an extensive dataset, named Eye Diseases Classification, that consists of a variety of eye fundus images that illustrate different conditions, including cataracts, glaucoma, and diabetic retinopathy. Expert ophthalmologists have painstakingly annotated every image in the dataset, offering precise ground-truth information that is essential for segmentation tasks.

The findings of the experiment demonstrate how well the suggested AI system performs in accurately classifying eye fundus diseases and recognizing impacted areas in the images. This study could potentially reduce the risk of blindness and severe vision impairment by revolutionizing the diagnosis of these diseases. The system expedites diagnosis by automating the classification process, enabling earlier intervention and treatment. Additionally, using AI lessens the workload for ophthalmologists, freeing up their time for more complicated cases and improving the effectiveness of healthcare as a whole. In the end, using AI to diagnose eye fundus illnesses is a huge step forward that will impact public and clinical health in many ways.

Keywords: Eye Fundus Disease · Artificial Intelligence · Deep Learning · Disease Classification · Segmentation · Convolutional Neural Networks · U-Net · Ophthalmology · Medical Imaging · Diagnostic System

1 Introduction

Early detection and treatment of diseases of the eyes like macular degeneration and diabetic retinopathy depend heavily on accurate diagnosis. Disparities in healthcare, medication, and therapy programs, and a lack of medical system integration all contribute to the underfunding of ocular disorders. Thanks to its sophisticated tools for segmenting and classifying eye fundus images, artificial intelligence (AI) has completely changed the field of ophthalmology. Large datasets are analyzed by machine learning algorithms,

P. D. Sivakumar et al. (Eds.): IRCCTSD 2024, CCIS 2360, pp. 125–133, 2025.
https://doi.org/10.1007/978-3-031-82389-3_11

which makes it possible to automatically identify and categorize minute pathological changes. By training on a variety of datasets, AI models for diagnosing eye fundus diseases identify patterns and characteristics suggestive of particular diseases, opening the door to reliable and precise automated diagnostics. Refraction error, cataracts, glaucoma, corneal opacities, macular degeneration, diabetic retinopathy, and presbyopia are among the conditions that, as defined by the World Health Organization, impact roughly 1.2 billion individuals globally.

Disruption to the veins of the retina which is located behind the human eye is the trigger of a condition called diabetic retinopathy It is the job of the retina to detect illumination and relay that information. Visual elements are seen by the brain via the decoding of messages. Both initial and progressed DR are phases of disease recurrence. Additionally, retinal veins begin to enlarge with uneven diameters.

Glaucoma causes impairment to the optical nerve, which connects the retina to the brain. The internal pressure is a force within the eye that causes damage to the nerve that signals vision. The risk of this disease, may cause sight impairment and visual loss if not caught in its initial stages, increases twice as insulin levels rise. When insulin levels are too elevated they cause protein to degenerate, leading to cataract, and subsequently causes the sight to get hazy and eventually causes poor sight. Individuals with diabetes are at a higher risk of getting cataracts and clouded optics at a younger age compared to those without the disease. The effective identification of the disease is impacted by disparities in health coverage, inadequate prevention services, treatment, and rehabilitation. According to the World Health Organization's 2019 report, deep learning techniques are essential for identifying, detecting, and categorizing these conditions [2, 4].

In this section, we will delve into the significance of employing artificial intelligence for eye fundus disease classification. ML is crucial for tasks like object recognition and picture interpretation. Numerous methods exist for accomplishing such goals [7]. Biomedical computing, visual argumentation, and many more fields rely heavily on visuals in today's digital environment. A plethora of characteristics may be utilized to convey each data record. However, not every trait is classified or analyzed with any degree of significance. As a result, there is a great deal of interest in studying choice of features and feature extraction. In image segmentation, distinct classes are assigned to each pixel in the image itself. One name for this pixel labeling activity is dense forecasting. Assuming that a picture contains a variety of items, such as shrubs, automobiles and animals. This means that picture segmentation will lump all shrubs into one category, while still assigning other categories to the appropriate groups. When it comes to picture segmentation, one crucial consideration is that it treats duplicate objects as just one class. Using instance segmentation, we can distinguish between objects of the same kind. Classifying images for pattern recognition and object identification is a common task for CNN.

2 Related Works

The paper [1, 16] suggests an approach for DR grading using a Federated Learning DL model. The extraction of local, global, and interim information from fundus pictures was accomplished via the use of a deep network. Training and testing the model on a broad collection of diverse consumers yielded better results than before. The issue of patient confidentiality arises from a suggested architecture that uses a centralized server to facilitate coordination across several medical facilities with a success rate of 98.6%.

In Elloumi, Y., Akil, M. and Boudegga, H., 2019 [4], identifying pathologies of the eyes from fundus image poses a significant challenge to medical care. In actuality, distinct severity phases for each pathology can be inferred by confirming the presence of particular lesions. Morphological characteristics define each lesion. Furthermore, a number of lessons from other disorders share characteristics. We observe that a patient may have multiple pathologies affecting them at the same time. As a result, the detection of ocular pathology offers a multiclass classification with a complicated resolution concept. Many techniques [3, 6] for classification of ocular diseases using fundus photos have been put forth. Because deep learning approaches may customize the network according to the detection purpose, they stand out for having superior performance detection [12]. The study [13] offers the method incorporating deep learning for detecting ocular pathologies. First, we examine the approaches that are currently in use for pathology categorization or lesion segmentation. Subsequently, we derive the fundamental processing processes and examine the suggested neural network architectures. Next, it is decided what software and hardware are required for employing the deep learning architecture. After that, we look at the principles of experimentation to assess the techniques and databases employed for the training and testing stages. Additionally presented and discussed are the execution timings and detection performance ratios.

Diabetes, also referred to as mellitus, or type 2 diabetes, disorder wherein an individual's body, either is unable in response to glucose produced through the pancreas, releases into the bloodstream or cannot create enough insulin. Over time, people with diabetes have an increased chance of developing a range of ocular disorders. Owing to developments in machine learning, it is now possible to identify diabetic eye disease early thanks to an automated method that has multiple benefits over manual detection. Numerous studies on the diagnosis of DR were just published. The automated methods for diagnosing diabetic retinopathy are thoroughly reviewed in Sarki R. et al., 2020 [5] from a number of angles, including: i) dataset; ii) image reprocessing methods; iii) ML models; and iv) performance indicators. Including cutting-edge techniques, this thorough overview of diabetes eye disease identification methods, the survey aims to educate diabetic patients, medical professionals, and research communities.

Artificial intelligence approaches have been the focus of medical health systems in order to facilitate rapid diagnosis [2]. Nonetheless, standardizing the way health data is recorded is still necessary to improve the accuracy as well as dependability machine learning by considering a number of variables. The aim of this undertaking is to create a foundational framework for the international standard format storage of diagnostic data, allowing algorithms for machine learning to use symptoms to predict a diagnosis of disease. An easy use interface was developed by Malik S. et al., 2019 [6] to ensure error-free data entry. Furthermore, several machine learning techniques, in addition to using neural network techniques for patient data analysis, taking into consideration a number of factors, include age, health history, and clinical findings. Medical practitioners used hierarchical hierarchies to structure the data, and the ICD-10 was used to make the diagnosis coding system of American Academies of Ophthalmology. The Random

Forest including decision tree algorithms outperform more sophisticated techniques such as neural networks along with a naive Bayes algorithm, with prediction rates of over 90%.

Glaucoma is a dangerous condition that is sometimes called the "unspoken thief of sight". When the disease is in its early stages, there are no outward signs or symptoms; however, blindness may result if treatment is not received. A lot of effort has gone into making glaucomatous changes in the eyes more accurate to spot and anticipate throughout the year as in Hagar, E., 2023 [11]. Fundus imaging modalities combined with artificial intelligence models are one of the most promising methods for accurately detecting and forecasting glaucoma [10].

3 Methodology

The groundbreaking CNN work derives its name from convolution, a mathematical operation that is linear between matrices. A CNN is made up of pooling layers, entirely connected layers, non-linearity layers, and convolutional layers. Pooling and non-linearity layers lack parameters as opposed to the fully-connected and convolutional layers' parameters. A CNN performs exceptionally well whenever applied to applications of machine learning [9]. The intelligent patterning classifier underwent construction, training, and evaluation using two distinct datasets. These were acquired online and were available without charge. To remove the flatness and retrieve the notable characteristics, the ReLU is used. Hard discharge, which are critical indicators of each disease e.g., diabetic retinopathy, can be detected by CNN [14, 15] (Fig. 1).

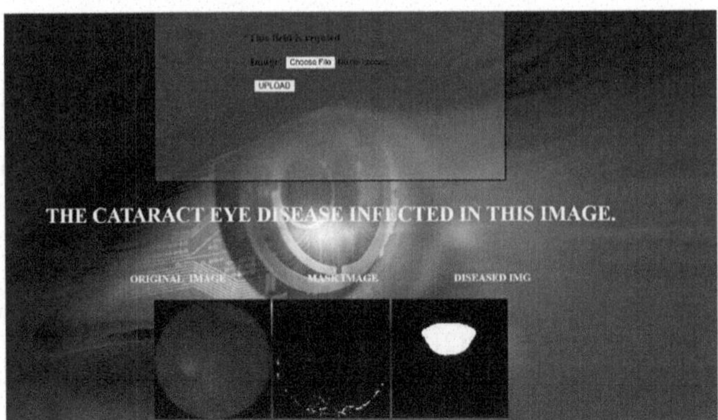

Fig. 1. Detection of Cataract in the Input eye image.

The detection of the rigid disc and the optic cup is characteristic of glaucoma. CNN training and testing may begin with these pre-processed pictures. This was accomplished by using the open-source Python neural network known as Keras using a pre-trained CNN is one of the benefits of this library [8, 15]. The numerical metrics presented in the paper are almost accurate. The metrics, including F1, recall, precision, and accuracy, are around 1.

The approach represents the benchmark in the realm of picture segmentation research. It is compatible with pictures that have three dimensions: height, breadth, and channel count. When there is a disparity between the images, picture enhancement is used. For an assortment with a lesser number of photographs compared to the other significant percentage of good retinal visuals, they are manipulated by mirroring, rotating, resizing, and cropping to create examples of the chosen images. Additional common way to prepare images for use is to scale them down. After determining whether the configuration is suitable, the picture is reduced to a decreased-resolution version. With improved lighting and clarity, system can pick up more significant details.

The picture sharpness is shown by the first two dimensions, while the variety of methods, or intensity ratings for red, green, and blue, is represented by the third one. In order to decrease computation time and prevent the issue of under fitting, it is common practice to compress the size of pictures before feeding them into the system. The aim of this is to standardize the raw images with respect to an analogue dataset in order to eliminate visual differences and create identical images. There is a label for each image in the collection that shows the number (zero to nine) it symbolizes. Our objective is to teach the convolution neural network model to correctly identify the number in every dataset through analyzing the structure of its pixels. The process of image segmentation involves breaking a picture to its component parts. The purpose of classification is to identify and retrieve individual pixel from a picture, which may include thousands of individual components. The ocular disc and cup may be segmented in fundus visuals.

A convolution layer named Convolution2D serves as the first hidden layer. Based on the intricate nature of a record, the convergence and clustering phases may be reiterated multiple times. This means that the characteristic maps decrease the number of details sent through successive component of the picture. It has the rectifier function and 32 feature maps that are 5×5 in size. The input layer looks like this. The MaxPooling2D pooling layer is responsible for taking the maximum output a 2×2 pool size configuration. The regularization process takes place at the layer of dropouts. To prevent over fitting, it is configured to arbitrarily remove twenty percent of the layer's connections. The fifth layer, flattened one is responsible for transforming the data from the two-dimensional matrix into a vector. A regular linked layer can analyze the output to its fullest extent. At last, the output layer uses activation function and 10 neurons per class to provide predictions that are similar to probabilities.

3.1 Proposed System

Our goal is to use artificial intelligence to create a quick and precise digital diagnostic for the categorization of various forms of eye fundus disease. According to studies, patients are at significant danger of losing their vision if they are not recognized early on. To address this, we strive to provide ophthalmologists with efficient and accurate support in recognizing and addressing various eye fundus disorders as soon as possible. We use artificial neural networks to evaluate the optimal strategy for predicting the discrete class of fresh input. The first stages include loading the dataset, data mining, and preprocessing.

Initially, the dataset is separated as train and test data. The findings are compared to the expected outcomes. The optimal model is finalized after careful consideration of

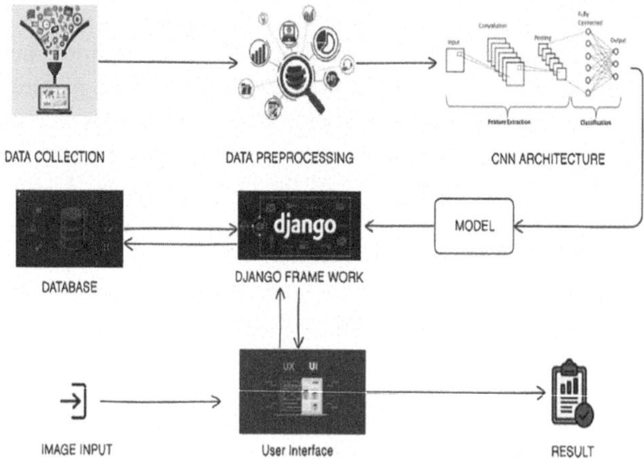

Fig. 2. Eye fundus disease classification architecture.

training data confirming with the validation dataset. This proposal intends to augment and increase the ability of eye specialists to diagnose the eye diseases through deep learning. Figure 2 displays the proposed architecture for classifying eye fundus illness with artificial intelligence.

4 Experimental Results

The experiments are carried out with Kaggle's Eye fundus disease public dataset. Figure 3 represents the accuracy comparison for classification of eye fundus disease. The line graph illustrates the efficacy in deploying the training data set for this proposal. It is worth noting that the accuracy begins at very low level and exponentially increases as the number of epochs increases.

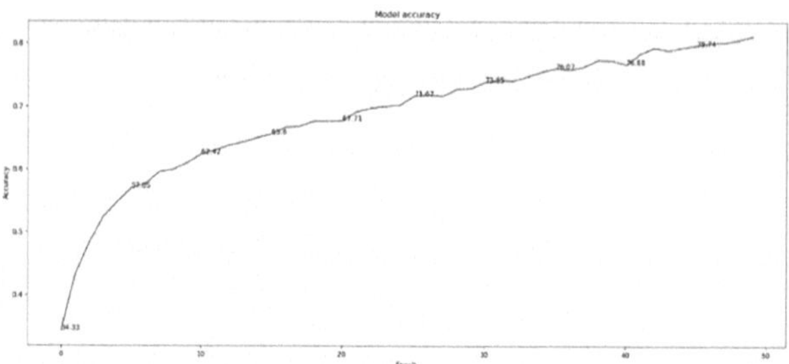

Fig. 3. Accuracy of LeNet.

Figure 4 shows a model loss graph for classifying eye fundus diseases. In contrast to model accuracy, model loss is high for the first few epochs and decreases exponentially as the value of the epoch is increased.

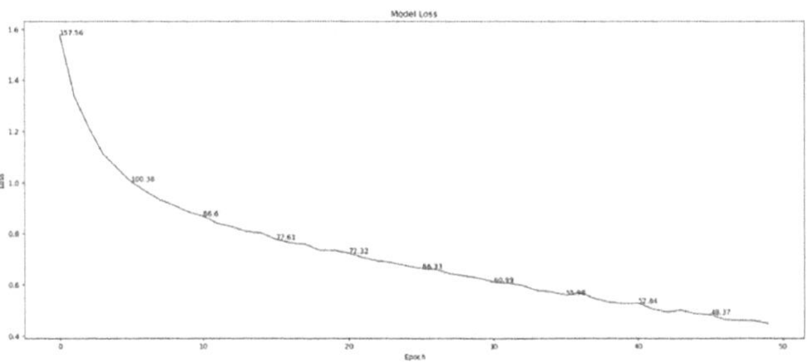

Fig. 4. Loss of LeNet.

The classification of multiple eye fundus diseases was analyzed using several machine learning models, including LnNet, LeNet, VGG, and ResNet. The LeNet model achieved the highest accuracy at 81.16%. LnNet followed with an accuracy of 65.74%, while VGG and ResNet attained 50.36% and 59.76% accuracy, respectively. Based on the comparison of these models, LeNet was identified as the most accurate approach for classifying eye fundus diseases.

5 Challenges

In order to effectively examine the progression of technology in healthcare, a threefold strategy is required. To begin with, there exists an urgent need for extensive and varied datasets that faithfully represent the intricacies of real-life situations. The effectiveness of ML models in the healthcare sector relies on the inclusiveness and accuracy of datasets, which guarantees resilient performance across diverse demographics and circumstances. Additionally, promoting enhanced collaboration among specialists becomes critical. This requires the collaborative effort by medical professionals and researchers to establish a consensus on disease-specific matters. A well-defined set of factors and a significant quantity of image evaluation result from this type of collaboration. These contributions enhance the precision and dependability of diagnostic models. Lastly, it is critical that medical professionals, patients, and data scientists collaborate more effectively. This process entails ongoing communication and iterative improvement of software interfaces throughout the development cycle. The objective is to supplement human interaction, which is fundamental to medical practice, rather than supplant it. This approach guarantees that technological progress enhances compassionate and individualized care, which is the essence of medicine, rather than substituting it.

6 Conclusion and Future Scope

Classification model embedded with CNN can contribute to the advancement of ophthalmology, particularly in regard to the worldwide issues of eye fundus diseases particularly glaucoma and diabetic retinopathy. Based on the study, convolution neural networks are among the most effective tools for picture categorization. Undoubtedly, in the future, the benefits that AI can reap great benefits to enhance the quality of life of the patients. Furthermore, it will conduct evaluations more rapidly and precisely than retina specialists can presently do in a sustainable manner, enabling them to devote more time to honing the skills as scientists and clinicians.

Deep Learning algorithms frequently require a substantial quantity of data for model training purposes; therefore, it is essential to establish adequate and labeled training datasets. When dealing with low-quality photos or instances of overlaying border pixels between divided objects, there is greater leeway to handle uncertainty. Critical jobs, such as tracking unexplained activity in constrained regions, need these processes to be accurate enough to be depended upon. Even a little error might have severe consequences.

In the near future, there will be a scarcity of doctors and other physicians, which will make medical inequality worse and make it harder to treat avoidable illnesses. Artificial Intelligence is a promising strategy for addressing these issues and enhancing medical treatment on a global basis and at the one-to-one basis via better, precise, and tailored management.

Achieving adequate accuracy becomes unattainable when the learning range is constrained. There are two possible approaches to this problem. Low-learning algorithms were initially employed to gather training data. To produce more accurate training data in order to develop Deep Learning models with more reliability and distinct traits, additional study is required. For instance, it is possible to create an algorithm that can both detect items and keep an eye on what is going on around it. Because of this, the usefulness of such methods will grow, and formerly labor intensive processes will become substantially more automated. Artificial intelligence will play a growing component within medicine in the years to come, enhancing the quality of care and enabling retinal experts to devote their time to being excellent doctors and researchers by performing examinations with greater efficiency and precision than is presently feasible.

References

1. Mohan, N.J., Murugan, R., Goel, T., Roy, P.: DRFL: federated learning in diabetic retinopathy grading using fundus images. IEEE Trans. Parallel Distrib. Syst. **34**(6), 1789–1801 (2023). https://doi.org/10.1109/TPDS.2023.3264473
2. Goutam, B., Hashmi, M.F., Geem, Z.W., Bokde, N.D.: A comprehensive review of deep learning strategies in retinal disease diagnosis using fundus images. IEEE Access **10**, 57796–57823 (2022). https://doi.org/10.1109/ACCESS.2022.3178372
3. Bernabé, O., Acevedo, E., Acevedo, A., Carreño, R., Gómez, S.: Classification of eye diseases in fundus images. IEEE Access **9**, 101267–101276 (2021). https://doi.org/10.1109/ACCESS.2021.3094649
4. Elloumi, Y., Akil, M., Boudegga, H.: Ocular diseases diagnosis in fundus images using a deep learning: approaches, tools and performance evaluation. In: Real-Time Image Processing and Deep Learning, vol. 10996, pp. 221–228. SPIE (2019)
5. Sarki, R., Ahmed, K., Wang, H., Zhang, Y.: Automatic detection of diabetic eye disease through deep learning using fundus images: a survey. IEEE Access **8**, pp. 151133–151149 (2020). ISSN 2169-3536
6. Malik, S., Kanwal, N., Asghar, M.N., Sadiq, M.A.A., Karamat, I., Fleury, M.: Data driven approach for eye disease classification with machine learning. Appl. Sci. **9**(14), 2789 (2019)

7. Daich Varela, M., et al.: Artificial intelligence in retinal disease: clinical application, challenges, and future directions. Graefe's Arch. Clin. Exp. Ophthalmol. **261**(11), 3283–3297 (2023)
8. Nemade, S.B., Sonavane, S.P.: Image segmentation using convolutional neural network for image annotation. In: 2019 International Conference on Communication and Electronics Systems (ICCES), pp. 838–843 (2019)
9. Nagula, J.M., Raman, M., Goel, T.: Role of machine and deep learning techniques in diabetic retinopathy detection. In: Chowdhary, C. (ed.) Multidisciplinary Applications of Deep Learning-Based Artificial Emotional Intelligence, pp. 32–46. IGI Global (2023). https://doi.org/10.4018/978-1-6684-5673-6
10. Hagar, E.: Detection of glaucoma using artificial intelligence in fundus image: a narrative review. Sci. Open Prepr. (2023). https://doi.org/10.14293/S2199-1006.1.SOR-.PP1AAS1.v1
11. Babaqi, T., Jaradat, M., Yildirim, A.E., Al-Nimer, S.H., Won, D.: Eye disease classification using deep learning techniques. arXiv preprint arXiv:2307.10501 (2023)
12. Khan, M.S., et al.: Deep learning for ocular disease recognition: an inner-class balance. Comput. Intell. Neurosci. (2022)
13. Ouda, O., Abdel Maksoud, E., Abd El-Aziz, A., Elmogy, M.: Multiple ocular disease diagnosis using fundus images based on multi-label deep learning classification. Electronics **11**, 1966 (2022)
14. Alyoubi, W.L., Shalash, W.M., Abulkhair, M.F.: Diabetic retinopathy detection through deep learning techniques: a review. Inform. Med. Unlocked **20**, 100377 (2020)
15. Nasajpour, M., Karakaya, M., Pouriyeh, S., Parizi, R.M.: Federated transfer learning for diabetic retinopathy detection using CNN architectures. In: SoutheastCon 2022, Mobile, AL, USA, pp. 655–660 (2022). https://doi.org/10.1109/SoutheastCon48659.2022.9764031
16. Bilal, A., Sun, G., Li, Y., Mazhar, S., Khan, A.Q.: Diabetic retinopathy detection and classification using mixed models for a disease grading database. IEEE Access **9**, 23544–23553 (2021). https://doi.org/10.1109/ACCESS.2021.30561

Deep Learning-Based Multi-disease Detecting Model

A. Vasuki(✉), A. Ragu Kaushik, and M. Darshan Kumar

Department of Electronics and Communication Engineering, SRM Institute of Science and Technology, Vadapalani, Chennai 600026, India
{vasukia,ak6433,dm1841}@srmist.edu.in

Abstract. Planning an effective treatment plan and managing patients with neurological and chronic diseases, such as Parkinson's, Alzheimer's, and diabetic retinopathy, requires early and accurate diagnosis. This work introduces a thorough method for multi-disease identification based on deep learning models, such as the well-known VGG-16, ResNet-50, and EfficientNet-B0 architectures that excel in computer vision applications. Integrating the predictions from multiple models using their varied architectures and learning capacities improves the system performance by using ensemble learning approaches. Based on comparative examination and intensive testing, the two best-performing models are determined based on diagnostic accuracy. This proposed work involves implementing a fusion method that integrates the advantages of the chosen models to enhance the overall precision and dependability of the multi-disease detection system.

Keywords: Deep learning · multi-disease detection · Ensemble learning · Feature fusion · Alzheimer's disease · Parkinson's disease · Diabetic retinopathy

1 Introduction

Because of their crippling consequences and difficult diagnosis, neurodegenerative and chronic diseases including Alzheimer's, Parkinson's, and diabetic retinopathy provide significant difficulties to healthcare systems around the world. The process of manually interpreting medical photographs to diagnose diseases is laborious and subjective, frequently leading to inconsistent results. Convolutional Neural Networks (CNNs), a type of deep learning, have become a potent tool for automating medical image interpretation and improving diagnostic precision. Consequently, this work proposes a comprehensive deep learning-based framework for the identification of diabetic retinopathy, Parkinson's disease, and Alzheimer's disease. Consequently, this suggested work makes use of the capabilities of many deep learning architectures, such as VGG-16, ResNet-50, and EfficientNet-B0, to extract various disease-related elements from medical pictures. Through the integration of feature fusion and ensemble learning approaches, this system seeks to improve diagnostic accuracy for a range of disorders. This suggested deep learning strategy performs tailored training of deep learning models for each target ailment using disease-specific datasets to optimise the learning process for precise diagnosis. The predictions of these models are then combined using ensemble learning, and

P. D. Sivakumar et al. (Eds.): IRCCTSD 2024, CCIS 2360, pp. 134–144, 2025.
https://doi.org/10.1007/978-3-031-82389-3_12

their characteristics are further integrated through feature fusion to improve diagnostic performance. Extensive testing and validation with truth-based labels validate that the suggested method is effective in correctly identifying diabetic retinopathy, Parkinson's disease, and Alzheimer's disease. In conclusion, this work introduces a thorough deep-learning framework for detecting various disorders, taking into account improvements in medical picture analysis. The suggested method enables early detection and individualised treatment plans for neurodegenerative illnesses, which may improve disease diagnosis and patient care. Our technique is shown to be effective in reliably diagnosing Alzheimer's disease, Parkinson's disease, and diabetic retinopathy through extensive tests and validation against ground truth labels. The suggested model introduces a unified deep-learning framework that can simultaneously detect different diseases, which significantly advances medical image analysis.

Therefore, using disease data, customised training of deep learning models is carried out for each target disease-specific dataset to discover patterns and traits unique to each disease. The predictions of various models are then integrated via ensemble learning, and their distinct features are combined via feature fusion to enhance overall diagnostic performance.

2 Neural Network Architectures

The basic concept and neural network details of the VGG-16, Resnet and EfficientNet-B0 are described in this section.

2.1 VGG-16

A major advancement in the field of deep learning was made when the Visual Geometry Group (VGG) at the University of Oxford presented the VGG-16 architecture [1], as shown in Fig. 1. The three completely linked layers and thirteen convolutional layers of the architecture give it its name. VGG-16 is characterised by a consistent architecture that uses max-pooling layers and small 3×3 convolutional filters across the network. Every convolutional layer in the VGG-16 architecture is followed by a max-pooling layer that down-samples the input feature maps. This structure improves the network's feature representation capabilities by allowing it to extract hierarchical features from input images gradually. Additionally, VGG-16's sensitivity to minute details in medical pictures is increased by using small convolutional filters, which allow it to collect local features with great spatial resolution. The characteristics that VGG-16 learns from medical picture datasets enable the network to identify patterns and abnormalities particular to a certain disease, which increases the network's usefulness for multi-disease detection tasks.

2.2 ResNet-50

Microsoft Research introduced ResNet-50, a deep learning architecture that stands for Residual Network with 50 layers [2], as seen in Fig. 2. By resolving the difficulties involved in training very deep architectures, it represents a milestone in the field of deep neural networks.

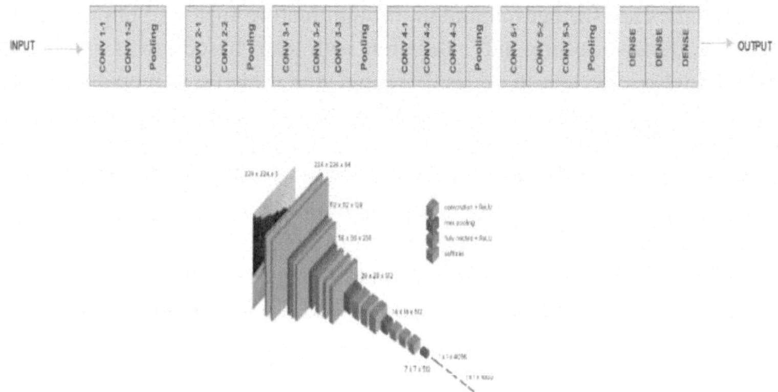

Fig. 1. VGG-16 Architecture

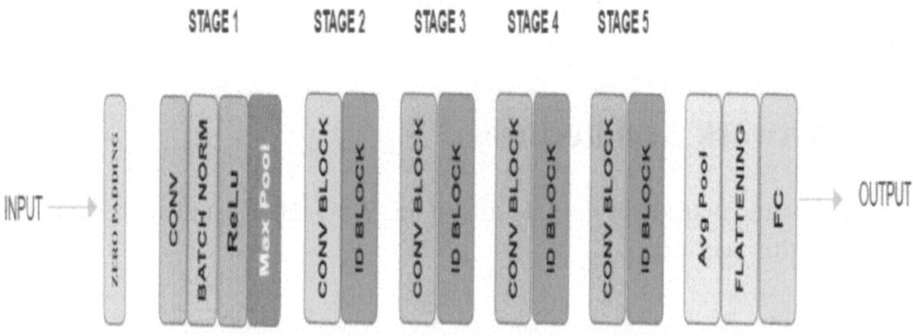

Fig. 2. Resnet-50 Architecture

ResNet-50 is equipped with a novel architecture that includes residual connections, often known as skip connections. By enabling the network to pick up residual features, these connections help to lessen the training process's vanishing gradient issue. Compared to conventional architectures, ResNet-50 can efficiently train considerably deeper networks. It consists of fifty convolutional layers arranged into blocks, where a shortcut connection comes after several convolutional layers in each block. ResNet-50 needs these residual blocks in order to extract hierarchical features from input images, such as minute details and intricate patterns that are important for jobs involving medical image analysis.

ResNet-50 has demonstrated remarkable performance in the field of medical image analysis in a number of applications, such as disease diagnosis, lesion recognition, and segmentation. Because of its ability to extract relevant features from medical pictures and the advantages of residual connections, ResNet-50 is an excellent choice for applications requiring precise and dependable feature extraction.

2.3 EfficientNet-B0

Figure 3 depicts the EfficientNet-B0 design. EfficientNet-B0, a convolutional neural network architecture created by Google Research, is a member of the EfficientNet family, which is intended to maximize accuracy and efficiency. It stands out in especially for being efficient and scalable, offering a well-balanced trade-off between performance and model size. The combined scaling approach of EfficientNet-B0 is a significant invention. It uses fewer parameters to achieve better performance by increasing the network's breadth, depth, and resolution proportionately. This method allows EfficientNet-B0 to maintain high accuracy while easily adapting to different computational budgets, which makes it appropriate for usage in situations with limited resources, including mobile and edge devices, for medical imaging applications [3]. With a core network structure that is scalable in width, depth, and resolution, the EfficientNet-B0 architecture produces a variety of models with varying computational costs and efficiencies. In the context of medical image analysis, EfficientNet-B0's streamlined architecture allows for the effective extraction of relevant features from medical images while minimizing computational resource usage. Its scalability allows to adapt the model to specific computational constraints without compromising accuracy, making it a versatile choice for multi-disease detection. In addition, EfficientNet-B0's efficient design facilitates rapid model training and optimization, allowing to perform efficient experiment with different architectures and frames. Using transfer learning techniques [4], EfficientNet-B0 can be refined to pre-train the models on medical image data, further improving performance and adaptability.

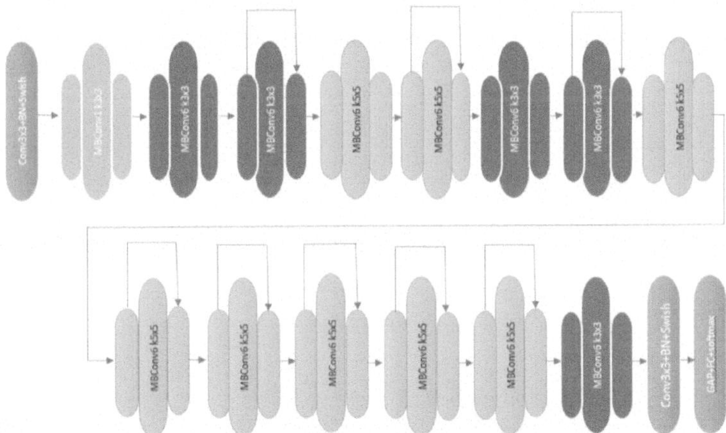

Fig. 3. EfficientnetB0 Architecture

2.4 Ensemble Learning

As shown in Fig. 4. Harnessing the power of multiple models to improve diagnosis, Ensemble learning is a powerful technique that uses the diversity and collective intelligence of several individual models to achieve better performance than any single model alone. In the context of medical image analysis, ensemble learning has emerged as a valuable approach to improve the accuracy and reliability of the diagnosis of various diseases, including Alzheimer's disease [5–7], Parkinson's disease [3, 8, 9] and diabetic retinopathy [2]. The basic idea behind ensemble learning is to combine predictions from multiple underlying models, everything getting trained on various variety subsets and datasets of the training data or using different machine-learning algorithms, to get a final prediction that is more reliable and accurate than any one model. Ensemble learning can be broadly classified into two main types [10, 11]: Averaging learning and Boosting learning. Averaging methods involve combining the predictions of several base models by taking their mean or weighted average. Common techniques include simple averaging, weighted averaging based on the per-performance of individual models, and geometric averaging.

2.5 Feature Fusion

Feature fusion as depicted as shown in Fig. 5, which is also known as multimodal fusion, is an important deep learning technique that combines features extracted from multiple categories or sources to create a unified and comprehensive representation. In the context of medical image analysis, feature fusion techniques play a crucial role in integrating multispecies information, clinical information, and genetic information from different imaging methods to improve diagnostic accuracy and reliability.

The main subjective of feature fusion is to exploit the complementary nature of information from different sources reducing their individual limitations, which ultimately improves the overall efficiency of the diagnostic system. Feature fusion is possible using a variety of methods, including early fusion, late fusion, and hybrid fusion [12, 13].

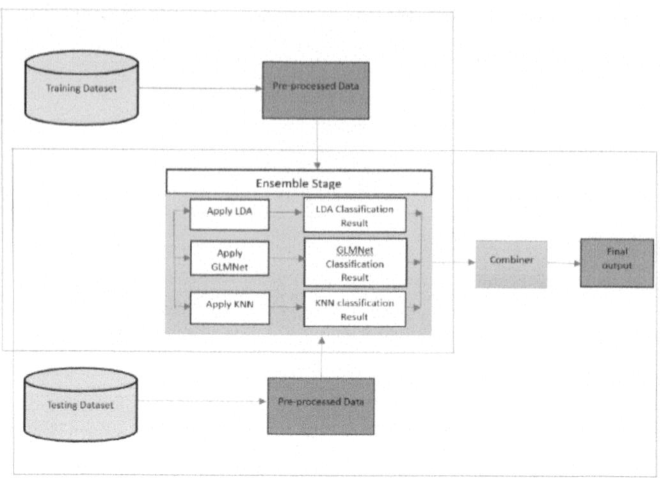

Fig. 4. Ensemble Learning Architecture

3 Proposed Deep Learning Model and Working Principle

The process involved and the working principle of the proposed method are discussed in this section.

3.1 Data Collection and Preprocessing

A diverse dataset encompassing medical images related to Alzheimer's disease, Parkinson's disease, and diabetic retinopathy from the Kaggle repository [19] is acquired. S The acquired dataset is standardized by ensuring uniformity in image size, resolution, and format to facilitate consistent model training. The preprocessing techniques [11, 14] such as normalization, cropping, and resizing are employed to enhance the quality and compatibility of the images. Then the dataset is augmented using techniques such as rotation, flipping, and zooming to increase its diversity and robustness, thereby improving the model's ability to generalize across different variations.

3.2 Model Development

Three process state-of-the-art deep learning architectures such as VGG-16, Res- Net-50, and EfficientNetB0 [15, 16] are implemented, leveraging established frameworks like TensorFlow or PyTorch [17]. The models are configured with appropriate point parameters, including the rate of learning, size of batch, and optimized algorithms. Then each model is trained on the preprocessed dataset using a portion of the data for training and a separate portion is divided for validation to monitor performance and prevent overfitting. Early stopping and checkpoints are applied to prevent the model from getting overfitted and preserve the better-performing models during the process of training.

3.3 Ensemble Learning

Employ ensemble learning [18] combines the predictions of the individual models VGG-16, ResNet-50, and EfficientNet-B0 to improve overall performance. Ensemble techniques such as model averaging or weighted averaging are implemented to aggregate the predictions of the individual models. The model's hyperparameters, such as the weights assigned to individual models are fine-tuned to optimize performance on the validation dataset.

3.4 Characteristic Fusion

The top-performing models are selected from the ensemble based on their performance on the validation dataset. The characteristic fusion techniques are implemented to integrate the distinct features extracted by the chosen models, leveraging their complementary information. The different fusion strategies such as feature concatenation, feature addition, or attention mechanisms are examined to effectively combine the features and enhance the model's discriminative power.

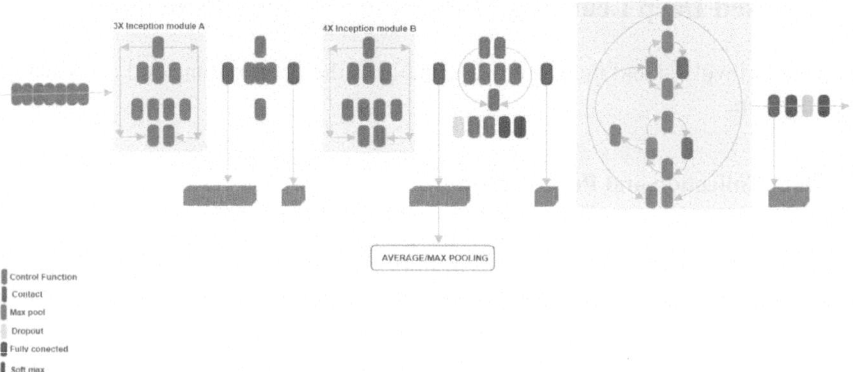

Fig. 5. Feature Fusion Architecture

3.5 Final Model Creation

The characteristic representations obtained from the selected models are fused to create the final deep-learning model for multi-disease detection. Then the final model is fine-tuned on the training dataset to optimize its performance further, adjusting hyperparameters such as regularization strength and dropout rates. Finally, the model performance is validated on a separate test dataset to assess its generalization and robustness.

3.6 Evaluation and Validation

The performance of the final model using standard evaluation metrics such as accuracy, precision, recall, and F1-score are evaluated. Assessing the model's performance across different disease categories ensures balanced and reliable detection. Statistical analysis and hypothesis testing are conducted to validate the significance of the model's performance improvements compared to baseline methods.

3.7 Cross-Validation and Interpretation

The cross-validation techniques are applied to assess the stability and consistency of the model's performance across multiple iterations and data splits. The model's predictions are interpreted using visualization techniques such as saliency maps or attention mechanisms to understand the features contributing to disease detection.

3.8 Documenting and Reporting

The proposed entire methodology, including data preprocessing steps, model architectures, hyperparameters, training procedures, and evaluation metrics is documented. A comprehensive report detailing the methodology, experimental setup, results, analysis and conclusions drawn from the study is prepared. Transparency and reproducibility are ensured by providing code repositories and supplementary materials.

4 Mathematical Model and Workflow

The proposed deep learning mathematical model workflow is illustrated as shown in Fig. 6. The model combines the features of VGG-16 and Efficientnet-B0 which are better performance in the prediction of diseases compared to Resnet architecture.

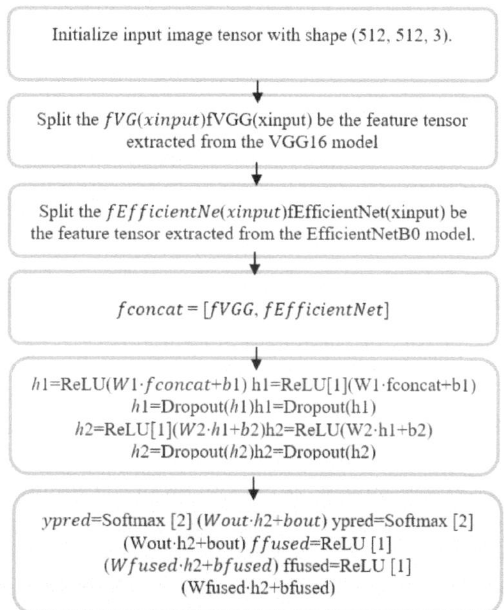

Fig. 6. Flow of Proposed Deep Learning Model

The different activation functions used in the proposed model are expressed. Equation 1 represents the activation function $f(x)$ for ReLU and expressed as

$$f(x) = \max(0, x) \tag{1}$$

The ReLU activation function is used in the hidden layers of the proposed integrated deep learning model as shown in Fig. 6. Also, the SoftMax activation function $\sigma(z)_i$ is employed at the output of the deep learning model and expressed in Eq. 2.

$$\sigma(z)_i = \frac{e^{z_i}}{\sum_{j=1}^{K} e^{z_j}} \tag{2}$$

The working principle of the proposed deep learning model is illustrated as shown in Fig. 7. The medical image datasets are gathered and preprocessed for uniformity and suitability in the model development phase.VGG-16, ResNet50 and EfficientB0 are trained using preprocessed data. The predictions of these three models are combined to enhance accuracy. Based on the predictions, features are fused to create a comprehensive final model. Then the performance of the combined deep learning model is fine-tuned by optimizing hyperparameters.

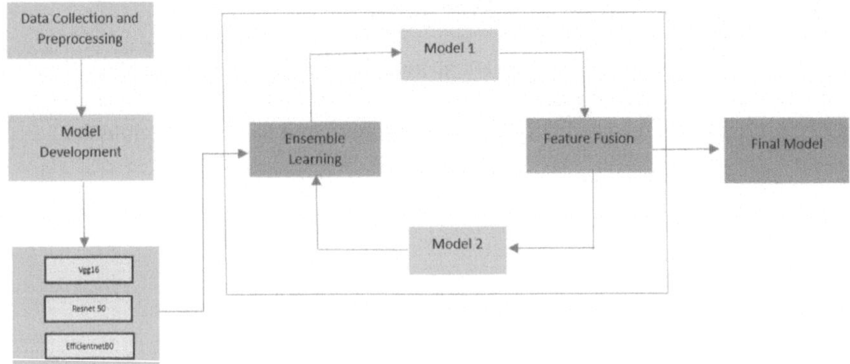

Fig. 7. Work Flow Diagram

5 Simulation Results

The simulation results of the proposed deep learning-based disease detection model are discussed in this section. The reference for the sample training images of Alzheimer's, Parkinson's, and Diabetes Retinopathy diseases [19] is, as illustrated in Fig. 7. The used simulation parameters are given in Table 1.

Table 2 lists the proposed models' accuracy. From the results, it has been shown that the VGG16 and Efficientnet-B0 models are better compared to other fusion models. Consequently, feature fusion is carried out on the VGG16 and Efficientnet-B0 models by taking into account both analysis aspects. This model differs from other models in that it combines three diseases: Alzheimer's, Parkinson's, and diabetic retinopathy. Additionally, aspects of VGG16 and EfficientB0 are fused and modelled obtaining better accuracy.

Table 1. Simulation Parameters

Parameter	Value
Library	TensorFlow
Platform	Python 3.10
Modulation	Feature Fusion
Number of images	3995
Activation function at the hidden layers	ReLu
Activation function at the output layer	Soft max
Optimizer	Adam
Loss function	Cross Entrophy

Table 2. Accuracy of the Deep Learning Architectures in Disease Prediction

MODEL	ALZHEIMER DISEASE	PARKINSON DISEASE	DIABETIC RETINOPATHY
VGG 16	59.50%	57.02%	55.88%
RESNET 50	49.85%	42.07%	49.25%
EFFICIENTNET-B0	57.88%	56.42%	60.35%

6 Conclusion

In this proposed work, a deep learning model is developed to detect three diseases, such as Alzheimer's disease, Parkinson's disease, and diabetic retinopathy. This work combines the significant features of VGG-16, ResNet50 and EfficientNetB0, using ensemble learning and signature fusion techniques. A deep learning model is designed to predict the multiple diseases for the given dataset. In future, a comprehensive evaluation of the proposed deep learning model could be carried out in terms of accuracy, precision, recall and F1 scores. The performance analysis can be done with model predictions and misclassification problems to enhance the accuracy further.

References

1. Bajwa, A., Nosheen, N., Talpur, K.I., Akram, S.: A prospective study on diabetic retinopathy detection based on modify convolutional neural network using Fundus images at Sindh institute of ophthalmology & visual sciences. Diagnostics **13**(3), 393 (2023)
2. Das, D., Biswas, S.K., Bandyopadhyay, S.: Detection of diabetic retinopathy using convolutional neural networks for feature extraction and classification (DRFEC). Multimedia Tools Appl. **82**(19), 29943–30001 (2023)
3. Shaban, M.: Deep learning for Parkinson's disease diagnosis: a short survey. Computers **12**(3), 58 (2023)
4. Wang, W., Lee, J., Harrou, F., Sun, Y.: Early detection of Parkinson's disease using deep learning and machine learning. IEEE Access **8**, 147635–147646 (2020)
5. Al-Shoukry, S., Rassem, T.H., Makbol, N.M.: Alzheimer's diseases detection by using deep learning algorithms: a mini-review. IEEE Access **8**, 77131–77141 (2020)
6. Al Shehri, W.: Alzheimer's disease diagnosis and classification using deep learning techniques. PeerJ Comput. Sci. **8**, e1177 (2022)
7. Joshi, S., Shenoy, D., Vibhudendra Simha, G.G., Rrashmi, P.L., Venugopal, K.R., Patnaik, L.M.: Classification of Alzheimer's disease and Parkinson's disease by using machine learning and neural network methods. In: 2010 Second International Conference on Machine Learning and Computing, pp. 218–222. IEEE (2010)
8. Pedrero-Sánchez, J.F., Belda-Lois, J.-M., Serra-Ano, P., Inglés, M., Lopez-Pascual, J.: Classification of healthy, Alzheimer and Parkinson populations with a multi-branch neural network. Biomed. Signal Process. Control **75**, 103617 (2022)
9. Templeton, J.M., Poellabauer, C., Schneider, S.: Classification of Parkinson's disease and its stages using machine learning. Sci. Rep. **12**(1), 14036 (2022)
10. Alsubaie, M.G., Luo, S., Shaukat, K.: Alzheimer's disease detection using deep learning on neuroimaging: a systematic review. Mach. Learn. Knowl. Extr. **6**(1), 464–505 (2024)

11. Pahuja, G., Nagabhushan, T.N.: A comparative study of existing machine learning approaches for Parkinson's disease detection. IETE J. Res. **67**(1), 4–14 (2021)
12. Dai, L., et al.: A deep learning system for detecting diabetic retinopathy across the disease spectrum. Nat. Commun. **12**(1), 3242 (2021)
13. Hamedani, A.G.: Deep learning, the retina, and Parkinson disease. JAMA Ophthalmol. **141**(9), 911–912 (2023)
14. Sharma, T., Shah, M.: A comprehensive review of machine learning techniques on diabetes detection. Vis. Comput. Ind. Biomed. Art **4**(1), 30 (2021)
15. Noor, M.B.T., Zenia, N.Z., Shamim Kaiser, M., Al Mamun, S., Mahmud, M.: Application of deep learning in detecting neurological disorders from magnetic resonance images: a survey on the detection of Alzheimer's disease, Parkinson's disease and schizophrenia. Brain Inform. **7**, 1–21 (2020)
16. Nancy Noella, R. S., Priyadarshini, J.: Machine learning algorithms for the diagnosis of Alzheimer and Parkinson disease. J. Med. Eng. Technol. **47**(1), 35–43 (2023)
17. Chaki, J., Thillai Ganesh, S., Cidham, S.K., Ananda Theertan, S.: Machine learning and artificial intelligence-based Diabetes Mellitus detection and self-management: a systematic review. J. King Saud Univ. Comput. Inf. Sci. **34**(6), 3204–3225 (2022)
18. Kaur, S., Aggarwal, H., Rani, R.: Diagnosis of Parkinson's disease using deep CNN with transfer learning and data augmentation. Multimedia Tools Appl. **80**(7), 10113–10139 (2021)
19. https://www.kaggle.com/datasets

Intelligent Posture Monitoring System with Real Time Notifications Using MediaPipe and OpenCV

Sneha Sathish Kumar(✉), S. Madesh, B. Nikitha, and Maheshwari

Department of Computer Science and Technology (Emerging Technologies), SRM Institute of Science and Technology , Vadapalani, Chennai, India
{ss8889,ms3478,nt3457,maheswas5}@srmist.edu.in

Abstract. Utilizing technological breakthroughs to address posture-related health issues and enhance general well-being is the goal of the "Posture Guard: Enhancing Well-being through Proactive Posture Management" initiative. In order to improve everyone's spinal health, this study explores the application of proactive posture management techniques in communities. In this work, a method for integrating Mediapipe and OpenCV technologies to achieve precise pose estimation is presented. Pre-trained models for human posture assessment are available through the machine learning framework Mediapipe, while a variety of picture and video processing functionalities are available through the popular computer vision library OpenCV. The user is then given real-time posture feedback and recommendations for corrective action.

Keywords: human pose estimation · convolution neural network · position estimation · landmark detection · real-time systems

1 Introduction

In a world where prolonged periods of sitting are becoming more popular, maintaining proper posture is essential for overall health and well-being. Many people struggle to maintain appropriate posture regularly. Especially when sitting or standing for an extended period. To address this issue, a unique solution has emerged: an Intelligent Posture Monitoring System with Real-Time Notifications that takes advantage of MediaPipe and OpenCV technology.

MediaPipe, an innovative framework developed by Google, offers a versatile suite of tools catering to diverse multimedia tasks, ranging from video processing to machine learning. Among its notable features is the capability to conduct real-time pose estimation, empowering developers with the ability to accurately discern and track human body movements in live video streams.

An extensive collection of tools and techniques for computer vision and image processing applications may be found in the robust open-source package OpenCV. Its ability to compute angles between objects or points in an image, including those between the nose, neck, and shoulders, is among its most amazing features.

P. D. Sivakumar et al. (Eds.): IRCCTSD 2024, CCIS 2360, pp. 145–155, 2025.
https://doi.org/10.1007/978-3-031-82389-3_13

The device can determine whether a person is maintaining proper posture or showing poor postural patterns, such as slouching or hunching over, by constantly tracking the position and alignment of various body parts.

When the system identifies the wrong posture, it immediately sends a notification to the user, suggesting they change their position and fix their posture. These real- time notifications provide users with small reminders to sit or stand up straight and maintain good posture throughout the day. By providing quick feedback, the technology supports users in developing a better understanding of their posture and promoting healthier habits over time.

The advantages of these systems are numerous. It minimizes the discomfort and pain associated with bad posture while reducing the possibility of developing musculoskeletal illnesses and injuries in the long run. Furthermore, the system promotes healthy posture methods, which improves users' general health, productivity, and quality of life.

In conclusion, the Intelligent Posture Monitoring System with Real-Time Notifications is a cutting-edge solution for maintaining proper posture. Using MediaPipe and OpenCV technologies, this solution enables users to take control of their posture and prioritize their physical well-being.

2 Related Works

Lately, there's been a lot of attention on making systems that can figure out how people are standing or moving in real-time. These systems are handy for things like helping with exercises, sports training, and keeping tabs on fitness progress. Let's dive into some research where scientists have used computer vision and machine learning to guess human posture.

In 2020, Zhang and colleagues developed a posture estimation system using a special camera that captures both color and depth information. This system used the depth map to figure out where the person's joints were, and then a kind of computer program called a convolutional neural network guessed the pose. It worked really well on a standard set of data and could do its job quickly, even in real-time [5, 7].

When it comes to position estimation, Mediapipe performs better than YOLO. The Mediapipe posture monitoring model has been created for this task, producing exact and consistent results in real time. While YOLO excels at object recognition accuracy, it may not perform as well at posture estimation due to its requirement for more intricate models and specific architectures [5, 7].

Working out with improper posture can lead to muscle damage and potential pain. Hence, a reliable automated posture assessment system is invaluable for detecting body misalignments. Essential body cues are utilized to assess posture accuracy. Ultimately, a classifier model is crafted to recognize human posture exercises in real-time, leveraging a basic web camera [2].

A different platform harnesses the power of MediaPipe to transform human-robot interactions. Tailored for developing a versatile machine learning pipeline, MediaPipe first employs a palm detection model to locate and outline hands within the video frame. Subsequently, it seamlessly transitions to a hand landmark model, capable of pinpointing 21 landmarks across the hand. These landmarks furnish intricate details about hand positions and gestures, streamlining interaction and fostering inclusivity by enabling even those unfamiliar with robotics to engage effectively [4].

By applying an OpenPose and MediaPipe-based algorithm, the system produced structured outputs that showed the pixel locations and coordinates of several bodily features, such as the eyes, necks, shoulders, elbows, and more. To identify each landmark point, a confidence score and pixel coordinates (x, y) were assigned. To precisely determine landmark positions and compare them with predefined poses, a live video feed was utilized.Users got input right away, which allowed them to correct their posture quickly [6].

In a separate study, the constraints of sensors have been mitigated through the utilization of MediaPipe, a computer vision library known for its lightweight computing capabilities, to discern the Range of Motion (ROM) on the human body. MediaPipe accomplishes this by tracking and forecasting the coordinates of an individual's body. This detection methodology aims to assess the precision of computer vision in measuring ROM [1].

A model for detecting squat motions is crafted by integrating the Mediapipe algorithm, subsequently enhanced by the YOLOv5 network. Known for its rapid detection speed and robust generalization capabilities, YOLOv5 further enhances the accuracy and operational efficiency of the algorithm [5].

Real-time posture feedback during exercise is facilitated by accurately detecting up to 33 body landmarks. OpenCV performs angle calculations and posture analysis while eliminating background interference. This approach excels over traditional methods by precisely evaluating posture across different conditions and accommodating complex movements. It assists users in enhancing form, minimizing injury risks, and efficiently reaching fitness objectives [3].

In conclusion, Several related studies have been employed in computer vision and machine learning to determine human posture in real time. The suggested system combines Mediapipe and OpenCV and develops an intelligent posture monitoring system. The system has various advantages, including the ability to provide real-time feedback to users and improve overall health.

3 Proposed Methodology

The proposed system aims to provide real-time notifications to laptop or computer users about their posture, specifically alerting them if they are slouching. Additionally, the system intends to track and report their most frequent posture.

The system utilizes a webcam or an external camera to capture video feed from the user as input. Once the input video is received, the system employs the MediaPipe Pose Landmark model to detect key landmarks such as the left shoulder blade, right shoulder blade, neck and nose.

Next, the system calculates the angles between the detected landmarks using trigonometric formulas such as the tangent formula. This calculation helps determine the posture of the user, including whether they are slouching or maintaining a correct posture.

The classification of angles for determining whether a posture is correct or slouched is established through an experiment involving 20 participants. The mean and standard deviation of these angles are calculated to define the range for each posture, as elaborated in the following sections.

The study identified a safe range for classifying posture as either correct or slouched. Individuals are then notified with the summary of their posture as an email alert based on this classification at the end of the session.

4 System Architecture

See Fig. 1.

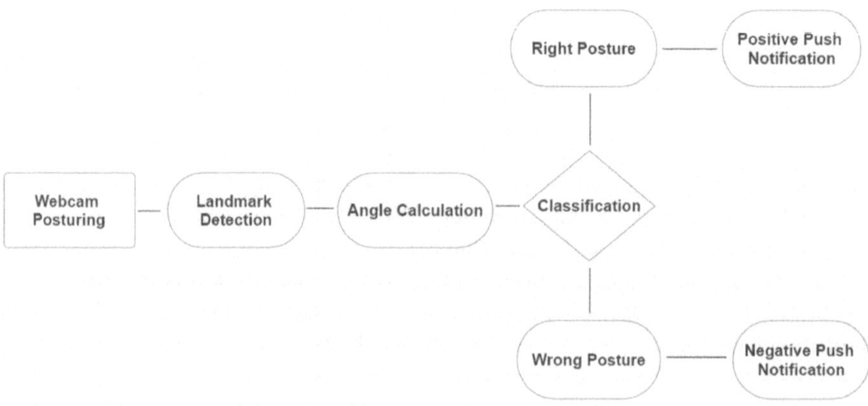

Fig. 1. System Architecture

5 Methodology

In this section, the methodology used in developing PostureGuard, a posture monitoring system utilizing MediaPipe and OpenCV is presented.

5.1 Data Collection and Preprocessing

The initial stages involved data collection and preprocessing, where human body postures were captured using a webcam or an external camera. Twenty volunteers, consisting of 10 males and 10 females, participated in the study. They were instructed to sit in a comfortable and upright posture, considered the correct posture, and then to slouch. Various postures commonly adopted while working on a computer were recorded, with consideration given to variations in lighting, camera angle, and posture to ensure the model's robustness in different conditions.

During these postures, experimental values were recorded for various points like the shoulder, neck, and nose (Fig. 2). Subsequently, mean values and standard deviations were calculated for each point, providing an overview of the participants' posture characteristics in both the correct and slouched positions.

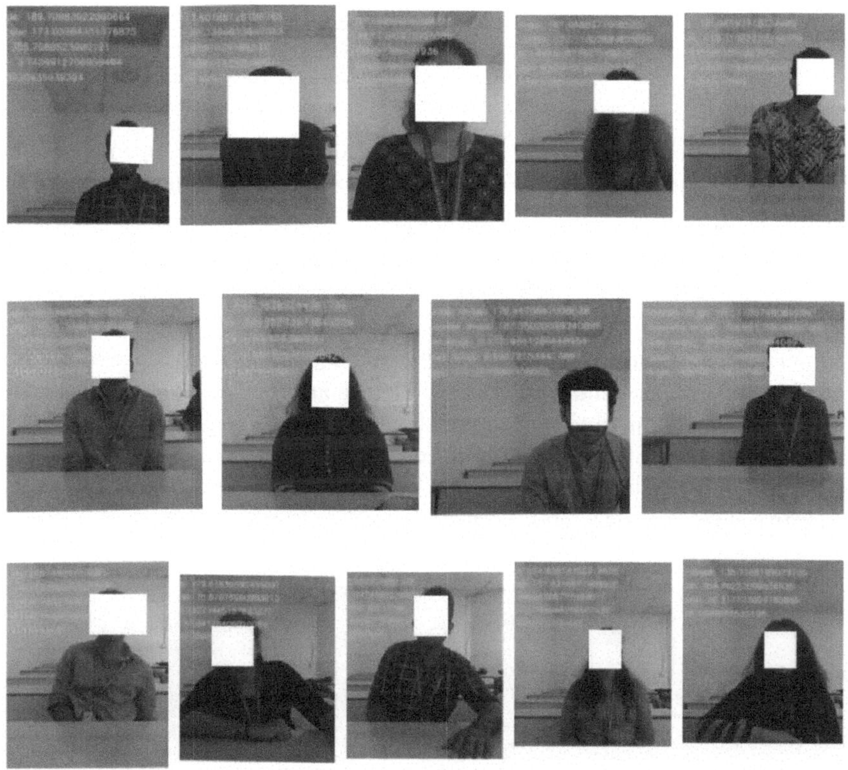

Fig. 2. Sample Dataset

Mediapipe

MediaPipe Pose Landmark is a component of the MediaPipe framework developed by Google, designed for real-time pose estimation tasks. It provides a set of pre-trained models and a unified pipeline for detecting key body landmarks, such as joints and key points, in human poses from input images or video streams. The Pose Landmark model uses a deep neural network (DNN) to predict the locations of these landmarks. It works by first detecting the human body within an image or frame and then estimating the precise locations of key body points. These points can include the wrists, elbows, shoulders, hips, knees, and ankles, among others, depending on the specific configuration of the model. Once the model has identified these landmarks, it can be used for a variety of applications, including fitness tracking, gesture recognition, augmented reality, and human-computer interaction. MediaPipe provides easy-to-use APIs for integrating the Pose Landmark model into your own applications, making it accessible to developers looking to incorporate pose estimation functionality into their projects.

In our study, we utilized the MediaPipe Pose Landmark component to detect key points such as the nose, neck, left shoulder, and right shoulder. MediaPipe Pose Landmark provides a convenient and accurate way to identify these key body landmarks in real-time from images or video streams.

Safe Angle Range Calculation

The determination of the safe angle involves detailed discussions on trigonometric formulas in the subsequent sections. We have worked on a novel calculation to determine the angles.

$$angle = \tan^{-1}\left(\frac{y2 - y1}{x2 - x1}\right) \quad (1)$$

In our study, we have used trigonometry to compute the angle created by three landmark points. We used the nose, neck, and left shoulder points to determine the left shoulder angle and similarly for the right shoulder. By applying the 'atan2' function from the Python math library, we determined the angle between the vectors created by these points (Fig. 3). Subsequently, we converted this angle from radians to degrees. For any negative angle, we added 360° to ensure a positive representation.

```
def calculateAngle(landmark1, landmark2, landmark3):
    x1, y1, _ = landmark1
    x2, y2, _ = landmark2
    x3, y3, _ = landmark3
    angle = math.degrees(math.atan2(y3 - y2, x3 - x2) - math.atan2(y1 - y2, x1 - x2))
    if angle < 0:
        angle += 360
    return angle
```

Fig. 3. Angle Calculation

For each posture (right and wrong), we observed the range of angles across the 20 participants. This provided insight into the variability in posture within the group. We then defined the ranges of angles for the right and wrong postures. Importantly, we found a clear distinction between safe and unsafe postures based on the measured angles in our study. There were no overlaps in the angles, which improves the accuracy of the model.

Classification

As part of the classification process, the angle is compared to the safe range. If the angle falls within the safe range, the correct posture counter is incremented; otherwise, the incorrect posture counter is incremented. Finally, the percentage of time a person is slouching versus sitting in the right posture in that specific interval is calculated and sent to the user.

Notification Module

Email is a type of communication that is sent between servers over the Simple Mail Transfer Protocol, or SMTP. It is an essential component of the internet's email infrastructure. In order to transmit emails safely and effectively, email clients and servers must communicate with one another according to the rules and processes defined by SMTP. Using the SMTP server of Gmail, the email notice is sent. It uses environment variables to get the application password and email address of the sender. After that, it creates an email message containing posture information and securely delivers it to the designated recipient (Fig. 4).

```
import smtplib
import os
from dotenv import load_dotenv

load_dotenv()


def send_mail(RECEIVER = "RECEIVER_MAIL_ID"):
    server = smtplib.SMTP_SSL('smtp.googlemail.com', 465)
    SENDER_MAIL_ID = os.getenv("SENDER_MAIL_ID")
    APP_PASSWORD = os.getenv("APP_PASSWORD")
    server.login(SENDER_MAIL_ID, APP_PASSWORD)
    message = f"""
    Subject: POSTURE ALERT

    Dear Madam/Sir,

    This is to inform you that the throughout the session, the posture details are as follows:
    Correct: {"{:.2f}".format(Correct_Percentage)}%
    Incorrect: {"{:.2f}".format(Incorrect_Percentage)}%

    """
    server.sendmail(SENDER_MAIL_ID, RECEIVER, message)
    server.quit()
    return True
```

Fig. 4. Email Notification

```
Subject: POSTURE ALERT

Dear Madam/Sir,

This is to inform you that the throughout the session, the posture details are as follows:
Correct: 98.48%
Incorrect: 1.52%
```

Fig. 5. Notification

Overall, the proposed methodology involves the collection and preprocessing of data, utilizing MediaPipe Pose Landmark for key body landmark detection, calculating safe angle ranges using trigonometry, and implementing a classification system to determine correct and incorrect postures. The methodology also includes a notification module that sends posture information to users (Fig. 5). Through these steps, the study aims to develop PostureGuard, a posture monitoring system that can effectively detect and classify posture, providing users with valuable feedback to improve their posture habits and overall health.

6 Results and Analysis

PostureGuard stands out as a novel posture monitoring system designed to combat poor sitting habits and promote better spinal health. Built upon the strengths of MediaPipe, a framework for real-time pose estimation, and OpenCV, a popular computer vision library, PostureGuard offers a practical solution. The researchers prioritized real-world applicability by collecting data from 20 participants engaged in various sitting postures, encompassing both proper form and slouching. They even factored in variations in lighting, camera angles, and individual posture differences to ensure the system's robustness in diverse situations.

PostureGuard's core function lies in its ability to calculate angles between key body landmarks like the nose, neck, and shoulders. By analyzing this geometric data, the system can differentiate between safe and unsafe postures. A key finding of the study was the clear distinction between the angle ranges for good and bad posture – they did not overlap. This non-overlap is critical for accurate posture detection.

Beyond simply identifying posture, PostureGuard actively monitors a user's posture over a chosen time interval. This allows it to track how frequently a person slumps compared to maintaining good posture. To provide feedback an raise awareness, PostureGuard sends email notifications summarizing the user's posture habits during the monitored period (Fig. 6).

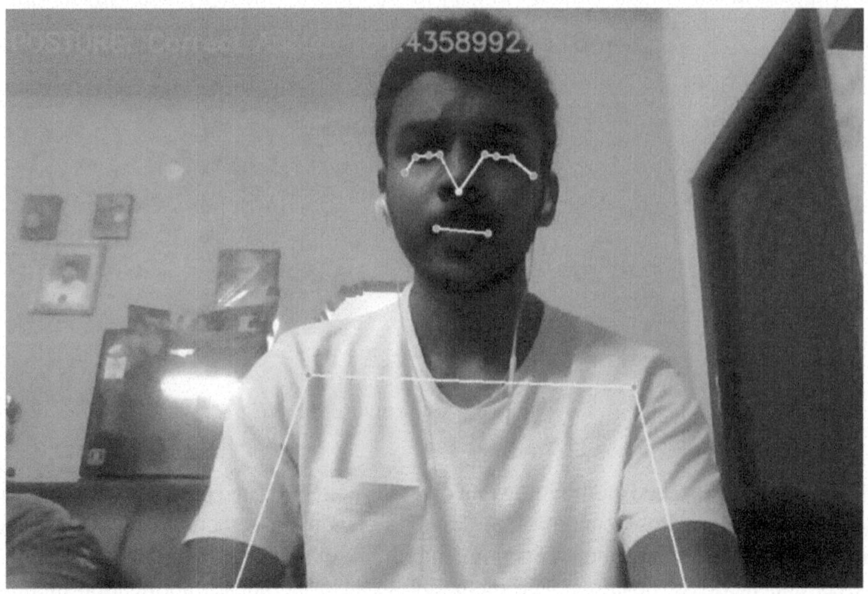

Fig. 6. Output

PostureGuard's utilization of MediaPipe for real-time pose estimation distinguishes it from other posture monitoring systems like OpenPose, PoseNet, MoveNet, and YOLO v7. While these systems also leverage computer vision techniques, their frameworks and methodologies vary, leading to differences in performance and applicability.

OpenPose, a well-established framework, excels in detecting multiple body keypoints and skeletal structures in real-time. However, its implementation requires more computational resources compared to MediaPipe, potentially impacting its suitability for resource-constrained environments. Additionally, OpenPose's focus on general pose estimation may not prioritize the specific angle calculations necessary for posture assessment like Mediapipe does.

Similarly, PoseNet, developed by Google, and MoveNet, known for its lightweight architecture, offer efficient pose estimation capabilities. Yet, their emphasis may lie more on general pose detection rather than the nuanced analysis of posture angles that Posture Guard specializes in. While they may offer real-time performance and low computational overhead, they lack the depth of posture assessment provided by Mediapipe.

YOLO v7, renowned for its object detection prowess, can also detect human figures but does not provide the detailed pose estimation required for posture analysis. Its primary focus on object detection might limit its suitability for applications demanding precise posture assessment like PostureGuard does.

PostureGuard's unique strength lies in its meticulous approach to data collection and analysis, ensuring robustness and accuracy in diverse real-world scenarios. By factoring in variations in lighting, camera angles, and individual posture differences, PostureGuard ensures its effectiveness across different environments. Its core function of calculating angles between key body landmarks enables precise posture differentiation, a feature crucial for accurate posture detection and correction.

In conclusion, the study effectively demonstrates PostureGuard's potential as a real-time posture monitoring system built upon readily available tools. The clear distinction between safe and unsafe posture angles, coupled with user feedback through email notifications, positions PostureGuard as a promising tool for promoting better posture and potentially reducing posture-related health issues.

7 Future Work

Various areas can be improved, including improving the posture analysis algorithm to detect more complicated posture concerns such as asymmetry, sway, and imbalance by leveraging machine learning techniques to classify and analyze various postural deviations. Consider integrating the posture monitoring system with VR or AR technologies to provide users with immersive experiences. This could include integrating visual signals or virtual avatars that demonstrate appropriate posture in real time. Implement accessibility features such as multilingual support, voice-guided directions, and screen reader compatibility to ensure equality and accessibility for users with varying needs. Collaborate with researchers and healthcare professionals to validate the system's performance in real-world contexts, and consider contributing to academic articles or clinical studies. Implement strong privacy and security controls to protect user data, particularly when integrating with third-party platforms or storing sensitive health information.

8 Conclusion

The Intelligent Posture Monitoring System with Real-Time Notifications offers an advanced solution to the common problem of poor posture. By utilizing MediaPipe and OpenCV technologies, this system continuously monitors posture and provides immediate feedback to users, prompting necessary adjustments. Through real-time notifications, it encourages users to develop healthier posture habits over time, ultimately reducing discomfort and the risk of musculoskeletal issues. Overall, this innovative system promotes well-being and enhances quality of life by seamlessly integrating into daily routines, representing a significant step towards advocating better posture and healthier lifestyles in today's sedentary world.

References

1. Mustar, M.Y., et al.: Range of motion detection system in humans based on MediaPipe special flexion-extension and abduction-adduction movements. In: 2023 3rd International Conference on Electronic and Electrical Engineering and Intelligent System (ICE3IS), Yogyakarta, Indonesia, pp. 486–491 (2023). https://doi.org/10.1109/ICE3IS59323.2023.10335332
2. Supanich, W., Kulkarineetham, S., Sukphokha, P., Wisarnsart, P.: Machine learning-based exercise posture recognition system using MediaPipe pose estimation framework. In: 2023 9th International Conference on Advanced Computing and Communication Systems (ICACCS), Coimbatore, India, pp. 2003–2007 (2023). https://doi.org/10.1109/ICACCS57279.2023.10112726
3. Bhamidipati, V.S.P., Saxena, I., Saisanthiya, D., Retnadhas, M.: Robust intelligent posture estimation for an AI gym trainer using Mediapipe and OpenCV. In: 2023 International Conference on Networking and Communications (ICNWC), Chennai, India, pp. 1–7 (2023). https://doi.org/10.1109/ICNWC57852.2023.10127264
4. Bensaadallah, M., Ghoggali, N., Saidi, L., Ghoggali, W.: Deep learning- based real-time hand landmark recognition with MediaPipe for R12 robot control. In: 2023 International Conference on Electrical Engineering and Advanced Technology (ICEEAT), Batna, Algeria, pp. 1–6 (2023). https://doi.org/10.1109/ICEEAT60471.2023.10426264
5. Zhang, S., Chen, W., Chen, C., Liu, Y.: Human deep squat detection method based on MediaPipe combined with YOLOv5 network. In: 2022 41st Chinese Control Conference (CCC), Hefei, China, pp. 6404–6409 (2022). https://doi.org/10.23919/CCC55666.2022.9902631
6. Negi, S., Garg, M., Maindola, H., Kansal, V., Jain, U., Bhatla, S.: Real-time human pose estimation: a MediaPipe and Python approach for 3D detection and classification. In: 2023 3rd International Conference on Technological Advancements in Computational Sciences (ICTACS), Tashkent, Uzbekistan, pp. 128–133 (2023). https://doi.org/10.1109/ICTACS59847.2023.10390506
7. Lugaresi, C., et al.: MediaPipe: a framework for building perception pipelines. arXiv preprint arXiv:1906.08172 (2019)
8. Mouthami, K., Yuvaraj, N., Navadharani, M.: A real-time drowsiness tracking framework using the single shot with Mediapipe. In: 2023 9th International Conference on Smart Structures and Systems (ICSSS), Chennai, India, pp. 1–6 (2023). https://doi.org/10.1109/ICSSS58085.2023.10407102
9. Cobb, B.S., Candan, F.: Improving workplace safety in the cargo industry through posture monitoring using Mediapipe and machine learning. In: 2023 International Conference on Innovations in Intelligent Systems and Applications (INISTA), Hammamet, Tunisia, pp. 1–6 (2023). https://doi.org/10.1109/INISTA59065.2023.10310609
10. Zhu, Y., Zeng, Y., Huang, D., Huang, J., Lu, H., Wang, W.: Occlusion- robust sleep posture detection using body rolling motion in a video. In: 2023 45th Annual International Conference of the IEEE Engineering in Medicine & Biology Society (EMBC), Sydney, Australia, pp. 1–5 (2023). https://doi.org/10.1109/EMBC40787.2023.10340050

11. Elavarasi, S.A., Ankit Kumar, P., Jayanthi, J.: Development of AI-based posture monitoring system to assist yoga training. In: 2023 International Conference on Research Methodologies in Knowledge Management, Artificial Intelligence and Telecommunication Engineering (RMKMATE), Chennai, India, pp. 1–4 (2023). https://doi.org/10.1109/RMKMATE59243.2023.10368735
12. Huu, P.N., Hong, P.D.L., Dang, D.D., Quoc, B.V., Bao, C.N.L., Minh, Q.T.: Proposing hand gesture recognition system using MediaPipe holistic and LSTM. In: 2023 International Conference on Advanced Technologies for Communications (ATC), Da Nang, Vietnam, pp. 433–438 (2023). https://doi.org/10.1109/ATC58710.2023.10318885
13. Renugadevi, R., Arul, P., Subashree, V., Sathi, G., Dharmateja, M., Indhumathi, G.: Deep learning-based GYM monitoring system using YOLOv5 and pose estimation algorithm. In: 2023 7th International Conference on Electronics, Communication and Aerospace Technology (ICECA), Coimbatore, India, pp. 697–702 (2023). https://doi.org/10.1109/ICECA58529.2023.10394921
14. Dedhia, U., Bhoir, P., Ranka, P., Kanani, P.: Pose estimation and virtual gym assistant using MediaPipe and machine learning. In: 2023 International Conference on Network, Multimedia and Information Technology (NMITCON), Bengaluru, India, pp. 1–7 (2023).https://doi.org/10.1109/NMITCON58196.2023.10275938
15. Tomar, Y., Himanshu, Devi, S., Kaur, H.: Human motion tracker using OpenCv and Mediapipe. In: 2023 3rd International Conference on Innovative Mechanisms for Industry Applications (ICIMIA), Bengaluru, India, pp. 1199–1204 (2023).https://doi.org/10.1109/ICIMIA60377.2023.10425865
16. Huang, Z., Li, J., Liang, J., Zen, B., Tan, J.: An IoT-oriented gesture recognition system based on ResNet-Mediapipe hybrid model. In: 2022 5th International Conference on Pattern Recognition and Artificial Intelligence (PRAI), Chengdu, China, pp. 1221–1228 (2022).https://doi.org/10.1109/PRAI55851.2022.9904258
17. Sharma, D., Gaur, M.: Deep learning-based diabetes detection and real- time exercise monitoring for enhanced healthcare. In: 2024 IEEE International Conference on Computing, Power and Communication Technologies (IC2PCT), Greater Noida, India, pp. 673–678 (2024). https://doi.org/10.1109/IC2PCT60090.2024.10486439
18. Ming, C., Yunbing, Y.: Perception-free calibration of eye opening and closing threshold for driver fatigue monitoring. IEEE Access **10**, 125469–125476 (2022). https://doi.org/10.1109/ACCESS.2022.3225453
19. Rimal, R., Herceg, M., Vranješ, M., Grbić, R.: Driver monitoring system using an embedded computer platform. In: 2023 International Conference on Software, Telecommunications and Computer Networks (SoftCOM), Split, Croatia, pp. 1–7 (2023). https://doi.org/10.23919/SoftCOM58365.2023.10271657
20. De La Cruz, P., Javier, K., Castañeda, P.: Patients rehabilitation with musculoskeletal sequel from Covid-19 using vision artificial in the city of Huancayo. In: 2022 4th International Conference on Advances in Computer Technology, Information Science and Communications (CTISC), Suzhou, China, pp. 1–5 (2022). https://doi.org/10.1109/CTISC54888.2022.9849767

Diabetes Prediction Using Deep Learning

B. Venkataramanaiah[1(✉)], R. Ramadoss[2], Kannan Ramakrishnan[3], K. Naga Harini[1], M. Kavya[1], and P. C. V. Saketh[1]

[1] Vel Tech Rangarajan Dr.Sagunthala R&D Institute of Science and Technology, Chennai, India
bvenkataramanaiah@veltech.edu.in
[2] SRM Institute of Science and Technology , Kattankulathur, Chengalpattu, India
rr5606@srmist.edu.in
[3] Saveetha Engineering College, Chennai, India

Abstract. The extend points to create a cutting-edge profound learning procedure for the early location of diabetes, moreover known as diabetes Mellitus. This malady postures a critical risk when a person's blood glucose or blood sugar levels gotten to be perilously tall, causing glucose to stay within the blood and not reach the body cells. Applying prevalent convolutional neural organize models to such data is constraining. In any case, ready to overcome these impediments by utilizing a neural organize design that combines convolutional neural systems and long short-term memory to offer an productive and precise early location of the malady.

Keywords: CNN · LSTM · diabetes

1 Introduction

Detecting diabetes early is crucial in mitigating its progression, and this model propose approach utilizing Deep learning technique. While traditional CNN models struggle with numerical data like the diabetes dataset is used in this research transforms Numerical data are transformed into graphs and images. This empowers strong. Three classification techniques are utilized on the coming about diabetes picture information. The think about illustrates the adequacy of this approach through the utilize of LSTM CNN models in recognizing diabetes at an early arrange. This consider recommends a new strategy leveraging profound learning procedures for early diabetes. Traditional convolutional neural network models encounter challenges with numerical medical data, such as the PIMA dataset in this research. To overcome this, numerical data is converted into images based on feature importance, enabling CNN models to effectively diagnose diabetes early on. Diabetes mellitus, a prevalent and serious health concern worldwide, poses significant challenges for early detection and management. Identifying diabetes at an early organize. It is crucial to ensure that this is done efficiently. Compelling treatment and anticipation of complications require appropriate consideration and care. Traditional strategies for diabetes discovery frequently depend on clinical markers and biochemical tests, which may have impediments in terms of precision and accessibility. In later

P. D. Sivakumar et al. (Eds.): IRCCTSD 2024, CCIS 2360, pp. 156–165, 2025.
https://doi.org/10.1007/978-3-031-82389-3_14

a long time, convolutional neural systems (CNNs) have appeared guarantee due to the headway of profound learning techniques in restorative picture investigation and malady conclusion.

This ponder points to investigate the potential of CNNs in identifying diabetes mellitus through the examination of restorative imaging information. By leveraging the control of CNNs, this investigate looks for to make strides the precision and proficiency of diabetes location, eventually contributing to way better persistent results and healthcare administration methodologies. Diabetes may be a constant condition that's characterized by elevated levels of glucose within the blood. It could be a critical worldwide wellbeing challenge, influencing the world's elderly populace in a extreme way. There were 463 million individuals diagnosed with diabetes around the globe. The Diabetes Federation expects this will continue rising and it will rise to 700 million individuals shortly. Traditional methods for diabetes detection involve manual analysis making them time-consuming and costly by utilizing "Deep learning techniques for detecting diabetes. Employing the LSTM and CNN offers promising avenues for efficient and accurate detection of disease.

2 Literature Review

The study employs three classification strategies on the generated diabetes image data according to the Diabetes Federation, the global diabetic population stood at 420 million, with projections indicating a rise to 635 million. Diabetes mellitus comprises a cluster of endocrine disorders marked by impaired glucose uptake, stemming from insufficient insulin hormone levels. It exhibits a chronic trajectory and disrupts various metabolic processes Diabetes is broadly categorized into Sort 1 diabetes (T1D), Sort 2 diabetes (T2D), gestational diabetes mellitus, and specific sorts due to other causes. Sort 1 diabetes (T1D), Sort 2 diabetes (T2D), gestational diabetes mellitus, and specific sorts due to other causes. Sort 1 diabetes (T1D), Sort 2 diabetes (T2D), gestational diabetes mellitus, and specific sorts due to other causes. Sort 1 diabetes (T1D), Sort 2 diabetes (T2D), gestational diabetes mellitus, and specific sorts due to other causes. Sort 1 diabetes (T1D), Sort 2 diabetes (T2D), gestational diabetes mellitus, and particular sorts due to other causes. T1D comes about from pancreatic beta cell annihilation, causing affront lack, whereas T2D emerges from ineffectual affront transportation into cells. T1D comes about from pancreatic beta cell annihilation, causing affront lack, whereas T2D emerges from ineffectual affront transportation into cells. Both sorts incline people to life-threatening complications such as strokes, heart assaults, and incessant renal disappointment, among others [4]. In this manner, opportune expectation and discovery of diabetes are vital for at-risk people. Manufactured insights (AI) strategies, especially convolutional neural systems (CNNs) have emerged as promising tools for disease diagnosis, exhibiting notable performance in classification tasks CNN, an extension of machine perception, employ multiple hidden layers to enhance feature extraction, facilitating robust model expression [7, 8].

Activation functions like Sigmoid, Softmax, tanh, ReLU, and Softplus introduce nonlinearity, crucial for model differentiation [9]. However, vanishing gradients can impede training efficacy due to signal saturation in deep networks. To mitigate overfitting, dropout regularization is commonly employed during training, randomly deactivating neuron nodes to improve generalization. In this setting, we proposed a novel CNN-based diabetes forecast show capable of not only forecasting future disease occurrence but also discerning between T1D and T2D. We integrate normalization layers to enhance model adaptability to unseen data, while dropout regularization combats overfitting. Additionally, we limit the accumulation window for past gradients to maintain efficient training Diabetes, a metabolic clutter that's characterized by tall levels of sugar within the blood is a significant health concern worldwide. With its prevalence on the rise, early detection and prediction of diabetes risk are imperative for effective prevention and management strategies. Deep learning techniques is used for predictive analytics in healthcare, offering the potential to uncover intricate patterns and relationships within vast datasets. Leveraging the capabilities of deep learning models, this consider points to create a strong prescient system for distinguishing people at tall hazard of creating diabetes. By harnessing the wealth of information available from diverse sources such as demographic data, medical records, and lifestyle factors, this research endeavors to contribute to the headway of personalized pharmaceutical and proactive healthcare intercessions within the battle against diabetes. The extend points to create a deep-learning method for early location of diabetes. "Profound learning procedures have ended up progressively prevalent in later a long time, demonstrating to be capable tools. There are a few numerous later methods have been created are various modern-day strategies have been set up with the advancement of modern innovation for the determination of diabetes mellitus.

Diabetes prediction is crucial for early diagnosis and effective management of the disease. Profound learning procedures have appeared guarantee in capturing complex designs in therapeutic information, making them important devices for prescient modeling. In this ponder, we conduct a comprehensive examination of diabetes forecast utilizing profound learning strategies. We investigate Different sorts of profound neural arrange designs, such as Convolutional Neural Network (CNNs), to develop accurate prediction models. We leverage diverse datasets, including demographic, clinical, and genetic information, "To train and evaluate the performance" is a complete sentence, but it lacks context. Could you please provide more information on what you're trying to achieve? That way, I can provide a more comprehensive and accurate response of our models. We employ preprocessing techniques to handle missing data and normalize features for improved model convergence. Additionally, we utilize dropout regularization to prevent overfitting and enhance model generalization. It illustrate the adequacy of profound learning approaches in diabetes expectation. We achieve high accuracy, sensitivity, and specificity in identifying individuals at risk of developing diabetes. Furthermore, we conduct comparative analyses with traditional machine learning algorithms to highlight the superiority of deep learning models in capturing intricate relationships in diabetes data.

3 Methodology

Our One way to detect diabetes is by using deep learning techniques. Unlike traditional machine learning methods that require explicit feature extraction, feature selection, and classification steps, deep learning networks embed these processes within themselves.

This means that the network can learn from the data in a self-directed manner, without the need for human intervention.

3.1 Deep Learning

Deep learning could be a sort of machine learning that employments counterfeit neural systems with a few layers to comprehend complex information. Not at all like ordinary machine learning approaches that require include building, profound learning calculations consequently extricate highlights from crude information, making them particularly valuable for dealing with gigantic amounts of unstructured information, such as pictures, content, and sound. At the center of profound learning are neural systems composed of interconnected layers of fake neurons, each layer preparing and progressively changing the information. The input layer gets crude information, such as pixel values in an image or words in a sentence, which is then passed through one or more hidden layers where the data is progressively abstracted and transformed. Finally, the output layer produces the model's prediction or classification Profound learning calculations are prepared utilizing expansive datasets to memorize complex patterns and relationships within the information.

This is often done through a prepare known as backpropagation, where the demonstrate iteratively alters its inner parameters to play down the contrast between anticipated and real yields. This optimization handle leads to tall levels of exactness in different errands such as picture acknowledgment, characteristic dialect preparing, discourse acknowledgment, and therapeutic determination.

3.2 Convolutional Neural Network (CNN)

Convolutional Neural Network (CNN) could be a profound learning calculation commonly utilized in computer vision like recognizing and classifying pictures. These systems they learn spatial pecking orders of highlights from input information. Profound learning may be a sort of machine learning impelled by the structure and work of the human visual cortex. CNNs have Convolutional layers slide little, learnable lattices over the input picture to detect patterns such as edges and surfaces. Usually done nearby pooling layers and completely associated layers. Pooling layers diminish the spatial measurements of the include maps, which makes a difference diminish the computational complexity of the organize whereas still Pooling operations such as max pooling and normal pooling downsample include maps by taking the most extreme or normal value each pooling window.

3.3 Long Short Term Memory (LSTM)

CNNs LSTM could be a sort of manufactured neural organize Neural Arrange outlined to address the issue of vanishing and detonating slopes. It Instead of simple RNN units, LSTM introduces memory blocks, which enables it to handle long-term dependencies better by remembering and connecting previous information to the present, even if the information lags far behind in time. An LSTM memory piece can be a complex dealing with unit that comprises of one or more memory cells. It utilizes a coordinate of multiplicative entryways, to be particular input and surrender entryways, to control the stream of data. Additionally, the operations of a memory square are supervised by a set of flexible multiplicative entryways. The input entryway chooses whether to allow or arrange of an input stream of cell incitation to a memory cell, though the abdicate entryway performs an allow or memory cell to a number of other centers. LSTMs are broadly utilized in common dialect preparing (NLP) applications due to their capacity to keep in mind data for long periods easily. They have Over the years, there have been noteworthy progressions in several areas of manufactured insights such as dialect modeling, computer vision, and speech acknowledgment, among others. In spite of the fact that Long Short-Term Memory (LSTM) was not broadly utilized within the past due to its complexity, it has become more prevalent in many areas due to its exceptional performance in language translation and speech processing. As research on LSTMs continues, new features such as forget gates and peephole connections have been introduced to make the network more efficient.

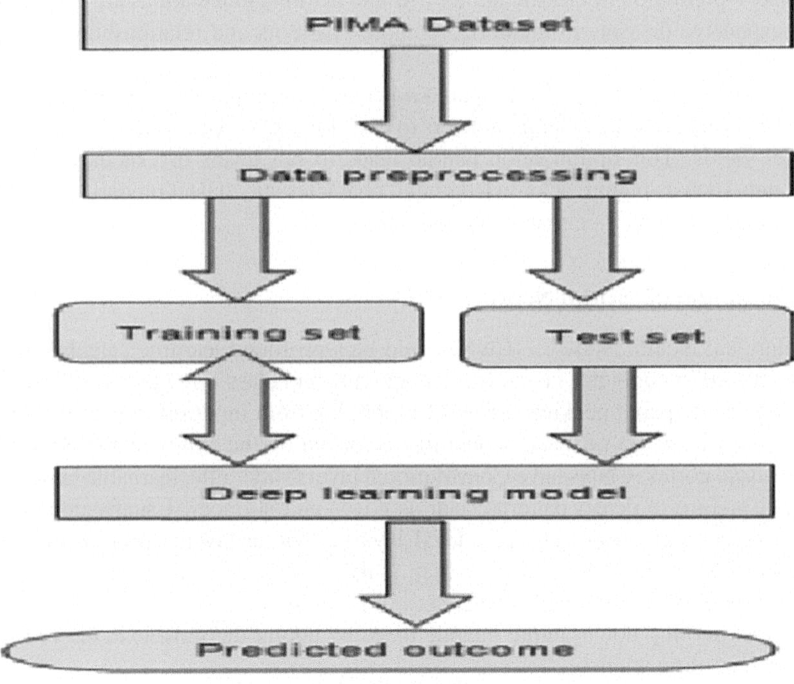

Fig. 1. Proposed method architecture

This enables the memory cell to forget or reset states. On the other hand, peephole connections connect the memory cell to all its gates, enabling it to learn the precise timing of the outputs as well as the internal state of a memory cell. In the prediction of diabetes using deep learning, the selection of the dataset is crucial for the performance and generalizability of the model. A well-curated dataset should encompass diverse features relevant to diabetes risk factors, including demographic information, clinical measurements, lifestyle habits, and genetic predispositions. Both systolic and diastolic blood pressure readings are crucial indicators of cardiovascular health. It's important to monitor both readings to get a complete picture of your heart's health which is closely linked to diabetes risk.

Blood Glucose Levels: Fasting plasma glucose levels or glycated hemoglobin levels provide direct measures of blood sugar regulation and are central to diabetes diagnosis and management.

Insulin Levels: Measures of insulin sensitivity or affront resistance can give experiences into the basic instruments of diabetes advancement.

Figure 1 shows proposed architecture which shows preprocessing, training followed by testing with the help of deep learning model. LSTM deep learning model is one of the highest accurate model used for prediction and classification.

3.4 PIMA Data Set

The Pima Indians Diabetes dataset consists of several numerical features Some examples of health metrics that can be tracked include incorporate Glucose levels, Blood weight, Skin thickness, Affront levels, and age, along with a binary target variable indicating whether each patient developed diabetes within five years. This dataset is often utilized for predictive modeling and exploring relationships between various health metrics and the onset of diabetes.

3.5 Pre-processing

Data preprocessing encompasses the tasks of refining, converting, and structuring unprocessed information to form it reasonable for assist investigation or model training. This process typically involves steps such as cleaning out irrelevant or inconsistent data points, transforming variables into appropriate formats, and organizing the data for efficient analysis. Data preprocessing involves cleaning, transforming, and organizing raw data for analysis into a format suitable for analysis.

3.6 Training Set

The preparing dataset comprises a expansive parcel of the accessible information and is utilized to prepare the deep-learning show Amid the preparing prepare, the show iteratively learns the basic designs and connections inside the information to form expectations. The preparing dataset ought to be differing and agent of the basic information dispersion to guarantee that the demonstrate learns vigorous and generalizable designs.

Each cycle, or age, includes nourishing clusters of preparing tests through the demonstrate, computing the misfortune (mistake) between the anticipated yields and the genuine targets, and overhauling the model parameters (weights and inclinations) to play down this misfortune utilizing optimization calculations like stochastic angle plummet (SGD) or Adam.

Table 1. Accuracy table

Authors	Year	Methods	Accuracy Obtained (in %)
Ref [17]	2013	Nonlinear	86.0
Ref\cite(j)	2021	Higher order spectrum	90.5
Ref [18]	2013	Discrete Wavelet transform	90.02
Ref [19]	2015	Empirical mode decomposition	95.63
Proposed method	2024	Deep learning (CNN-LSTM)	80

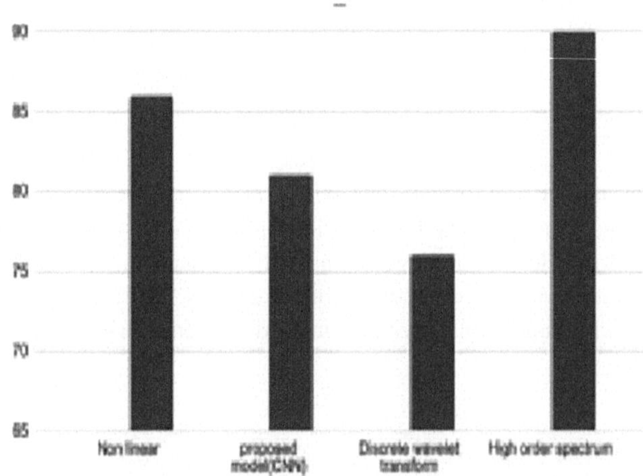

Fig. 2. Comparison of proposed method with existing works

4 Result

The cross-validation exactness and the nitty gritty test exactness are given in Table 1.

We utilized a few standard assessment measurements to assess the system's execution, counting precision, accuracy, review, and Fl-score. The accuracy of classification is the extent of accurately recognized tests to the complete number of tests inside the dataset. Figure 2 shows comparison of proposed method with existing works

$$\text{Accuracy} = (\text{TN} + \text{TP})/(\text{FP} + \text{FN} + \text{TN} + \text{TP}) \tag{1}$$

TP talks to the number of veritable positive tests, TN talks to the number of honest to goodness negative tests, FP talks to the number of unfaithful positive tests, and FN talks to the number of unfaithful negative tests. The number of unfaithful negative tests of off-base negative tests. Precision is the extent of veritable positive tests to the generally number of positive tests.

Accuracy expected by the appear:

$$\text{Sensitivity} = \text{TP} /(\text{FP} + \text{TP}) \tag{2}$$

The survey is the extent of honest to goodness positive tests to the in general number of positive tests inside the dataset:

$$\text{Specificity} = \text{TN}/(\text{FP} + \text{TN}) \tag{3}$$

$$\text{Recall} = \text{TP}/(\text{FN} + \text{TP}) \tag{4}$$

$$\text{Precision} = \text{TP}/(\text{FP} + \text{TP}) \tag{5}$$

We calculated these measurements for each category of diabetes expectation in Dataset and the model's in general execution. Figure 3 represents min-max Data set.

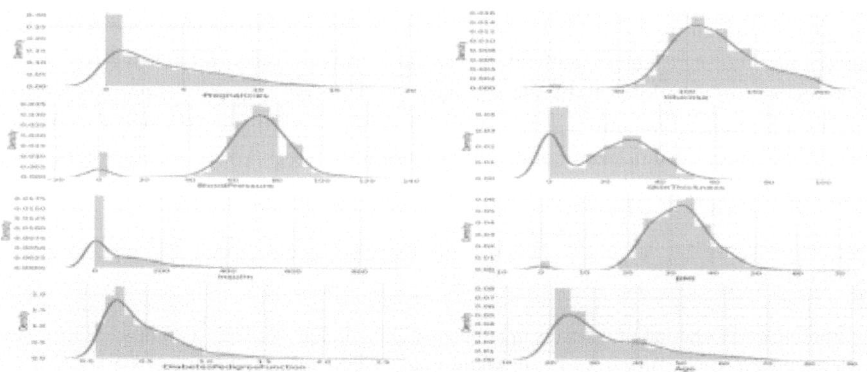

Fig. 3. Min-max Data set

5 Conclusion

In this term paper, we look at the utilize of profound learning methods in diabetes prediction holds immense promise for enhancing early detection, personalized intervention, and improved patient outcomes. By leveraging diverse datasets encompassing demographic, clinical, lifestyle, and genetic factors, deep learning models can effectively capture complex relationships and patterns underlying diabetes risk. Through meticulous model training, validation, and testing procedures, deep learning practitioners can develop robust and accurate predictive models capable of identifying individuals at risk of developing diabetes and distinguishing between different types of the disease. The integration of advanced Profound learning models, such as Long Short-Term Memory (LSTM) systems and convolutional neural systems (CNNs) further enhances the predictive capabilities of diabetes prediction models, enabling the extraction of temporal and spatial features from sequential and image data, respectively. Additionally, techniques such as dropout regularization, batch normalization, and hyperparameter optimization contribute to improving model generalization and performance. The successful deployment of deep learning-based diabetes prediction models in clinical practice has the potential to revolutionize diabetes care by enabling early intervention, personalized risk assessment, and targeted preventive strategies. By identifying high-risk individuals and providing of diabetes-related complications, Making strides persistent results and decreasing healthcare costs are two basic objectives that can be accomplished through the utilize of profound learning approaches. In any case, it is fundamental to recognize the challenges and confinements related with these strategies. One of the greatest challenges is the require for huge and assorted datasets to prepare the models successfully. Additionally, the interpretability of complex models is another limitation of deep learning approaches. Ethical considerations surrounding data privacy and algorithm bias also need to be taken into account. Therefore, future research efforts should focus on addressing these challenges to ensure that deep learning approaches can be used effectively in healthcare. Refining model architectures, and conducting prospective clinical studies to validate the real world effectiveness of deep learning-based diabetes prediction models. In summary, deep learning offers a powerful framework for advancing diabetes prediction research, with the potential to revolutionize preventive healthcare and contribute to the global fight against diabetes. Through interdisciplinary collaboration and continuous innovation, deep learning practitioners can pave the way for a future where early detection and personalized intervention transform the landscape of diabetes management.

References

1. Bhoi, S.K.: Prediction of diabetes in females of pima Indian heritage: a complete supervised learning approach. Turk. J. Comput. Math. Educ. (TURCOMAT) **12**, 3074–3084 (2021)
2. Mellitus, D.: Diagnosis and classification of diabetes mellitus. Diabetes Care **28**, S5–S10 (2005)
3. Madan, P., et al.: An optimization-based diabetes prediction model using CNN and bi-directional LSTM in real-time environment. Appl. Sci. **12**, 3989 (2022). https://doi.org/10.3390/app12083989
4. Pippitt, K., Li, M., Gurgle, H.E.: Diabetes mellitus: screening and diagnosis. Am. Fam. Physician **93**, 103–109 (2016)
5. Xu, G., et al.: Prevalence of diagnosed type 1 and type 2 diabetes among US adults in 2016 and 2017: population based study. BMJ **362**, k1497 (2018). https://doi.org/10.1136/bmj.k1497
6. Chiefari, E., Arcidiacono, B., Foti, D., Brunetti, A.: Gestational diabetes mellitus: an updated overview. J. Endocrinol. Investig. **40**, 899–909 (2017). https://doi.org/10.1007/s40618-016-0607-5
7. Mirghani, D.A., Doupis, J.: Gestational diabetes from A to Z. World J. Diabetes **8**, 489–511 (2017). https://doi.org/10.4239/wjd.v8.i12.489
8. Sarwar, A., Ali, M., Manhas, J., Sharma, V.: Diagnosis of diabetes type-II using hybrid machine learning based ensemble model. Int. J. Inf. Technol. **12**, 419–428 (2020). https://doi.org/10.1007/s41870-018-0270-5
9. Zhou, H., Myrzashova, R., Zheng, R.: Diabetes prediction model based on an enhanced deep neural network. EURASIP J. Wirel. Commun. Netw. **2020**, 148 (2020). https://doi.org/10.1186/s13638-020-01765-7
10. Cho, N., et al.: IDF diabetes atlas: global estimates of diabetes prevalence for 2017 and projections for 2045. Diabetes Res. Clin. Pract. **138**, 271–281 (2018). https://doi.org/10.1016/j.diabres.2018.02.023

11. Azrar, A., Ali, Y., Awais, M., Zaheer, K.: Data mining models comparison for diabetes prediction. Int. J. Adv. Comput. Sci. Appl. **9**, 320–323 (2018). https://doi.org/10.14569/IJACSA.2018.090841
12. Larabi-Marie-Sainte, S., Aburahmah, L., Almohaini, R., Saba, T.: Current techniques for diabetes prediction: review and case study. Appl. Sci. **9**, 4604 (2019). https://doi.org/10.3390/app9214604
13. Jahani, M., Mahdavi, M.: Comparison of predictive models for the early diagnosis of diabetes. Healthc. Inform. Res. **22**, 95–100 (2016). https://doi.org/10.4258/hir.2016.22.2.95
14. Ayon, S.I., Islam, M.: Diabetes prediction: a deep learning approach. Int. J. Inf. Eng. Electron. Bus. **11**, 21 (2019)
15. Naz, H., Ahuja, S.: Deep learning approach for diabetes prediction using PIMA Indian dataset. J. Diabetes Metab. Disord. **19**, 391–403 (2020). https://doi.org/10.1007/s40200-020-00520-5
16. Acharya, U.R., et al.: An integrated diabetic index using heart rate variability signal features for diagnosis of diabetes. Comput. Methods Biomech. Biomed. Eng. **16**(2), 222–234 (2013)
17. Jian, L.W., Lim, T.-C.: Automated detection of diabetes by means of higher order spectral features obtained from heart rate signals. J. Med. Imaging Health Inform. **3**(3), 440–447 (2013)
18. Acharya, U.R., Faust, O., Kadri, N.A., Suri, J.S., Yu, W.: Automated identification of normal and diabetes heart rate signals using nonlinear measures. Comput. Biol. Med. **43**(10), 1523–1529 (2013)
19. Acharya, U.R., Vidya, K.S., Ghista, D.N., Lim, W.J.E., Moli-nari, F., Sankaranarayanan, M.: Computer-aided diagnosis of diabetic subjects by heart rate variability signals using discrete wavelet transform method. Knowl. Based Syst. **81**, 56–64 (2015)
20. Pfeifer, M.A., et al.: Quantitative evaluation of cardiac parasympathetic activity in normal and diabetic man. Diabetes **31**(4), 339–345 (1982)

Efficient Information Extraction from Medical Records

G. Paavai Anand(✉), Rishi Sundaram, S. Karthikk Raja, and P. Kalpana

Department of Computer Science and Engineering, SRM Institute of Science and Technology, Vadapalani, Chennai, India
{paavaiag,rs9164,ks2724,kc3560}@srmist.edu.in

Abstract. Over the years, information extraction has undergone numerous significant changes. This also holds true for the medical field. Medical transcripts are valuable resources filled with information that can help medical professionals and patients with insights. However, extracting this data manually is not efficient. Automating or minimalizing the process is required to perform efficiently. We aim to achieve No Code data extraction from medical transcripts to achieve inclusivity. To achieve this, we fine-tune the model using curated datasets. LLMs are powerful learners, but lack specific knowledge. By feeding them medical data like research papers and clinical documents, we adjust their internal understanding towards the medical domain. This lets them recognize medical terms, grasp relationships between diseases and treatments, and understand the context of medical queries. So, when we ask a question, the fine-tuned LLM can search through the data it learned to give us an accurate and relevant answer [5, 6]. We evaluate the model's performance using RAG metrics for data retrieval. Compared to Mixtral 7B, our proposed model shows 8% improvement in MMLU performance. Translation being the only weak section of our proposed model as it is trained for data retrieval and not translation. The fine-tuning process is carried out with fixed parameters listed in the PEFT section. Methodology of training is PEFT and for learning, reinforcement learning with a learning rate of 1e−4. This value is experimental and the relation between learning rate and results vary due to difference in nature and quality of the training data.

Keywords: Medical Transcripts · Data extraction · Large Language Model (LLM) · Low-Code/No-Code · Finetuning · Prompt Engineering

1 Introduction

Medical transcripts are a treasure trove of information about patients' health, but manually extracting this data can be a slow and laborious task. This study delves into the potential of Large Language Models (LLMs), a rapidly evolving technology in Artificial Intelligence (AI), to automate this process. Researchers and healthcare professionals, for instance, could pose a question like "Identify all patients diagnosed with a specific condition within a defined timeframe". The LLM would then scour vast databases of transcripts and retrieve the relevant data. This capability has the potential to significantly accelerate research efforts, allowing scientists to unlock new insights from healthcare data at an unprecedented pace.

P. D. Sivakumar et al. (Eds.): IRCCTSD 2024, CCIS 2360, pp. 166–179, 2025.
https://doi.org/10.1007/978-3-031-82389-3_15

An LLM trained on general text might misinterpret medical jargon, leading to inaccurate data retrieval. Medical transcripts often contain unstructured or semi-structured text, requiring LLMs to be adept at handling different document formats. We address these challenges by exploring techniques like fine-tuning, which involves specifically training the LLM on medical language and terminology.

Data extraction techniques have undergone a remarkable transformation. We have progressed from the laborious process of manual extraction to leveraging the power of Natural Language Processing (NLP) techniques, initially requiring code-based implementations. The rise of web technologies introduced scraping as a means of data retrieval. Today, deep learning models stand at the forefront, offering even greater capabilities.

This proposal advocates for the integration of deep learning models, specifically Large Language Models (LLMs), to achieve no-code data extraction. This would democratize the process, making it accessible to a broader range of users, regardless of their technical expertise [9, 12].

This approach would significantly enhance the ease and efficiency of data extraction, fostering a wider range of data-driven applications and insights.

2 Related Works

Large language models (LLMs) have the potential to minimalize the workload in Information Extraction. Challenges in this field include ensuring the safety, reliability, efficacy, and privacy of the technology. Addressing these issues, many studies and experiments have been conducted in the potential usage of LLMs in this area [22]. Before this, Scraping, NLP techniques such as Named Entity Recognition, Deep learning models were used in Information extraction [21]. LLMs perform better than the previous techniques and is far easier to use with minimal technical knowledge [4, 7]. Additionally, there is a need for more research to develop LLMs that are aligned with the medical domain and that can avoid hallucinations [21, 23]. That is why we try to solve this by fine tuning models to understand medical domain conversations. The training methods used are similar to [11, 19]. We curated datasets from popular public datasets with quality in mind. Dataset curation methodology is referenced from the papers [13, 14]. LLMs could potentially assist in various areas of medicine, given their capability to process complex concepts, as well as respond to diverse requests and questions (prompts) However, these models also raise concerns about misinformation, privacy, biases in the training data, and potential for misuse [22]. LLMs can sift through vast quantities of scientific literature to identify promising drug targets and accelerate the drug discovery process. Additionally, they can help researchers by summarizing complex scientific concepts and analysing research data [5, 23].

3 Methodology

Creating a framework to extract data from medical transcripts is done using various technologies.

We create the flow diagram and structural diagram for the model. The technologies used have various substitutes; we use these as they are open sourced. They are as follows:

3.1 Dataset Curation

To facilitate the LLM to understand medical jargon, we fine-tune the LLM with a curated dataset of medical Q&A. We put together quality Q&A from various public and exclusive datasets to get the training data.

3.2 Fine-Tuning

To achieve efficient and effective fine-tuning of our model, we have leveraged a combination of techniques. Reinforcement Learning offers a powerful approach that simplifies the training process while delivering strong performance. Additionally, Parameter-Efficient Fine-Tuning (PEFT) [1] methods like Low-Rank Adaptation (LoRA) significantly reduce the dimensionality of model parameters, leading to faster training times and lower computational costs. This combination allows for cost-effective and time-efficient LLM fine-tuning (Fig. 1).

$$f(s)_i = \frac{e^{s_i}}{\sum_j^C e^{s_j}} \qquad CE = -\sum_i^C t_i log(f(s)_i)$$

Fig. 1. SoftMax computation Cross-Entropy Loss function

3.2.1 Parameter Efficient Fine-Tuning (PEFT)

Training an LLM from scratch is computationally expensive and can be inefficient. To address this, we employ selective parameter update techniques. These techniques focus training on a subset of the model's parameters with a higher impact on the desired task. This approach, compared to full model retraining, significantly reduces training time and resource consumption without compromising the model's performance on the target task [2, 3].

Figure 3 below shows the parameters we used for training for 2 epochs. Resulting model is stored in "results" folder.

```
prompt_template="""
You are a virtual assistant providing information from the given medical transcripts with context.
Provide information only available in the input context.\n
If answer is not available, just say "Please consult medical professional. I am not able to give the answer".\n
Always print a warning message at the end after 5 lines of space that says
" Always consult Medical Professional for treatment and diagnosis.
\n I am an assistant in training. \nTake Care... and remember that You only have one Life.\n"
Context:\n {context} \n
Question:\n {question} \n
Answer:
"""
model=ChatGoogleGenerativeAI(model="gemini-pro",temperature=0.3)
prompt=PromptTemplate(template=prompt_template,input_variables=["context","question"])
chain=load_qa_chain(model,chain_type="stuff",prompt=prompt)
```

```
training_params = TrainingArguments(
    output_dir="./results",
    num_train_epochs=2,
    per_device_train_batch_size=4,
    gradient_accumulation_steps=2,
    optim="paged_adamw_32bit",
    save_steps=20,
    logging_steps=20,
    learning_rate=1e-4,
    weight_decay=0.001,
    fp16=False,
    bf16=False,
    max_grad_norm=0.4,
    max_steps=-1,
    warmup_ratio=0.03,
    group_by_length=True,
    lr_scheduler_type="constant",
    report_to="tensorboard"
)
```

Fig. 2. PEFTWorkflow

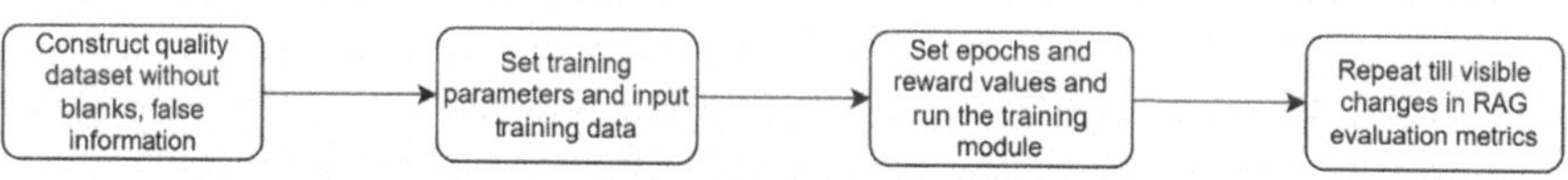

Fig. 3. Prompt Engineered Template

4 Workflow Diagram

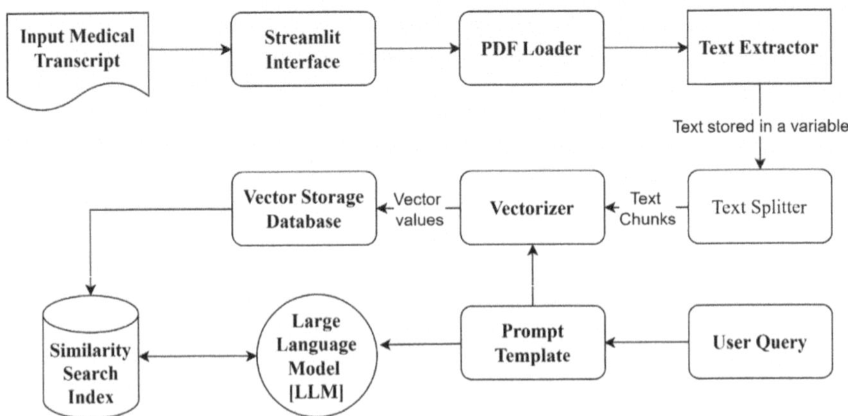

5 Module Descriptions

Streamlit Interface: The frontend interface is created using Streamlit and is connected through LangChain. Takes the medical transcripts as input and feeds it to the Document loader. The user query is also taken as the other input. The query is passed to the Prompt template module.

Document Loader: PyPDF2 module is used to load the input medical transcript into the framework. Takes input transcript from interface and extracts the textual data into a variable which is passed to the Text Splitter module.

Text Splitter: This module takes the variable containing the extracted text from the document loader module and splits it into smaller batches for easier processing. Batch size and overlap can be set in this module. Gives the output as text batches of the given sizes.

Vectorizer: We use GoogleGenerativeAI embeddings model "001" as the transformer to vectorize the input data. Inputs are the text chunks from the text splitter module, output is given as vectors of the input texts.

Vector Storage: The vectors from vectorizer module is store in the vector storage either locally or via cloud. We use FAISS as the vector database to compare the vectors. This module also functions in creating the similarity search index for the input document and the user query.

Prompt Template: The user query is inserted in the prompt template along with the instruction prompts. The template contains instructions, question, answer and the LLM model. Having a proper prompt template gives the model awareness on the boundaries we configure. This is set by lowering the temperature and adding stop words to filter out unwanted content. The prompt template is then fed to the LLM.

LangChain: We use LangChain to connect the Q&A chains (Question Answer pairs) and the interface created using Streamlit. New question answer pairs are generated for each new document uploaded. The previous interactions are not kept as residual memory to avoid hallucinations and breach of data.

LLM: The Large Language Model (LLM) is placed at the centre of the framework between the input document and user query. The LLM has access to the Vector database and the Similarity search index is created with the vectors of the input document and the user query.

Fine-Tuning Module: The fine tuner has the training dataset and training parameters as the input data. Epoch, learning rate, reward value, batch size are the main training parameters, The LLM is inserted to start the training process [17]. Evaluation metrics are the RAG metrics used to evaluate data retrieval processes. The training module is run for a maximum of 2 epochs due to limited resources. Variations of training parameters are used for many iterations to arrive at the best possible RAG score.

6 Testing and Documentation

Our model's performance is rigorously evaluated using F1 scores calculated across a diverse set of test datasets. This comprehensive evaluation ensures the model's generalizability and robustness. Additionally, meticulous documentation practices are implemented throughout the training process. This includes capturing all updates, modifications, and hyperparameter tuning decisions. This thorough documentation facilitates reproducibility, analysis, and future improvements to the model.

7 Results

This model facilitates a user-friendly interface for medical transcript information retrieval. Users can interact with the model by posing natural language queries. The model then leverages its understanding of the transcripts to locate relevant data points and present them in a clear and concise manner. Additionally, the model can rephrase complex medical terminology or simplify findings for improved user comprehension.

This model demonstrates a measurable improvement over the previous iteration (2022–2023) [15–17]. However, there is still significant potential for further advancement. Two key areas for exploration include increasing the model size and exploring novel fine-tuning methodologies.

7.1 RAG Performance Metrics

Proposed model is compared with GPT 3 and Llama2 base models. Proposed model is built on Gemini experimental model available at OpenAI (Figs. 4 and 5).

Metric	Description	Proposed Model	GPT 3	Llama 2
ROUGE [Recall Oriented Understudy for Gisting Evaluation]	Calculated with the count of overlapping n-grams between the system output and the reference summaries.	**0.769**	0.751	0.784
BLEU [BiLingual Evaluation Understudy] *Our model is not trained for translation	Calculated with n-gram precision	0.268	0.799	0.826
RECALL	Identified Correct Answer divide by Total Correct answers	**0.826**	0.835	0.793
PRECISION	Accuracy measure	**0.756**	0.729	0.762
F1 Score	Harmonic Mean of Precision and Recall	**0.797**	0.772	0.886
HIT RATE	Successful attempts divided by Total Attempts	0.689	0.795	0.719

Fig. 4. MMLU metric compared with Mixtral models and LLaMA models with base parameters ranging from 7B to 70B. Our proposed model's size is 12B parameters.

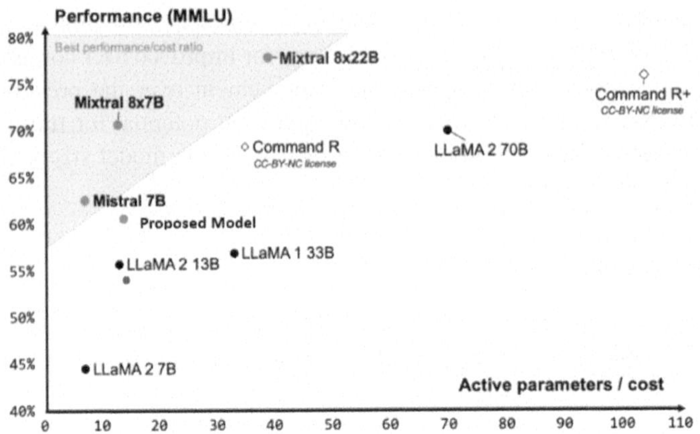

Fig. 5. Sample Training Data used for fine tuning the model.

8 Output

See Figs. 6, 7, 10, 11, 12 and 13

###24494281

OBJECTIVE To observe the difference in the clinical efficacy on oculomotor impairment between electroacupuncture and acupuncture and explore the best therapeutic method in the treatment of this disease .

METHODS Sixty cases of oculomotor impairment were randomized into an electroacupuncture group and an acupuncture group , 30 cases in each one .

METHODS In the electroacupuncture group , the points were selected on extraocular muscles , the internal needling technique in the eye was used in combination of electroacupuncture therapy .

METHODS In the acupuncture group , the points and needling technique were same as the electroacupuncture group , but without electric stimulation applied .

METHODS The treatment was given 5 times a week , 15 treatments made one session .

METHODS After 3 sessions of treatment , the clinical efficacy , palpebral fissure size , pupil size , oculomotor range and the recovery in diplopia were compared before and after treatment in the two groups .

RESULTS In the electroacupuncture group , the palpebral fissure size was (9.79 + / -2.65) mm and the eyeball shifting distance was (18.12 + / -1.30) mm , which were hig-her than (8.23 + / -2.74) mm and (16.71 + / -1.44) mm respectively in the acupuncture group .

RESULTS In the electroacupuncture group , the pupil diameter was (0.44 + / -0.42) mm , which was less than (0.72 + / - 0.53) mm in the acupuncture group , indicating the significant difference (all P < 0.05) .

RESULTS The cured rate was 63.33 % (19/30) and the total effective rate was 93.33 % (28/30) in the electroacupuncture group , which was better than 36.67 % (11/30) and 83.333 (25/30) in the acupuncture group separately , indicating the significant difference (all P < 0.05) .

CONCLUSIONS Electroacupuncture presents the obvious advantages in the treatment of oculomotor impairment , characterized as quick and high effect , short duration of treatment and remarkable improvements in clinical symptoms , there are important significance for the improvement of survival quality of patients .

Fig. 6. Sample Input Transcript that will be used as the input from the user.

Description: Patient presented to the bariatric surgery service for consideration of laparoscopic roux en Y gastric bypass surgery.

Speciality: Bariatrics

Sample: Gastric Bypass Discussion – 3

Transcription: PAST MEDICAL HISTORY:, Significant for hypertension. The patient takes hydrochlorothiazide for this. She also suffers from high cholesterol and takes Crestor. She also has dry eyes and uses Restasis for this. She denies liver disease, kidney disease, cirrhosis, hepatitis, diabetes mellitus, thyroid disease, bleeding disorders, prior DVT, HIV and gout. She also denies cardiac disease and prior history of cancer.,PAST SURGICAL HISTORY: , Significant for tubal ligation in 1993. She had a hysterectomy done in 2000 and a gallbladder resection done in 2002.,MEDICATIONS: , Crestor 20 mg p.o. daily, hydrochlorothiazide 20 mg p.o. daily, Veramist spray 27.5 mcg daily, Restasis twice a day and ibuprofen two to three times a day.,ALLERGIES TO MEDICATIONS: , Bactrim which causes a rash. The patient denies latex allergy.,SOCIAL HISTORY: , The patient is a life long nonsmoker. She only drinks socially one to two drinks a month. She is employed as a manager at the New York department of taxation. She is married with four children.,FAMILY HISTORY: , Significant for type II diabetes on her mother's side as well as liver and heart failure. She has one sibling that suffers from high cholesterol and high triglycerides.,REVIEW OF SYSTEMS: , Positive for hot flashes. She also complains about snoring and occasional slight asthma. She does complain about peripheral ankle swelling and heartburn. She also gives a history of hemorrhoids and bladder infections in the past. She has weight bearing joint pain as well as low back degenerating discs. She denies obstructive sleep apnea, kidney stones, bloody bowel movements, ulcerative colitis, Crohn's disease, dark tarry stools and melena.,PHYSICAL EXAMINATION: ,On examination temperature is 97.7, pulse 84, blood pressure 126/80, respiratory rate was 20. Well nourished, well developed in no distress. Eye exam, pupils equal round and reactive to light. Extraocular motions intact. Neuro exam deep tendon reflexes 1+ in the lower extremities. No focal neuro deficits noted. Neck exam nonpalpable thyroid, midline trachea, no cervical lymphadenopathy, no carotid bruit. Lung exam clear breath sounds throughout without rhonchi or wheezes however diminished. Cardiac exam regular rate and rhythm without murmur or bruit. Abdominal exam positive bowel sounds, soft, nontender, obese, nondistended abdomen. No palpable tenderness. No right upper quadrant tenderness. No organomegaly appreciated. No obvious hernias noted. Lower extremity exam +1 edema noted. Positive dorsalis pedis pulses.,ASSESSMENT: , The patient is a 56-year-old female who presents to the bariatric surgery service with a body mass index of 41 with obesity related comorbidities. The patient is interested in gastric bypass surgery. The patient appears to be an excellent candidate and would benefit greatly in the management of her comorbidities.,PLAN: , In preparation for surgery will obtain the usual baseline laboratory values including baseline vitamin levels. Will proceed with our usual work up with an upper GI series as well as consultations with the dietician and the psychologist preoperatively. I have recommended six weeks of Medifast for the patient to obtain a 10% preoperative weight loss.

Fig. 7. Conclusion of a study conducted on 150 patients created by giving the prompt of "conclusion? why?"

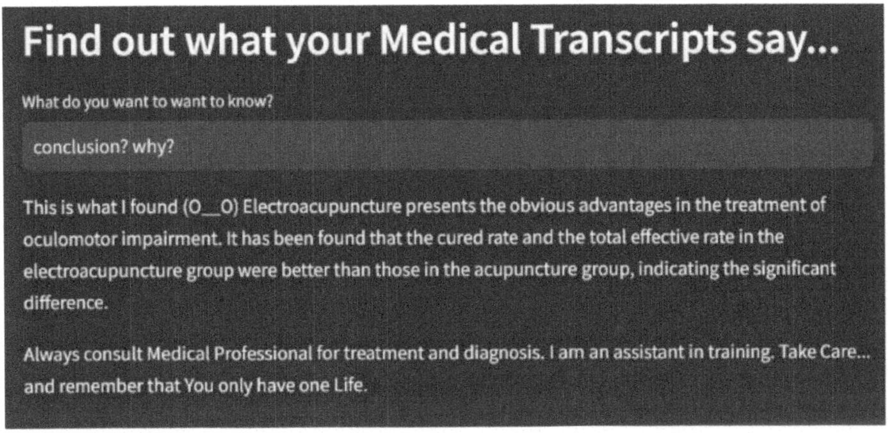

Find out what your Medical Transcripts say...
What do you want to want to know?
conclusion? why?
This is what I found (O__O) Electroacupuncture presents the obvious advantages in the treatment of oculomotor impairment. It has been found that the cured rate and the total effective rate in the electroacupuncture group were better than those in the acupuncture group, indicating the significant difference.
Always consult Medical Professional for treatment and diagnosis. I am an assistant in training. Take Care... and remember that You only have one Life.

Fig. 8. Summary of Fig. 2 generated with the constraint of 5 lines.

Give me a summary of the documnet in 5 lines
Reply: - A 56-year-old female with a BMI of 41 and obesity-related comorbidities presents for consideration of laparoscopic roux en Y gastric bypass surgery.
The patient has a history of hypertension, high cholesterol, dry eyes, and has undergone tubal ligation, hysterectomy, and gallbladder resection.
She takes Crestor, hydrochlorothiazide, Veramist spray, Restasis, and ibuprofen.
She is a lifelong nonsmoker, drinks socially, and is employed as a manager.
She has a family history of type II diabetes, liver and heart failure, high cholesterol, and high triglycerides.

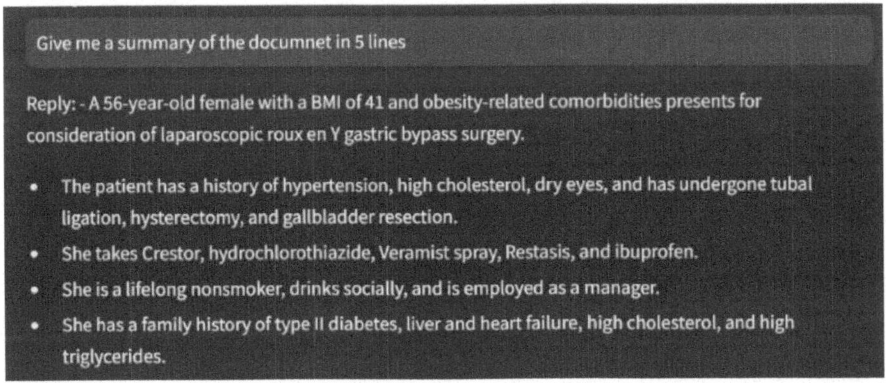

Fig. 9. Interface of the model with which user will interact. Caution message is always displayed as this is an experimental model and should not be used as replacement for a doctor.

Welcome!
Upload your Trancripts and Click Submit.
Drag and drop files here
Limit 200MB per file
Browse files
Submit and Process
Find out what your Medical Transcripts say...
What do you want to want to know?

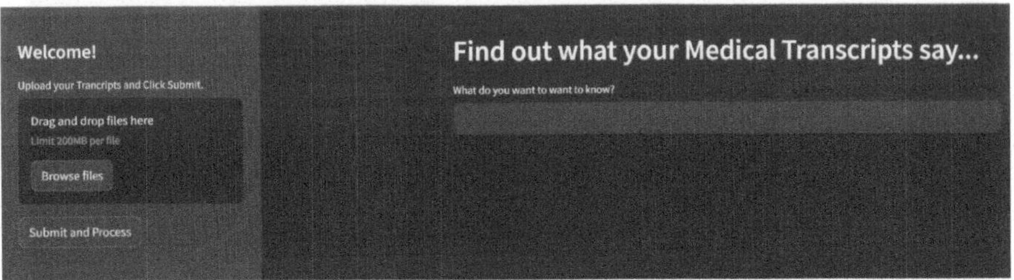

Fig. 10. This output retrieves the procedure conducted on the patient, using natural language input and not as a technical command.

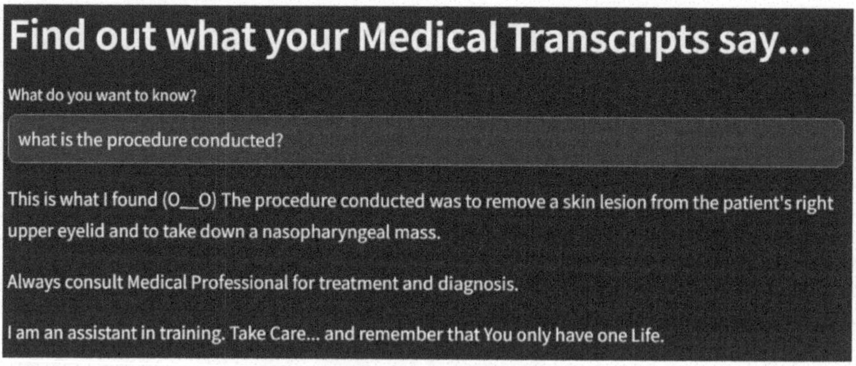

Find out what your Medical Transcripts say...

What do you want to know?

what is the procedure conducted?

This is what I found (O__O) The procedure conducted was to remove a skin lesion from the patient's right upper eyelid and to take down a nasopharyngeal mass.

Always consult Medical Professional for treatment and diagnosis.

I am an assistant in training. Take Care... and remember that You only have one Life.

Fig. 11. Figure 8 Patient History Retrieval

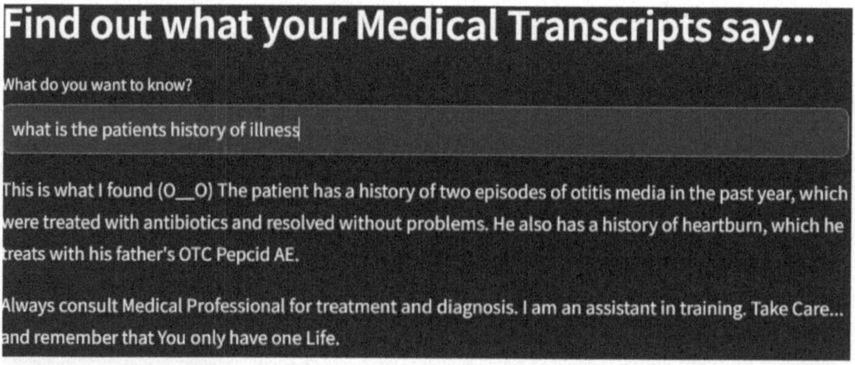

Find out what your Medical Transcripts say...

What do you want to know?

what is the patients history of illness

This is what I found (O__O) The patient has a history of two episodes of otitis media in the past year, which were treated with antibiotics and resolved without problems. He also has a history of heartburn, which he treats with his father's OTC Pepcid AE.

Always consult Medical Professional for treatment and diagnosis. I am an assistant in training. Take Care... and remember that You only have one Life.

Fig. 12. Figure 9 Insufficient Information Handling - Shows how the model responds when the data requested is not available in the given document. Displays a message stating that the information is not available.

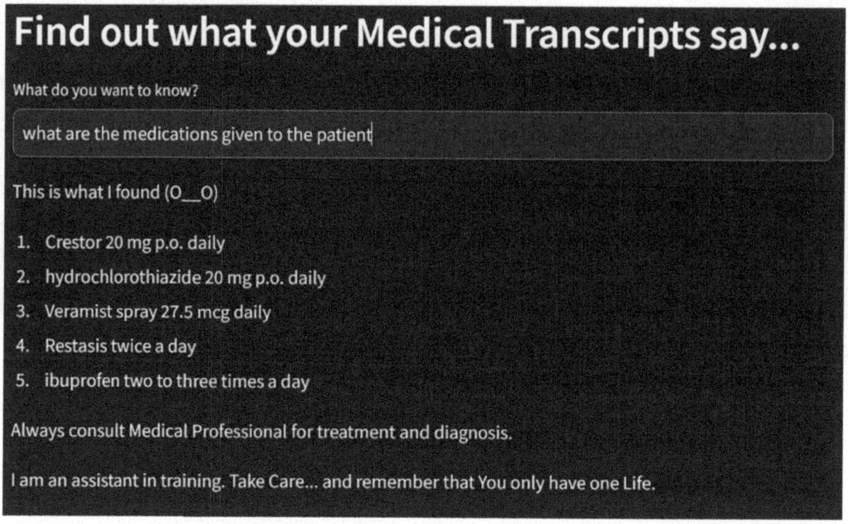

Find out what your Medical Transcripts say...

What do you want to know?

what are the medications given to the patient

This is what I found (O__O)

1. Crestor 20 mg p.o. daily
2. hydrochlorothiazide 20 mg p.o. daily
3. Veramist spray 27.5 mcg daily
4. Restasis twice a day
5. ibuprofen two to three times a day

Always consult Medical Professional for treatment and diagnosis.

I am an assistant in training. Take Care... and remember that You only have one Life.

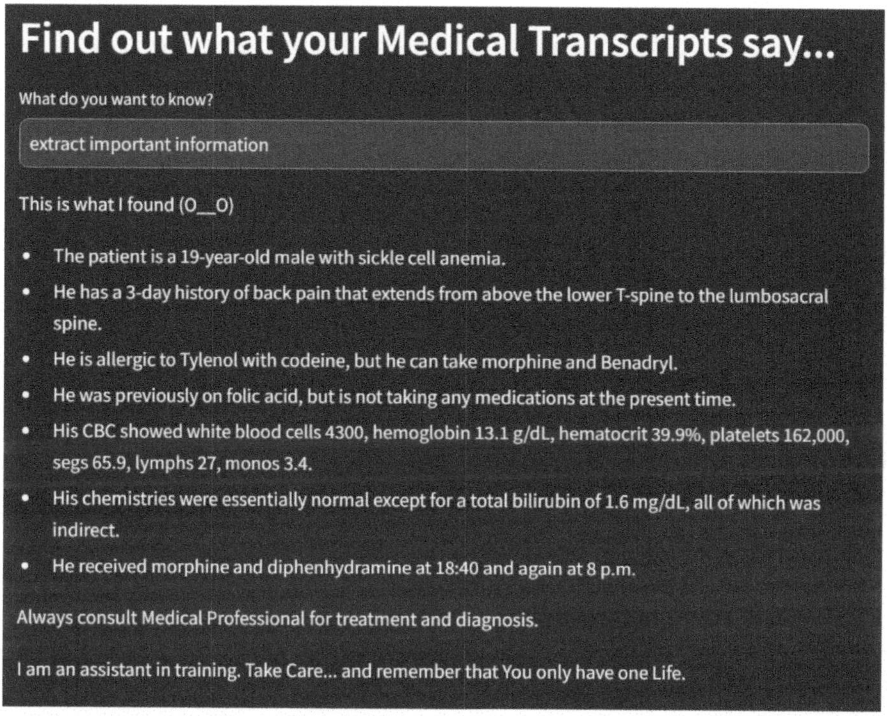

Find out what your Medical Transcripts say...

What do you want to know?

extract important information

This is what I found (O__O)

- The patient is a 19-year-old male with sickle cell anemia.
- He has a 3-day history of back pain that extends from above the lower T-spine to the lumbosacral spine.
- He is allergic to Tylenol with codeine, but he can take morphine and Benadryl.
- He was previously on folic acid, but is not taking any medications at the present time.
- His CBC showed white blood cells 4300, hemoglobin 13.1 g/dL, hematocrit 39.9%, platelets 162,000, segs 65.9, lymphs 27, monos 3.4.
- His chemistries were essentially normal except for a total bilirubin of 1.6 mg/dL, all of which was indirect.
- He received morphine and diphenhydramine at 18:40 and again at 8 p.m.

Always consult Medical Professional for treatment and diagnosis.

I am an assistant in training. Take Care... and remember that You only have one Life.

Fig. 13. Medication History of patient is retrieved using a single simple input. This finds and returns the tablets and drugs the patient has been taking.

9 Conclusion

Large Language Models (LLMs) are poised to revolutionize healthcare by empowering patients with virtual health assistants for self-care, fostering better communication across languages, and improving efficiency for clinicians. However, careful development and implementation are crucial to ensure the accuracy and transparency of LLM outputs, paving the way for a future of more efficient, effective, personalized, and accessible healthcare for all. LLMs are the building base on which the artificial human will be created one day. Finetuning will be efficient in improving the model's performance. The base model is a general-purpose model and we have fine-tuned it and made it precise using medical transcripts and Q&A pairs, enabling the new proposed model to have a competitive edge in handling medical data when compared to the generic LLMs. This enables the model to be used in the medical field for research and with further advancements, can be deployed to support patients as well.

10 Future Work

Decision making: LLMs are still under development but the impact they can have in the medical field is undeniably crucial. Further refining this, decision making is where LLMs are currently lacking. We can overcome this only through extensive research and experimentation in supervision of medical professionals.

Diagnosis: We can train the LLMs to diagnose patients through descriptions and other input methods like sensors, patient history etc. These are highly sensitive grounds and these systems can be used suggestively and not autonomously.

Patient Support: LLMs can be enabled to translate medical jargon into simple terms so that patients can understand exactly what they are going through. Curated assistance and monitoring can be done using LLMs.

References

1. Landolsi, M.Y., Hlaoua, L., Ben Romdhane, L.: Information extraction from electronic medical documents: state of the art and future research directions. Knowl. Inf. Syst. **65**, 463–516 (2023). https://doi.org/10.1007/s10115-022-01779-1
2. Cohen, K.B., et al.: Coreference annotation and resolution in the Colorado richly annotated full text (craft) corpus of biomedical journal articles. BMC Bioinform. **18**, 372 (2017)
3. Kreuzthaler, M., Schulz, S.: Detection of sentence boundaries and abbreviations in clinical narratives. BMC Med. Inform. Decis. Mak. **15**, S4 (2015)
4. Uzuner, O., Solti, I., Cadag, E.: Extracting medication information from clinical text. J. Am. Med. Inform. 17, 514–518 (2010)
5. XLNet: generalized autoregressive pretraining for language understanding (2019)
6. LLMP: exploiting LLDP for latency measurement in software-defined data centred networks. Comput. Sci. Technol. (2023)
7. Llama 2: open foundation and fine-tuned chat models. GenAI META (2023)
8. PaLM 2 technical report 2023. Google Brain
9. Reasoning about physical commonsense in natural language. In: AAAI Conference on Artificial Intelligence (2022)
10. Jang, D., Yun, T.R., Lee, C.Y., Kwon, Y.K., Kim, C.E.: GPT-4 can pass the Korean national licensing examination for Korean medicine doctors. PLOS Digit. Health **21**, e0000416 (2023)
11. Attention is all you need. In: 31st Conference on Neural Information Processing Systems (NIPS 2017) CA, USA (2017)
12. Long short-term memory-networks for machine reading (2016)
13. MRN: a locally and globally mention-based reasoning network for document-level relation extraction. In: Findings of Association for Computational Linguistics (2021)
14. Towards expert-level medical question answering with LLMS (2023)
15. Jiang, X., Cheng, Y., Zhang, S., Wang, J., Ma, B.: APIE: an information extraction module designed based on the pipeline method. Array **21**, 100331 (2023)
16. Transformers. Or as i like to call it attention on steroids. Data Sci. (2020)
17. Understanding attention mechanism: NLP. Analytics Vidhya (2022)
18. NLP: zero to hero. Medium (2023)
19. When scaling meets LLM finetuning: the effect of data, model and finetuning method. In: ICLR (2024)
20. Towards expert-level medical question answering with large language models. Google Deep Mind (2024)

21. Li, J., Sun, Y., Johnson, R.J., Sciaky, D., Wei, C.-H., Leaman, R., et al.: BioCreative V CDR task corpus: a resource for chemical disease relation extraction. Database (2016). https://doi.org/10.1093/database/baw068
22. Wu, Y., Luo, R., Leung, H.C.M., Ting, H.F., Lam, T.W.: RENET: a deep learning approach for extracting gene-disease associations from literature. In: Cowen, L. (eds.) RECOMB 2019. LNCS, vol. 11467, pp. 272–284. Springer, Cham (2019). https://doi.org/10.1007/978-3-030-17083-7_17
23. Beltagy, I., Lo, K., Cohan, A.: SciBERT: a pretrained language model for scientific text (2019). https://doi.org/10.48550/arXiv.1903.10676

Decoding the Mind: Translating Human Thought with EEG Signals

Neenu Francis(✉) and G. Vadivu

SRM Institute of Science and Technology, Kattankulathur, Chengalpattu, India
{nf5767,vadivug}@srmist.edu.in

Abstract. The electroencephalogram (EEG) signal has several uses in biomedicine, including diseases diagnosis, rehabilitation, brain-computer interfaces, and sleep research, due to its complexity and lack of invasiveness. Researchers have proposed several advanced methods for pre-processing and feature extraction to tackle the complexity of EEG data analysis. This study highlights the importance of EEG to serve the purpose of decoding humans' thoughts by exploring current methods that help structure brain signals and extract significant information by implementing various effective processing approaches. The discourse encompasses the model architecture of the EEG signal processing, commencing with the initial recording and progressing through data extraction and classification. The study discusses various techniques like Independent Component Analysis, Canonical Correlation Analysis, Discrete Wavelet Transform, and Empirical Mode Decomposition have been developed for various use cases to denoise the EEG signals for accurate data analysis. It also discuss the metrics that evaluates denoising the EEG signals. The studies shows the ability to classify EEG signals using traditional models and deep learning models across multiple domains with the accuracies ranging from 84.2% to 99.3%. Futhermore, the paper confronts the challenges associated with the present technologies and looks at upcoming progressions. Additionally, it provides several suggestions for future research initiatives within this domain.

Keywords: BCI-Brain Computer Interface · electroencephalography · signal processing · thought translation · machine learning · deep learning

1 Introduction

Brain science has emerged as a critical discipline in unveiling the enigmas of existence, propelled by progress in medical technology and an expanding comprehension of the brain. The investigation of the brain's complex mechanisms has been a subject of scientific interest since the mid-20th century [1]. EEG is an essential tool for many brain-related studies [2], as examining the electrical signals of the human brain is a significant area of research.

Non-invasiveness and safety are elements contributing to EEG's widespread use. Placing electrodes on the scalp makes it easier to capture electrical signals coming

P. D. Sivakumar et al. (Eds.): IRCCTSD 2024, CCIS 2360, pp. 180–190, 2025.
https://doi.org/10.1007/978-3-031-82389-3_16

from the brain, which gives important information about how the brain functions [3]. These signals assist in the diagnosis of neurological illnesses as well as the study of cognitive functions like memory. Combining EEG with other imaging modalities, such as functional near-infrared spectroscopy (fNIRS) and also magnetic resonance imaging (MRI), can provide a greater understanding of the structure and function of the brain [4, 5].

The potential of EEG transcends the realm of research. Additional developments in brain-computer interactions, the facilitation of recognizing emotions, and the rehabilitation of individuals with partial paralysis are all potential benefits of such technology [6, 7]. Alzheimer's, epilepsy, and cognitive impairment are among the conditions for which clinicians and researchers utilize EEG to diagnose dysfunction of the brain [8, 9]. As a result of their inherent complexities, and susceptibility to noise, however, meaningful information extraction from EEG signals necessitates meticulous analysis [10].

The paper's organization is intended to provide a framework for this investigation. In Sect. 2, we look at the EEG signal process's pipeline architecture. This includes signal acquisition, denoising methods, and feature engineering methods like time-frequency, spectral, and non-linear dynamic analysis. Furthermore, EEG signal classification is investigated using conventional and deep learning methodologies. Future implications and constraints are addressed in Sect. 3, whereas an overview of the research endeavours and contributions of the current study is provided in Sect. 4.

2 Eeg Signal Analysis Process

The architecture model with EEG Signal processing for classification or regression is illustrated in Fig. 1. The process begins by acquiring and recording the raw EEG signal from the partipants by fixing the electrodes on the scalp. The next segment is preprocessing which is necessary to improve the quality of signal by filtering and removing unwanted noise and artifacts. Subsequently, extract the relevant features or patterns for classification or recognition of different mental states or events. Lastly, add a feedback mechanism that enables researchers to enhance and strengthen the dynamic nature of experimental protocols that leads to more reliable and valid results.

2.1 EEG Signal Acquisition

Electrodes are placed either non-invasively or invasively to the cranium to assess the neural activity in various spatial regions of the brain. This approach has undergone tremendous growth since its invention in 1924. These electrodes identify and capture tiny voltage fluctuations while a neural transmission occurs are then digitised and amplified for further investigation. Critical insights into cognitive processes can be obtained by researchers through the identification of activity in specific region of brain [11].

The EEG signals exhibit discrete wave oscillations [12] with respect to the frequency spectrum, is measured in cycles per second and quantified in Hertz (Hz). The brain waves, namely Delta (δ) wave, Theta (θ) wave, Alpha (α) wave, Beta (β) wave, and Gamma (γ) wave, have been listed in Table 1 in accordance with their corresponding cerebral activities.

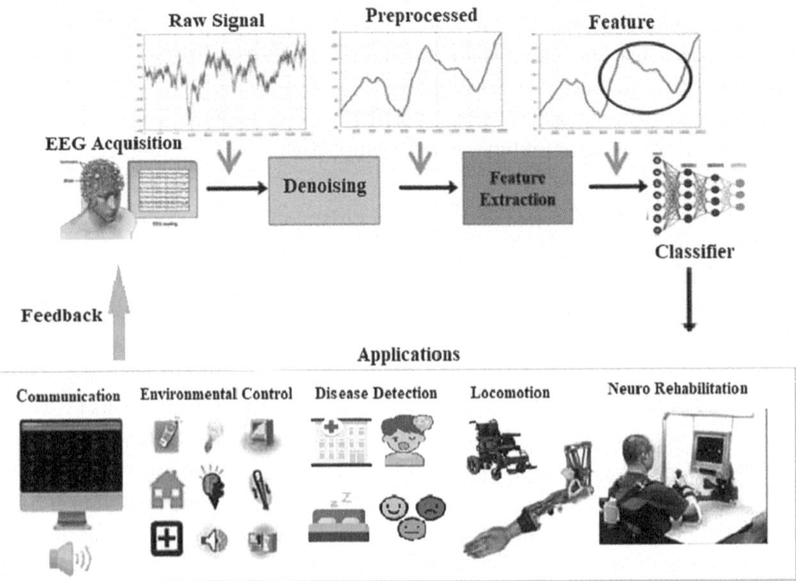

Fig. 1. EEG Signal Process architecture

Table 1. EEG brain waves and its characteristics [8]

Brainwave Type	Cognitive States	Associated tasks and behaviours
Delta (δ) (0.5–4 Hz)	Deep sleep, unconscious	not moving, not attentive, low level of arousal
Theta (θ) (4– 8 Hz)	Light sleep, imaginary, dreamlike, and drowsy	creative and intuitive, but may also be distracted or unfocused
Alpha (α) (8–12 Hz)	relaxed, not agitated, but not fatigued, conscious	silent, soothing, and therapeutic during meditation
Beta (β) (12–30 Hz)	calm but focused, integrated, thoughtful, conscious of self and environment	Actively vigilant while remaining calm
Gamma (γ) (30–100 Hz)	thinking, integrated thoughts	associated with information-rich task processing

2.2 Denoising of EEG

Due to the error caused from the biological and environmental noise, the quality of the EEG signal may worsen while capturing it with scalp electrodes [13]. Examples of such artifacts that have the potential to significantly impact outcomes are physiological artifacts, including eye movements, blinking, and muscular activity [14]. Researchers have devoted considerable effort to the denoising of EEG signals in recognition of this difficulty. Implementing this technique to remove undesired anomalies is critical for ensuring the accuracy of the characteristics extracted from the EEG data. Various denoising techniques (refer to Table 2) and the metrics to evaluate (refer to Table 3) have been devised to address this issue.

2.3 EEG Feature Engineering

For extracting pertinent information from the denoised EEG, feature engineering [21] plays a key role in machine learning (ML) as well as deep learning (DL) approaches. This process bridges the gap between raw EEG data and the algorithms, allowing them to leverage the underlying brain activity. Common techniques include transformation (segmenting [22] the denoised data into manageable parts depending on task needs or brain activity, feature extraction), and feature selection.

To understand brain activity from EEG signals, researchers extract features using techniques [23] like time domain analysis (analyzing basic statistics like mean, variance, and signal amplitude within specific time windows), frequency domain analysis (analyzing frequency components using techniques like Fast Fourier Transform (FFT) to identify brainwave activity), methods that combine both (e.g., wavelet transform) to capture how frequency changes over time, and nonlinear feature analysis (using techniques like approximate entropy [24], which can be helpful for tasks like seizure detection). Once they have been extracted from the brain signals, it is essential to select the most revealing factors (called feature selection). This shortens processing times while also increasing model accuracy. Common techniques include using feature significance ranking, training models with various feature combinations, and integrating selection into deep learning training.

2.4 EEG Classification

Classification of EEG signals is the most critical aspect of brain activity analysis. A wide variety of processing methodologies can be utilized for classifying EEG signals, encompassing statistical analysis, machine learning, deep learning, and additional strategies. This section will center on the classification methods that are implemented across diverse domains of EEG application, placing specific emphasis on approaches that rely on machine learning. A variety of both machine learning and deep learning methodologies for the intent of categorization have been shown in Table 4 and it is illustrated in Fig. 2 and Fig. 3.

Deep learning (DL) models offers more substantial benefits in the domain of EEG signal classification and prediction, despite the simplified architecture of traditional machine learning (ML) models. By taking inspiration from the structure and functioning of the human brain, deep learning (DL) excels at discerning complex patterns within extensive datasets. Many researchers have analysed these EEG signals for epilepsy, emotion classification, motor imagery, and other purposes using deep learning models.

Table 2. Varıuos De-Noısıng Technıques

Technique	Description	Benefits	Drawbacks	Mathematical Representation
Regression Analysis [15]	Removes eye artifacts using EOG recordings	Effective for eye blinks/movements	Can remove some desired EEG data, struggles with other artifacts	$eeg(t) = EEG(t) - \sum_{g=0}^{T} \beta_g eog(t-g)$ eeg(t) - eeg at time t, eog(t-g) - eog at time t-g, EEG(t) unaltered EEG at time t, βg measures *EOG* on *eeg*(*t*) at time (*t*-*g*)
Independent Component Analysis (ICA) [16]	Separates mixed EEG into source signals (brain and artifacts)	Powerful for various artifacts	Traditional ICA may struggle with muscle activity	X = AS where **X** - contains raw EEG data, **A** contains combination of EEG and artifacts, matrix **S** contains independent components
Principal Component Analysis (PCA) [17]	Reduces noise by eliminating components with small eigenvalues	Simpler method, good for general noise reduction	May remove weak brain signals	$XX^T\omega_i = \lambda_i\omega_i$ where λ - the eigenvalue, ω - the eigenvector, XXT decomposing the eigenvalues of the matrix
Canonical Correlation Analysis (CCA) [18]	Effective for muscle artifact removal	Handles low correlation of muscle activity	More complex than other techniques	$\rho = \frac{u^T R_{xy} v}{\sqrt{u^T R_{xx} u}\sqrt{v^T R_{yy} v}}$ Where ρ - canonical correlation, u and v are the canonical vectors, R_{xy}, R_{xx}, R_{yy} - covariance matrix, u^T, n v^T - transposes of *u*, v

(*continued*)

Table 2. (*continued*)

Technique	Description	Benefits	Drawbacks	Mathematical Representation
Discrete Wavelet Transform (DWT) [19, 20]	Decomposes signal for targeted artifact removal	Preserves clean brain activity	Lacks perfect translation invariance	$DWT(m, n) = \int_{-\infty}^{+\infty} x(t)\psi_{m,n}(t)dt$ Where: DWT(m, n) - wavelet coefficient at scale m and translation n x(t) - input signal $\psi_{m,n}$ (t)dt - is the wavelet function, scaled and translated
Empirical Mode Decomposition (EMD) [14]	Decomposes signal into IMFs to isolate artifacts	Handles non-stationary and non-linear data	Sensitive to spike noise	$x(t) = \sum_{i=1}^{n} C_i(t) + r_N(t)$ Where: x(t) - original signal $C_i(t)$ - component signals $rN(t)$ - residual signal after n components have been extracted

3 Future Prospects and Obstacles in EEG

While traditional methods for EEG data processing require large datasets, deep learning simplifies the pipeline and opens doors to new research areas. EEG data remains valuable for understanding brain activity with both traditional and machine-learning approaches. EEG data analysis struggles with inconsistency across datasets due to different recording devices. Domain adaptation techniques [30], help reduce this data variation for improved performance. Data privacy laws restrict access to raw EEG data due to privacy concerns. Federated learning [31] offers 91% accuracy on sleep data, a solution by training models on local data without sharing it. EEG research benefits from various databases storing electrical brain activity recordings. These databases cover diverse topics, from seizure development to brain responses to sounds. Researchers can leverage this data to improve our understanding of the brain.

Processing of EEG signals exposes a complex challenge due to several inherent characteristics of the data. It is challenging to extract the true brain activity from EEG signals because noise and artifacts from different sources frequently corrupt them. Furthermore, because of their non-stationary nature, certain methods are needed to collect the data that

Table 3. Metrics For Evaluating De-Noising

Sl. No	Metrics	Description	Mathematical Representation
1	Mean Squared Error (MSE)	Evaluates similarity between the actual and denoised signals. Lower MSE indicates a better denoising process	$MSE = \frac{1}{n}\sum_{i=1}^{n}\left(Y_i - \widehat{Y}_i\right)^2$ Where Y_i - original signal $\widehat{Y}_i$ - denoised signal i – time n - total number of values
2	Root Mean Squared Error (RMSE)	Square root of MSE, which signifies the average differences between the actual and denoised signals. Lower RMSE signifies better denoising	$RMSE = \sqrt{MSE}$
3	Signal-to-noise ratio (SNR)	Compares the signal's strength to the noise level. Higher SNR indicates better denoising	$SNR = 10\,log_{10}\frac{\sum_{i=1}^{n}{y_i}^2}{\sum_{i=1}^{n}\left(y_i - \widehat{y}_i\right)^2}$ Where y_i – signal values $\widehat{y}_i$ – reconstructed signal values n - number of samples

is always changing. Further complicating analysis, volume conduction causes signals from different brain regions to overlap, making it hard to pinpoint the exact origin of specific activity. The choice of reference electrode also impacts the data, adding another layer of complexity. Furthermore, the interpretability of deep learning models used for EEG analysis remains a hurdle, potentially limiting patient trust in AI-based diagnoses. EEG signals also exhibit significant variability between individuals due to anatomical differences, requiring specialized methods for comparison. Finally, interpreting these indirect measures of neural activity demands expertise in both neurology and signal processing. Scientists working hard to come up with new ways to deal with these problems. Some of these ideas are signal processing methods powered by AI, source localization methods, and better electrode designs.

Table 4. Machine Learning And Deep Learning Approaches To Classification

Machine Learning Approaches[25]

Domain	Technique	Accuracy
Brain Activity Classification	Naïve ayes	87%
Emotion Classification	SVM	96.83%
Motor Imagery Classification	KNN, SVM	88.7%
Epilepsy Classification	ANN	99.3%

Deep Learning Approaches

Domain	Technique	Accuracy
Motor Imagery Classification	CNN (with adaptive transfer learn) [26]	84.19%
Motor Imagery Classification	DBN[27]	97.69%
Epilepsy	RNN-LSTM[28]	96.1%
Emotion Classification	GCN[29]	98.64%

SVM -Support Vector Machine, **KNN**- K- Nearest Neighbour, **ANN** – Artificial Neural Network, **CNN**- Convolutional Neural Network, **DBN**- deep belief networks, **RNN** -Recurrent Neural Network, **LSTM** – Long short-term Memory, **GCN**- Graph Convolutional Network.

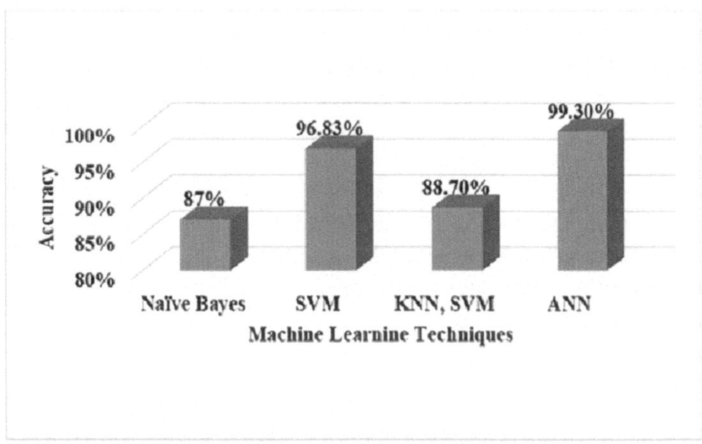

Fig. 2. Machine Learning Classification Accuracy Representation

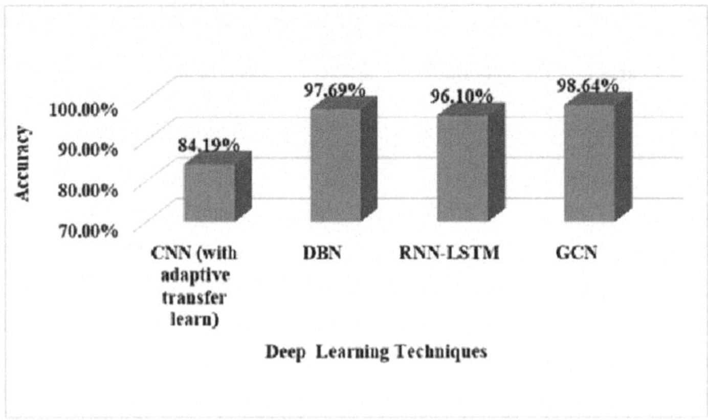

Fig. 3. Deep Learning Classification Accuracy Representation

4 Conclusion

Brain activity research may be effectively conducted with the use of EEG analysis have a major role in understanding the brain's intricate mechanism. Cognitive research may assist in the diagnosis of neurological diseases and give useful information about thinking processes. An overview of the EEG signal processing pipeline has been given in this paper, consists of signal capturing, denoising, feature engineering, and classification. EEG signals from scalp electrodes can be influenced by biological and environmental noise, causing artifacts like eye movements. By eliminating these abnormalities using several denoising strategies as well as including metrics like MSE and SNR, the quality of data or the data accuracy will improve. An essential tool in deep learning and machine learning for EEG data processing is feature engineering, which makes use of methods such as nonlnear analysis, time domain, and frequency domain. The deep learning model gave the accuracy ranging from 84.19% to 99.3% in recognising complex patterns from massive dataset, offer a world of possibilities for their inclusion in ML and DLbased brain-based monitoring applications.

The difficulty in accurately detecting brain activity from EEG signals is due to factors such as interpretability, reference electrode choice, noise, artifacts, non-stationarity, volume conduction, and EEG signals. It is necessary to use specific techniques for comparing EEG data since they also differ across people. These indirect metrics need a high level of expertise in signal processing and neurobiology to understand. In response to these difficulties, researchers are developing new approaches to signal processing that are driven by artificial intelligence, new ways to localize sources, and better electrode designs.

References

1. Lee, D.: Brain for learning. Birth Intell., 127–154 (2020). https://doi.org/10.1093/oso/9780190908324.003.0007
2. da Silva, F.L.: EEG: origin and measurement. In: Mulert, C., Lemieux, L. (eds.) EEG – fMRI, pp. 23–48. Springer, Cham (2022). https://doi.org/10.1007/978-3-031-07121-8_2
3. Islam, Md.K., Rastegarnia, A.: Editorial: recent advances in EEG (non-invasive) based BCI applications. Front. Comput. Neurosci. **17** (2023). https://doi.org/10.3389/fncom.2023.1151852
4. Schirner, M., Ritter, P.: Integrating EEG–fMRI through brain simulation. In: Mulert, C., Lemieux, L. (eds) EEG – fMRI, pp. 745–777. Springer, Cham (2022). https://doi.org/10.1007/978-3-031-07121-8_30
5. Chaddad, A.: Brain function evaluation using enhanced fNIRS signals extraction. In: 2014 48th Annual Conference on Information Sciences and Systems (CISS) (2014). https://doi.org/10.1109/ciss.2014.6814079
6. Orban, M., Elsamanty, M., Guo, K., Zhang, S., Yang, H.: A review of brain activity and EEG-based brain-computer interfaces for rehabilitation application. Bioengineering **9**(12), 768 (2022). https://doi.org/10.3390/bioengineering9120768
7. Liu, M., Ushiba, J.: Brain–machine interface (BMI)-based neurorehabilitation for post-stroke upper limb paralysis. Keio J. Med. **71**(4), 82–92 (2022). https://doi.org/10.2302/kjm.2022-0002-oa
8. Jiao, B., et al.: Neural biomarker diagnosis and prediction to mild cognitive impairment and Alzheimer's disease using EEG technology. Alzheimer's Res. Ther. **15**(1) (2023). https://doi.org/10.1186/s13195-023-01181-1
9. Shir, D., et al.: Analysis of clinical features, diagnostic tests, and biomarkers in patients with suspected Creutzfeldt-Jakob disease, 2014–2021. JAMA Netw. Open **5**(8), e2225098 (2022). https://doi.org/10.1001/jamanetworkopen.2022.25098
10. Chaddad, Wu, Y., Kateb, R., Bouridane, A.: Electroencephalography signal processing: a comprehensive review and analysis of methods and techniques. Sensors **23**(14), 6434 (2023). https://doi.org/10.3390/s23146434
11. Choi, H., Park, J., Yang, Y.-M.: A novel quick-response eigenface analysis scheme for brain-computer interfaces. Sensors **22**(15), 5860 (2022). https://doi.org/10.3390/s22155860
12. Wang, J., Wang, M.: Review of the emotional feature extraction and classification using EEG signals. Cogn. Robot. **1**, 29–40 (2021). https://doi.org/10.1016/j.cogr.2021.04.001
13. Akuthota, S., Kumar, K.R., Janapati, R.: Artifacts removal techniques in EEG data for BCI applications: a survey. In: Computational Intelligence and Deep Learning Methods for Neuro-rehabilitation Applications, pp. 195–214 (2024). https://doi.org/10.1016/b978-0-443-13772-3.00004-2
14. Ghosh, R., Phadikar, S., Deb, N., Sinha, N., Das, P., Ghaderpour, E.: Automatic eyeblink and muscular artifact detection and removal from EEG signals using k-nearest neighbor classifier and long short-term memory networks. IEEE Sens. J. **23**(5), 5422–5436 (2023). https://doi.org/10.1109/jsen.2023.3237383
15. Wu, Q., Zhang, W., Wang, Y., Zhang, W., Liu, X.: Research on removal algorithm of EOG artifacts in single-channel EEG signals based on CEEMDAN-BD. Comput. Methods Biomech. Biomed. Eng. **24**(12), 1368–1379 (2021). https://doi.org/10.1080/10255842.2021.1889525
16. Al-Qazzaz, N.K., Aldoori, A.A., Ali, S.H.B.M., Ahmad, S.A., Mohammed, A.K., Mohyee, M.I.: EEG signal complexity measurements to enhance BCI-based stroke patients' rehabilitation. Sensors **23**(8), 3889 (2023). https://doi.org/10.3390/s23083889
17. Gu, L., Jiang, J., Han, H., Gan, J.Q., Wang, H.: Recognition of unilateral lower limb movement based on EEG signals with ERP-PCA analysis. Neurosci. Lett. **800**, 137133 (2023). https://doi.org/10.1016/j.neulet.2023.137133

18. Downey, R.J., Ferris, D.P.: ICanClean removes motion, muscle, eye, and line-noise artifacts from phantom EEG. Sensors **23**(19), 8214 (2023). https://doi.org/10.3390/s23198214
19. Narmada, A., Shukla, M.K.: A novel adaptive artifacts wavelet denoising for EEG artifacts removal using deep learning with meta-heuristic approach. Multimedia Tools Appl. **82**(26), 40403–40441 (2023). https://doi.org/10.1007/s11042-023-14949-2
20. Alyasseri, Z.A.A., Khader, A.T., Al-Betar, M.A., Abasi, A.K., Makhadmeh, S.N.: EEG signals denoising using optimal wavelet transform hybridized with efficient metaheuristic methods. IEEE Access **8**, 10584–10605 (2020). https://doi.org/10.1109/access.2019.2962658
21. García-Ponsoda, S., García-Carrasco, J., Teruel, M.A., Maté, A., Trujillo, J.: Feature engineering of EEG applied to mental disorders: a systematic mapping study. Appl. Intell. **53**(20), 23203–23243 (2023). https://doi.org/10.1007/s10489-023-04702-5
22. Li, Y., Liu, A., Yin, J., Li, C., Chen, X.: A segmentation-denoising network for artifact removal from single-channel EEG. IEEE Sens. J. **23**(13), 15115–15127 (2023). https://doi.org/10.1109/jsen.2023.3276481
23. Chen, D., Huang, H., Bao, X., Pan, J., Li, Y.: An EEG-based attention recognition method: fusion of time domain, frequency domain, and non-linear dynamics features. Front. Neurosci. **17** (2023). https://doi.org/10.3389/fnins.2023.1194554
24. Chen, W., et al.: An automated detection of epileptic seizures EEG using CNN classifier based on feature fusion with high accuracy. BMC Med. Inform. Decis. Mak. **23**(1) (2023). https://doi.org/10.1186/s12911-023-02180-w
25. Hosseini, M.P., Hosseini, A., Ahi, K.: A review on machine learning for EEG signal processing in bioengineering. IEEE Rev. Biomed. Eng. **14**, 204–218 (2020)
26. Zhang, K., Robinson, N., Lee, S.-W., Guan, C.: Adaptive transfer learning for EEG motor imagery classification with deep convolutional neural network. Neural Netw. **136**, 1 (2021). https://doi.org/10.1016/j.neunet.2020.12.013
27. Cheng, L., Li, D., Yu, G., Zhang, Z., Li, X., Yu, S.: A motor imagery EEG feature extraction method based on energy principal component analysis and deep belief networks. IEEE Access **8**, 21453–21472 (2020). https://doi.org/10.1109/access.2020.2969054
28. Kumar, S., Sharma, R., Sharma, A.: OPTICAL+: a frequency-based deep learning scheme for recognizing brain wave signals. PeerJ Comput. Sci. **7**, e375 (2021). https://doi.org/10.7717/peerj-cs.375
29. Fan, Z., Chen, F., Xia, X., Liu, Y.: EEG emotion classification based on graph convolutional network. Appl. Sci. **14**(2), 726 (2024). https://doi.org/10.3390/app14020726
30. Wang, Y., Qiu, S., Li, D., Du, C., Lu, B.-L., He, H.: Multi-modal domain adaptation variational autoencoder for EEG-based emotion recognition. IEEE/CAA J. Autom. Sin. **9**(9), 1612–1626 (2022). https://doi.org/10.1109/jas.2022.105515
31. Sun, L., Wu, J.: A scalable and transferable federated learning system for classifying healthcare sensor data. IEEE J. Biomed. Health Inform. **27**(2), 866–877 (2023). https://doi.org/10.1109/jbhi.2022.3171402

Detection of Thyroid Stages Classification by Using Convolutional Neural Network Techniques

V. PrasannaKumari(✉), P. Sanjay, and M. G. Raja Ramanan

Department of IT, Rajalakshmi Engineering College, Chennai, TamilNadu, India
prasannakumari.v@rajalakshmi.edu.in

Abstract. Thyroid nodules present a diagnostic challenge because of their variable nature and possible cancerous nature. The process of manually classifying nodules from ultrasound pictures might be inconsistent and lead to incorrect diagnoses. By creating a deep learning-based system for the automatic classification of thyroid nodules using ultrasound images, this study seeks to address this difficulty. To improve classification accuracy, we built a multi-scale, multi-channel architecture by utilizing convolutional neural networks (CNNs). Our system's accuracy rate in differentiating between thyroid nodule kinds was a promising 94%. The incorporation of sophisticated deep learning methodologies enabled accurate recognition of radiographic characteristics suggestive of several nodule categories. Our results show that deep learning has the potential to increase the precision and effectiveness of thyroid nodule categorization, providing invaluable assistance to medical practitioner.

Keywords: Thyroid nodules · Ultrasound images · Deep learning · Convolutional neural networks · Automated classification

1 Introduction

Thyroid diseases pose significant challenges to global health systems, necessitating accurate and prompt diagnosis for effective treatment. While traditional diagnostic methods rely on manual interpretation of medical images, recent years have seen the rise of deep learning techniques, notably convolutional neural networks (CNNs), offering promising automation solutions. However, current approaches face challenges like limited datasets, computational complexity, and the need for improved accuracy and generalizability. Pioneering work by Sharma Dua and Singh in 2017 laid the foundation for CNN-based thyroid disease diagnosis, showcasing its potential to automate diagnostic tasks. Wang, Yan, and Gong's 2018 deep learning approach combining CNNs and Recurrent Neural Networks (RNNs) demonstrated significant accuracy improvement, yet challenges regarding computational complexity and transfer learning mechanisms persisted. In 2019, Reddy, Raju, and Rao proposed a hybrid CNN-SVM model, achieving exceptional accuracy, albeit with concerns regarding data requirements and model

P. D. Sivakumar et al. (Eds.): IRCCTSD 2024, CCIS 2360, pp. 191–201, 2025.
https://doi.org/10.1007/978-3-031-82389-3_17

complexity. Subsequent research by Rakas Rahman and Rahman in 2020 introduced transfer learning and ensemble learning, addressing some challenges but highlighting the need for further real-world application development. Dash's 2008 approach combining multi-level CNN architecture and an alerting mechanism achieved remarkable accuracy, yet computational complexity and hyperparameter tuning challenges persisted. Rao's 2022 proposal of a lightweight CNN architecture prioritized computational efficiency without sacrificing accuracy, signaling progress while underscoring the need for continued refinement for efficacy across diverse clinical scenarios. Given these challenges and advancements, this paper aims to contribute to the advancement of thyroid disorder diagnosis. The subsequent sections provide a literature review, experimental implementation, results, and conclusions, outlining future research directions.

2 Literature Survey

Thyroid Disease Diagnosis Using Convolutional Neural Networks [1].
A. S. Sharma Dua and S. P. Singh in their 2017 article, "Thyroid Disease Diagnosis Using Convolutional Neural Networks," present an innovative approach to thyroid disease classification using convolutional neural networks (CNNs). The study focuses on a carefully curated dataset of thyroid images from various clinical conditions, with the goal of improving the accuracy and efficiency of diagnosing thyroid disease. A CNN-based system is adapted to classify thyroid images into three basic states: normal, hyperthyroid, and hypothyroid. In particular, the study achieved an impressive 92.3% accuracy, highlighting CNN's ability to detect complex patterns and features in thyroid images. This breakthrough represents a major advance in the development of automated diagnostic tools, offering the potential benefit of speeding up the identification of thyroid disorders and helping clinicians make informed decisions about patient care.

Drawbacks
Despite its achievements, the document acknowledges some limitations that require attention. A major drawback is the relatively modest material used in the study. Although the accuracy achieved is significant, the generalizability of the system to a wider population is a concern due to the size of the dataset. A larger and more diverse dataset is crucial to improve overall assessment and reliability in real-world scenarios. Furthermore, research is limited to the three main thyroid diseases – normal, hyperthyroid and hypothyroid – excluding other common disorders such as goiter and thyroid cancer. Fixing this limitation is essential for the continued development of the system. Future iterations should consider expanding the scope to include a wider range of thyroid disorders, thus increasing versatility and relevance in clinical practice. In conclusion, although the paper makes an important contribution to the diagnosis of thyroid diseases using CNNs, addressing the recognized limitations is necessary to improve the efficiency and practical usability of the system in all health settings.

A Deep Learning Approach for Thyroid Disorder Diagnosis [2].
In their 2019 paper, "Classification of Gonadal Disorders Using a Hybrid CNN-SVM Model," A. R. P. Reddy, K. Raju, and S. V. S. L. S. Rao presents an innovative approach

combining Convolutional Neural Network (CNN) and Support Vector Machine (SVM) techniques for thyroid disease classification. This hybrid model aims to improve diagnostic accuracy and efficiency by exploiting the strengths of both CNN and SVM. With an impressive 96.1% accuracy, the hybrid model shows the potential synergy between deep learning and classical machine learning methods, enabling accurate thyroid image classification and distinguishing between normal, hyperthyroid and hypothyroid.

Drawbacks
Despite significant achievements, the paper acknowledges the limitations of the system, especially its high computational cost, which prevent practical use in clinical settings where efficiency is critical. Solving this problem is critical to ensure that the system is suitable for routine diagnostics. Another limitation highlighted is the lack of transfer learning, which can improve adaptability and efficiency, especially with limited labeled data. Incorporating transfer learning can improve efficacy and generalizability across clinical scenarios. In conclusion, although the deep learning proposed by Wang, Yan and Gong for the diagnosis of thyroid disorders shows promising accuracy, considering the computational cost and exploring the benefits of transfer learning are important for continuous improvement and practical application in real clinical environments.

Thyroid Disorder Classification Using a Hybrid CNN-SVM Model [3].
In their 2019 paper, "Classification of Gonadal Disorders Using a Hybrid CNN-SVM Model," A. R. P. Reddy, K. Raju, and S. V. S. L. S. Rao presents an innovative approach combining Convolutional Neural Network (CNN) and Support Vector Machine (SVM) techniques for thyroid disease classification. This hybrid model aims to improve diagnostic accuracy and efficiency by exploiting the strengths of both CNN and SVM. With an impressive 96.1% accuracy, the hybrid model shows the potential synergy between deep learning and classical machine learning methods, enabling accurate thyroid image classification and distinguishing between normal, hyperthyroid and hypothyroid.

Drawbacks
Although the CNN-SVM hybrid model seems promising, it has significant limitations. Its reliance on extensive training data presents challenges in clinical settings where data acquisition is difficult. In addition, the complexity of the model can hinder its implementation, especially in health services that prioritize simplicity. Simplifying the architecture without sacrificing accuracy is crucial for practical feasibility. Despite improvements in accuracy, consideration of data requirements and model complexity is crucial to integrate the model into real clinical settings. Ongoing research to optimize these aspects will contribute to the development of practical tools for the diagnosis of hypothyroidism.

3 Implementation

A. Data Collection
We collected throat X-rays for thyroid nodule analysis by downloading the Kaggle dataset. The data set will likely consist of annotated laryngeal radiographs showing the presence or absence of thyroid nodules. To ensure robust model training and evaluation,

we divide the dataset into training and validation subsets using the standard 70% training and 30% validation ratio. This approach allows the model to learn from sufficient data while keeping a separate section for estimating unseen samples. Additionally, we ensured consistent class distribution across both subsets to avoid bias and promote generalization to new information.

B. Data Preprocessing

Data preprocessing is crucial to prepare collected data for model training [4–6], which involves various tasks to improve the quality and usability of machine learning algorithms. These steps include grayscale conversion to ensure consistent pixel intensity, resizing images to a fixed 36 × 36 pixel size for standardized processing, converting images to NumPy matrices for numerical processing, resizing the array to match the input format of the model, and pixel normalization. Values for the region. 0–1 for effective model training.

C. Model Implementation

The accompanying Convolutional Neural Network (CNN) model is designed to predict thyroid nodules on throat radiographs. It includes convolutional layers, batch normalization, maximum pooling, removal and fully connected layers for classification. With 634,370 parameters, input format (36, 36, 1) and binary classification output format (2), the CNN architecture automatically distinguishes the automatic input data from relevant features, captures spatial hierarchies [7–9] and adapts to different images - related tasks, including medical image analysis.

D. Load a Trained Model

Loading a trained model involves downloading a saved model file containing the architecture, weights and specifications from storage. The file is then deserialized using machine learning libraries such as TensorFlow, the CNN architecture is reconstructed with the stored data, and the weights and biases are initialized with the optimized parameters learned during training.

E. Testing and Evaluation

Thyroid nodule prediction algorithms improve diagnostic capabilities by analyzing image features to identify potential signs of thyroid nodules. An alarm mechanism notifies health professionals of abnormalities, facilitating timely diagnosis and treatment. Rigorous testing validates system functionality, performance, and reliability, including accuracy, response time, and integration testing under various conditions.

Thyroid Detection Algorithm

A Convolutional Neural Network (CNN) architecture that has been tuned for real-time analysis of throat X-ray images powers the project's thyroid nodes prediction algorithm. The purpose of this specialized CNN architecture is to identify and categorize thyroid nodes-related patterns in X-ray pictures that the imaging system has taken. The following operations are performed by the algorithm:

A. Convolution Operation

One of the basic operations in image processing is convolution. To extract features from an input image, a filter is used [10, 11]. Sliding the filter over the image and computing the element-wise product and sum of the overlapped regions allows you to do this. In mathematical notation, it is expressed as:

$$(I * K)(i, j) = \sum m \sum n I(m, n) \cdot K(i - m, j - n)$$

Where I represents the input image, K is the filter (also known as kernel), and (i, j) are the pixel coordinates.
Implementation

- Slide the filter over the input image.
- Calculate the element-wise product between the filter and the overlapped region of the image in each position.
- Add the results to determine the output value for that position.
- Repeat the procedure for each position in the image. *Algorithm for 2D*

Convolution operation 1.
Initialize an empty feature map F of size $(M - m + 1) \times (N - n + 1)$.

For $x = 0$ to $M - m$:
For $y = 0$ to $N - n$:

- Extract a region R of size $m \times n$ from I starting at position (x, y).
- Compute the element-wise product of R and Kernel matrix W.
- Sum the element-wise products to get a single value Assign this value to F(x, y).
  Return F

Explanation

- The algorithm performs a sliding window operation on the input image I, with each window having the same size as the convolution kernel W.
- At each position (x, y), subtract a region (R) from I and multiply by the kernel (W).
- The sum of these element-wise products is assigned to the corresponding position in the output feature map (F) (Fig. 1).

B. Activation Function (ReLU). The Rectified Linear Unit (ReLU) is an activation function that introduces non-linearity to the network. It computes the maximum between zero and the input value. Mathematically, it is represented as:

$$f(y) = max(0, y)$$

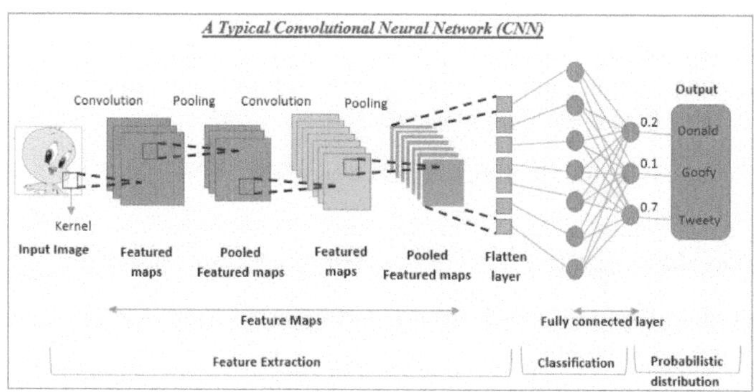

Fig. 1. Convolution operation

Implementation
For each element in the feature map, apply the ReLU function. If the element is negative, set it to zero. Otherwise, leave it unchanged.
Algorithm

1. If y is greater than or equal to zero: Return y as the output.
2. If y is less than zero: Return zero as output

Explanation
The ReLU activation function is defined as f(y) = max (0, y). When the input x is non-negative, the output is equal to the input. When the input x is negative, the output is set to zero. This introduces non-linearity, making it easier for the network to learn complex relationships (Fig. 2).

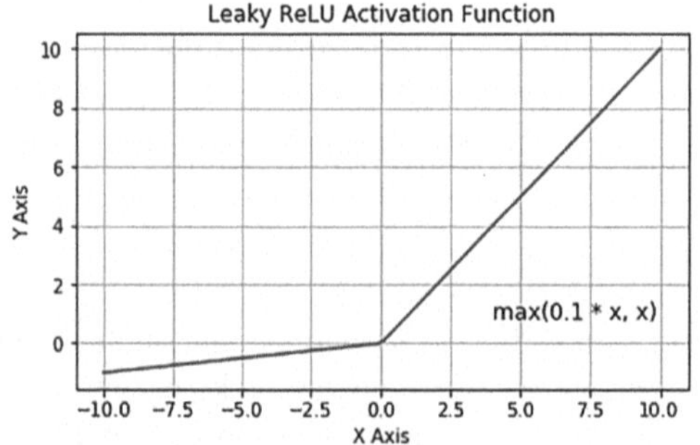

Fig. 2. ReLU Activation Function

C. Max-Pooling Operation
Max-pooling is a downsampling operation that reduces [12, 13] the spatial dimensions of the feature maps. It does this by selecting the maximum value from a specified window. Mathematically, it is represented as:

$$MaxPooling(x) = max_{(i,j)\in Window} x(i,j)$$

Implementation
Divide the feature map into non-overlapping regions. For each region, find the maximum value. Create a new downsampled feature map with these maximum values.
Algorithm for the 2D Max-Pooling Operation

1. Initialize an empty output feature map O of size M/k × N/k
2. For i = 0 to M/k − 1:

   For j = 0 to N/k − 1:

- Extract a region R of size k × k from I starting at position (i × k, j × k). Compute the maximum value within R.

   Assign this maximum value to O(i, j). Return O.

Explanation

The algorithm divides the input feature map into non-overlapping regions of size k × k. For each region, it finds the maximum value and assigns it to the corresponding position in the output feature map. The result is a downsampled feature map with reduced spatial dimensions (Fig. 3).

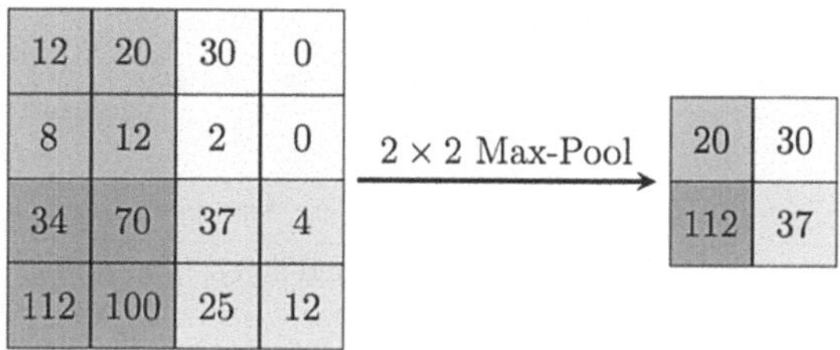

Fig. 3. Max pooling operation

D. Fully Connected Layer

A fully connected layer is a neural network layer where each neuron is connected to every neuron in the previous layer. It applies a linear transformation followed by an activation function. Mathematically, it is represented as:

$$h = \sigma(Wx + b)$$

Where h is the output, W is the weight matrix, x is the input vector, b is the bias vector, and σ is the activation function

Implementation

Multiply the input vector x by the weight matrix W. Add the bias vector b to the result. Apply the activation function σ to obtain the output h.

Algorithm

1. Compute the weight of the inputs: $Z = Wx + b$
2. Apply an activation function f element-wise:

$$h = f(z)$$

Explanation

The algorithm calculates the weighted sum of the inputs x by multiplying with the weight matrix W and adding the bias b. The activation function f introduces non-linearity to the output of the layer. Common activation functions include ReLU, sigmoid, tanh, etc. (FIg. 4).

E. Categorical Cross-Entropy Loss

Categorical cross-entropy loss [14, 15] is a measure of the difference between the predicted and true distributions. It is commonly used in multi-class classification tasks.

Mathematically, it is represented as:

$$L(y, \hat{y}) = -\Sigma_i y_i \, log\,(\hat{y}_i)$$

Where y is the true table distribution and ŷ is the predicted label distribution

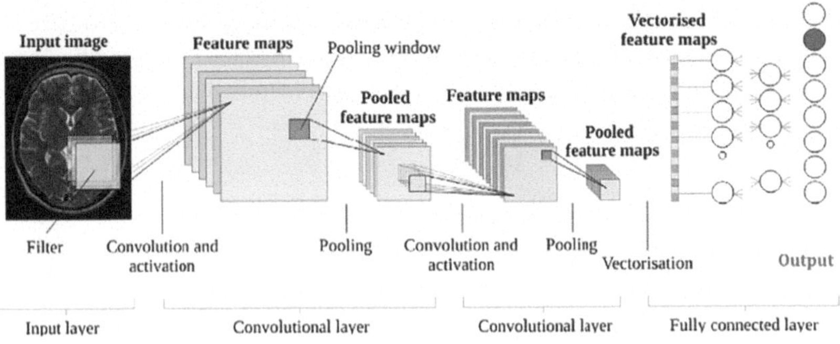

Fig. 4. Fully-connected layer

Implementation
Compute the softmax probabilities for each class in the output layer. Take the natural logarithm (log) of the softmax probabilities. Utilize one-hot encoding to represent the true labels. Sum the element-wise product of the encoded labels and the log probabilities. Take the negative of this sum. Divide the obtained loss by the number of samples in the batch for normalization.
Algorithm

1. Initialize the loss L to 0.
2. For each class i:

   - Compute the contribution of class i to the loss: $L_i = -y_i.log(p_i)$
   - Add L_i to the total loss L.

3. Return L.

Explanation
For each class, the loss term L_i measures the discrepancy between the true label y_i and the predicted probability p_i using the logarithmic scale. When the true label y_i is 1, the loss term simplifies to $-\log(p_i)$, which penalizes low predicted probabilities. When the true label y_i is 0, the loss term becomes 0, as there is no contribution from that class. The total loss L is the sum of the individual losses for each class (Fig. 5).

F. Gradient Descent Update Rule. Gradient descent is an optimization algorithm used to adjust the model's weights during training. The update rule is based on the derivative of the loss function with respect to the weights. Mathematically, it is represented as:

$$W \leftarrow W - \alpha\, \partial L/\partial W$$

Where α is the learning rate.
Implementation
Calculate the gradient of the loss with respect to the model parameters using backpropagation. Adjust the model parameters using the computed gradients and a predefined learning rate [17, 18]. Repeat steps 1 and 2 for a specified number of iterations or until convergence criteria are met. Check convergence criteria, such as a minimum change in loss or a maximum number of iterations reached [19, 20].
Algorithm

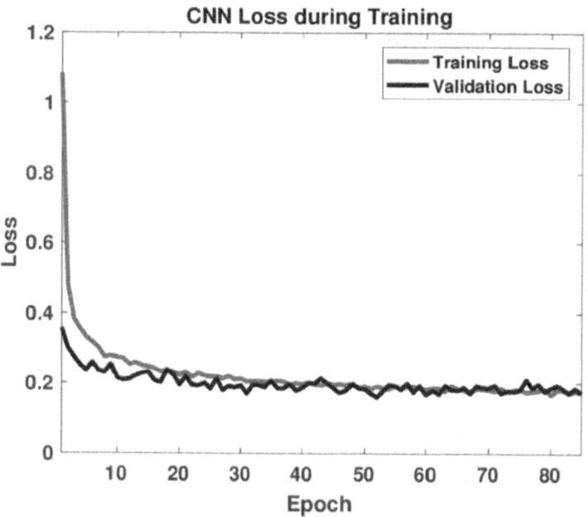

Fig. 5. Categorical Cross-Entropy Loss

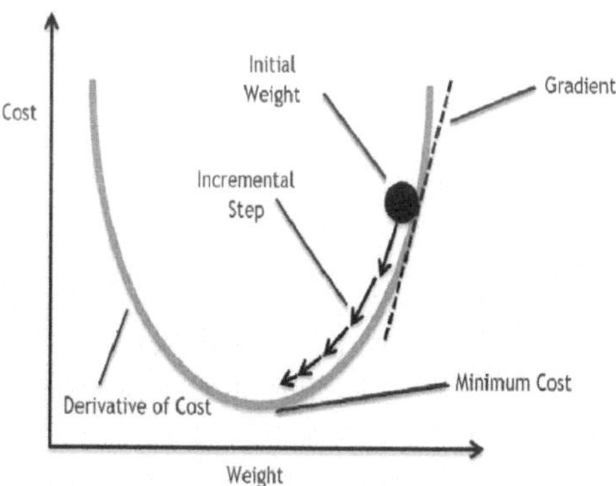

Fig. 6. Gradient Descent Algorithm working

1. Initialize the model parameters θ randomly or with some predefined values.
2. For each iteration (epoch) until convergence or a specified number of epochs:
   a. Compute the gradient of the loss function with respect to the parameters:

$$\nabla L = \partial L/\partial \theta$$

   b. Update the parameters using the gradient and the learning rate: $\theta' = \theta - \alpha \cdot \nabla L$
   c. Repeat steps (a) and (b) for a predefined number of iterations or until convergence criteria are met.

Explanation

- The algorithm starts with an initial set of model parameters θ. In each iteration, the gradient ∇L is computed, which represents the direction of steepest ascent of the loss function. The parameters θ are then updated by moving in the opposite direction of the gradient, scaled by the learning rate α. The learning rate controls the step size of the updates. A smaller learning rate leads to smaller steps, potentially requiring more iterations for convergence, while a larger learning rate may lead to overshooting the minimum (Fig. 6).

4 Conclusion

This project used advanced machine learning techniques to develop an accurate hypothyroidism-specific classification system with a specific focus on hypothyroidism. Using a custom convolutional neural network (CNN) architecture, the project achieved an impressive 94.25% accuracy, a major milestone in medical image classification. This success highlights the potential of CNNs to detect complex patterns in medical images, which is central to accurate diagnosis in clinical settings. Despite its achievements, the project faced major challenges, mainly related to computational requirements and model interpretability. The resource intensity of deep learning models such as the developed CNN presents practical obstacles to their application in clinical settings. Balancing accuracy and computational efficiency continues to be a problem that requires exploration of optimization strategies for resource-constrained settings. Furthermore, the need for model interpretability in a healthcare context highlights the need to address the black box of complex neural network architectures, thus increasing transparency and strengthening trust between healthcare providers and patients. In the future, future iterations of the project aim to deepen the inclusion of different data sources, refine the model architecture and explore common learning strategies. Expanding the dataset to include a variety of thyroid images and expanding the model's ability to classify a wider spectrum of thyroid disorders will improve its clinical relevance and utility. In addition, efforts to optimize hyperparameters and streamline deployment models are essential to ensure practical implementation in real-world healthcare scenarios. By continuously improving the model and addressing existing challenges, this project is a significant step towards improving the diagnosis of thyroid diseases using advanced machine learning techniques, which will ultimately improve patient outcomes and revolutionize medical diagnostic health practices.

References

1. Zhang, J.: Thyroid nodule classification using deep learning from ultrasound images (2018)
2. Li, W.: A novel deep learning approach for thyroid nodule classification using transfer learning (2019)
3. Wang, S.: Multi-view convolutional neural network for thyroid nodule classification using ultrasound images (2020)
4. Wang, Z.: Ensemble of deep learning models for thyroid nodule classification from ultrasonography (2021)
5. Zhao, J.: Deep learning-based thyroid nodule classification using texture features and convolutional neural networks (2022)
6. Zhang, L.: Attention-based deep learning for thyroid nodule classification from ultrasonic images (2023)
7. Sun, H.: An efficient deep learning approach for thyroid nodule classification using ultrasound images with few-shot learning (2023)

8. Chen, Y.: A novel deep learning framework for thyroid nodule classification using ultrasound images with data augmentation (2023)
9. Liu, X.: Multi-task learning for thyroid nodule classification and feature extraction from ultrasound images (2023)
10. Li, C.: A hierarchical CNN-based deep learning model for thyroid nodule classification from ultrasound images (2023)
11. Ma, J., Wu, F., Zhu, J., Kong, D.: Cascade convolutional neural networks for automatic detection of thyroid nodules in ultrasound images. Med. Phys. **44**(5), 1678–1691 (2017)
12. Poudel, P., Ataide, E., Illanes, A., Friebe, M.: Linear discriminant analysis and k-means clustering for classification of thyroid texture in ultrasound images. Proc. IEEE Eng. Med. Biol. Soc. (2018)
13. Poudel, P., Illanes, A., Friebe, M.: Evaluation of commonly used algorithms for thyroid ultrasound images segmentation and improvement using machine learning approaches. J. Healthc. Eng. (2018)
14. Chang, C.-Y., Chung, P.-C., Hong, Y.-C., Tseng, C.-H.: A neural network for thyroid segmentation and volume estimation in CT images. IEEE Comput. Intell. Mag. **6**(4), 43–55 (2011)
15. LeCun, Y., Bengio, Y., Hinton, G.: Deep learning. Nature **521**(7553), 436–444 (2015)
16. Schmidhuber, J.: Deep learning in neural networks: an overview. Neural Netw. **61**(1), 85–117 (2015)
17. Ker, J., Wang, L., Rao, J., Lim, T.: Deep learning applications in medical image analysis. IEEE Access **6**(1), 9375–9389 (2018)
18. Shen, D., Wu, G., Suk, H.-I.: Deep learning in medical image analysis. Annu. Rev. Biomed. Eng. **19**(1), 221–248 (2017)
19. Ma, L., Ma, C., Liu, Y., Wang, X., Xie, W.: Diagnosis of thyroid diseases using SPECT images based on convolutional neural network. J. Med. Imaging Health Inform. **8**(8), 1684–1689 (2018)
20. Liu, T., Xie, S., Xu, J., Niu, L., Sun, W.: Classification of thyroid nodules in ultrasound images using deep model based transfer learning and hybrid features. In: IEEE International Conference on Acoustics, Speech and Signal Processing (ICASSP), New Orleans, LA, USA, pp. 919–923, March 2017

Evaluating Biglycan as a Biomarker in Breast Cancer Detection: A Custom CNN Architecture

Anish Samantaray, M. Poonkodi(✉), and Anita Christaline Johnvictor

School of Computer Science Engineering, Vellore Institute of Technology, Chennai, India
poonkodi.m@vit.ac.in

Abstract. Breast cancer ranks as a significant health threat to women globally, making timely and accurate diagnosis critical. In response to the delays experienced in traditional laboratory analyses, a novel approach has been suggested leveraging the Biglycan protein as a biomarker for breast cancer detection. This method employs image data of the biomarker, which undergoes initial preprocessing before being analyzed through a custom Convolutional Neural Network (CNN) architecture. This innovative approach is further evaluated by comparing its performance against established pretrained models, including ResNet, MobileNet, and EfficientNet. The comparison reveals that proposed model leads in accuracy, achieving a 91.68% success rate in detecting breast cancer. Remarkably, EfficientNet closely follows, registering an 89.57% accuracy rate. This finding underscores the potential of using the Biglycan protein as a reliable biomarker in the development of advanced AI-driven diagnostics of breast cancer.

Keywords: Machine Learning · Breast Cancer · Biglycan · Diagnosis · Longitudinal Analysis · Accuracy

1 Introduction

Breast cancer primarily originates from the cells lining the milk ducts or the lobules responsible for milk production in the breast. It's one of the most prevalent types of non-skin cancers among women, ranking as the second most common cancer following lung cancer, and it stands as the fifth leading cause of cancer-related deaths globally. In 2004, it was responsible for 519,000 deaths worldwide, making up about 1% of all cancer deaths and 7% of total mortality rates. While breast cancer is significantly more common in women—with their risk being nearly 100 times higher than in men—men tend to experience poorer outcomes due to later diagnoses [1].

The conventional method for detecting Breast Cancer involves a biopsy, a procedure in which a small sample of breast tissue is extracted for laboratory analysis [2]. This biopsy provides crucial information about the presence of cancerous cells, contributing to the diagnosis of the disease, this process however takes a bit of time i.e. around 3–14 days [3]. However, advancements in technology, particularly in the realm of Machine Learning and computer vision, offer a promising alternative. Instead of relying solely on traditional biopsy methods, leveraging these technologies allows for the analysis of cell

P. D. Sivakumar et al. (Eds.): IRCCTSD 2024, CCIS 2360, pp. 202–215, 2025.
https://doi.org/10.1007/978-3-031-82389-3_18

images with greater efficiency and accuracy. By employing sophisticated algorithms and image recognition techniques, Machine Learning (ML) can identify patterns and anomalies within cellular structures, aiding in the ear detection of Breast Cancer. This approach not only has the potential to streamline the diagnostic process but also offers a non-invasive and potentially less resource-intensive means of identifying cancerous cells, marking a significant stride towards more accessible and efficient breast cancer detection methods [4].

Using the machine learning (ML) algorithms for the diagnosis of breast cancer and prognosis has been the subject of a sizable corpus of research. Studies have looked into various biomarkers, particular cellular characteristics, and digital mammography images to improve the precision of cancer identification. While digital mammography examinations offer detailed insights into cellular abnormalities, machine learning algorithms have found specific markers suggestive of breast cancer in the analysis of cell structures [5]. Beyond diagnosis, machine learning has been essential in forecasting the disease's course, with an emphasis on characteristics of tumour cells [6].

Breast cancer stem cells represent a subset within tumors characterized by heightened metastatic potential. Through our investigation, we have identified the biglycan protein as a prospective molecular target within Breast cancer stem cells, playing a regulatory role in the aggressive phenotypes displayed by these cells. This discovery provides a foundational basis for the development of therapeutics aimed at targeting biglycan, with the goal of eliminating Breast cancer stem cells and preventing the onset of metastatic breast cancer. The implications of these findings open up promising avenues for the development of targeted treatments that could significantly impact the management and prevention of metastatic breast cancer [7].

The dataset comprises of photomicrographs that show Biglycan (BGN) immunohistochemistry expression in breast tissue using 3-3$'$ diaminobenzidine (DAB) staining [8], in both cancerous and non-cancerous cases. This dataset serves as a valuable resource for feature extraction and the application of Deep Learning algorithms. By leveraging this dataset, we can classify cells into cancerous and non-cancerous categories, thereby contributing to our research on the detection of breast cancer.

2 Related Work

In a 2017 study conducted by Yun Liu and colleagues, a Convolutional Neural Network (CNN) was applied to the CAMELYON16 Gigapixel microscopy image dataset for the purpose of breast cancer diagnosis. The results showcased the model's high efficiency, with an accuracy rate of 96.7% and an Area Under the Curve (AUC) score of 97% [9].

In 2019, Ripon Patgiri explored the utilization of the Support Vector Machine (SVM) algorithm on the WDBC dataset [10], which includes data from mammographic images. His research achieved a peak accuracy of 99.41%, with the lowest observed accuracy being 92.39%. Patgiri also compared the performance of the SVM with various other algorithms such as Decision Trees (D-Tree), Random Forest, Artificial Neural Networks (ANN), k-Nearest Neighbors (k-NN), and Naïve Bayes, highlighting the SVM's effectiveness in this context [11].

In their 2019 research, Aman Kumar and his team investigated the WDBC dataset, employing techniques such as Principal Component Analysis (PCA), Linear Discriminant Analysis (LDA), and a stacked ensemble approach that integrated various algorithms. The study found that after applying PCA, the kernel-based Support Vector Machine (SVM) showed outstanding performance in classifying breast cancer data, recording a high accuracy rate of 98.24%. This achievement underscores the efficacy of combining dimensionality reduction and advanced machine learning models in medical data analysis [12].

In 2022, Viswanatha Reddy Allugunti conducted research on a segmented thermographic image dataset consisting of images of more than 150 patients. Three algorithms were used in the study, with the deep learning convolution neural network performing best with an accuracy of 99.67%, followed by Random forest with 90.55% and SVM with 89.84% [13].

Research on the interaction between biglycan (BGN) and breast cancer stem cells (BCSCs) was carried out in 2022 by Kanakaraju Manupati et al. Building on earlier research that showed BGN to be a regulatory factor in BCSCs, Manupati et al. found that BGN enhances these cells' unique characteristics, highlighting its involvement in determining the aggressive phenotypes of breast cancer. This new information adds to our growing understanding of the complex mechanisms regulating BCSCs and emphasises the potential importance of biglycan targeting in treatment plans intended to reduce the risk of metastatic breast cancer [7].

Researchers examined the behaviour of dormant cancer cells in a 2023 study by Ashley Sunderland et al. with the goal of illuminating their activation, which frequently results in cancer recurrence. Using a brain metastasis model from triple-negative breast cancer, the researchers found that these quiescent cells had a significant overexpression of the BGN gene. Remarkably, in real patient comparisons, brain tumours showed significantly lower BGN levels than the initial breast tumours. By highlighting the possible importance of comprehending and targeting genes like BGN to stop the reawakening of dormant cancer cells and reduce the chance of cancer relapse, Sunderland's study expands on earlier [14].

The study by Ana Paula Thiesen et al. from 2023 compares the expression of the Biglycan (BGN) protein in the group of cells in the breast when cancer is present and absent using immunohistochemical methods and supervised deep learning neural networks (SDLNN). The results of the study, which comprised 24 formalin-fixed, paraffin-embedded tissues, showed that BGN expression was lower in DAB units in both tissues when cancer is present and absent. The D-HScore showed a mean of 6.2 (0.8 to 12.4) and 27.31 (5.3 to 81.7) DAB units, respectively. The SDLNN classification accuracy of 85.3% indicates that breast cancer tissue has lower levels of BGN protein expression than normal tissue. According to this, BGN could be a helpful biomarker for more precise tumour detection and treatment. [15].

A collection of 336 photomicrographs demonstrating Biglycan (BGN) immunohistochemical expression in breast tissue, both in cancerous and non-cancerous conditions, was published in 2023 by Pedro Clarindo da Silva Neto et al. 3-3′ diaminobenzidine staining and a colour deconvolution plugin were used to process the photos. Two categories were created out of the dataset: (I) those with cancer and (II) those without. Based on BGN colour intensity, this dataset helps train and validate the CNN (deep learning) models for breast cancer diagnosis, recognition, and classification [8].

In the work of Yun Liu et al. [9], the focus was on utilizing lymph node-based images for analysis. Following this, Ripon Patgiri [11] concentrated on analyzing features extracted from breast masses. Similarly, Aman Kumar and colleagues [12] also delved into examining features derived from breast masses. Lastly, Viswanatha Reddy [13] employed thermographic imagery in his research. Given the extensive theoretical research on BGN's involvement in the distinct traits of BCSCs and its potential impact on cancer latency and recurrence [14], a deep learning model dedicated to examining BGN expression could significantly enhance the existing research on its utility as a biomarker for breast cancer. This targeted exploration might provide groundbreaking insights into the function of BGN in oncology, which lead us to the identification of new therapeutic avenues.

Among these, only the research conducted by Ana Paula and her team applied deep learning techniques to underscore the relevance of BGN in breast cancer [15] where she has achieved a maximum of 85.3% accuracy with another dataset collected by her team. So we saw an opportunity to outperform this accuracy and do better than the existing model.

3 Proposed Work

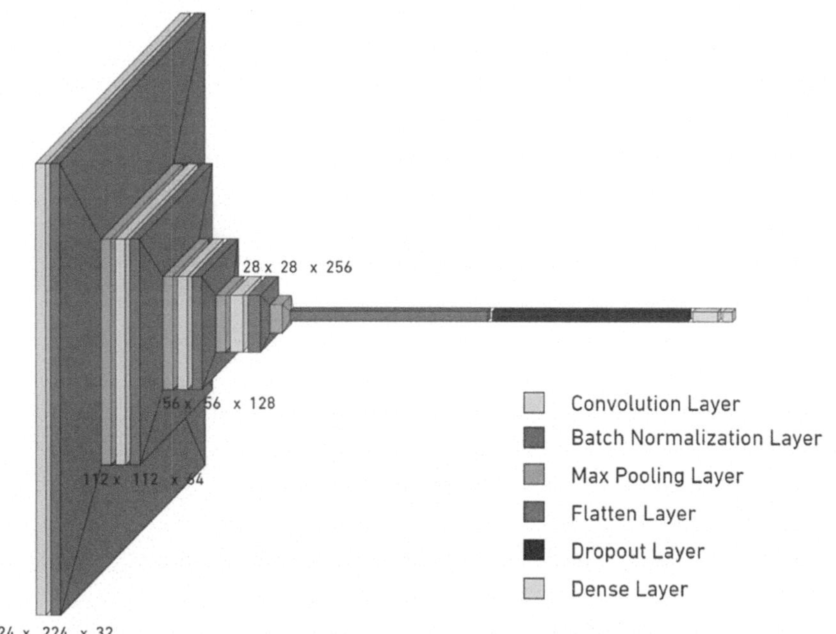

Fig. 1. Architecture of the proposed DL Model

The image depicted in Fig. 1 outlines the proposed CNN architecture for the categorization of breast cancer imagery into non-cancerous and cancerous-classes. In the preliminary phase, image dataset is amassed, to apply the deep learning model. Subsequent to acquisition, the datasets are subjected to a preprocessing, which includes standardizing image dimensions to ensure uniformity for CNN input. The images are then transformed into a tensor structure—a format amenable to neural network processing. Pixel values are regularized to fall within the 0 to 1 range (originally spanning from 0 to 255), a practice that accelerates training by maintaining minimal neural network weight magnitudes. The datasets are partitioned into two subsets: one for model training and the other for gauging model accuracy.

An innovative CNN framework as shown in Fig. 1 is introduced and its efficacy is gauged against three established models: ResNet, MobileNet, and EfficientNet. Subsequently, the models undergo a hyperparameter optimization process, tweaking the learning rate—which dictates the adjustment magnitude of the model's weights during training—and the batch size—the quantity of samples evaluated prior to updating the model. Following this, a 10-Fold cross-validation method is employed to ascertain the model's proficiency on unseen data. This technique involves dividing the dataset into 10 segments. Each segment is alternately used in as a test set, while the remaining segments serve as the training set. The resulting performance metrics are averaged across all 10 iterations, yielding a more reliable measure of the model's performance.

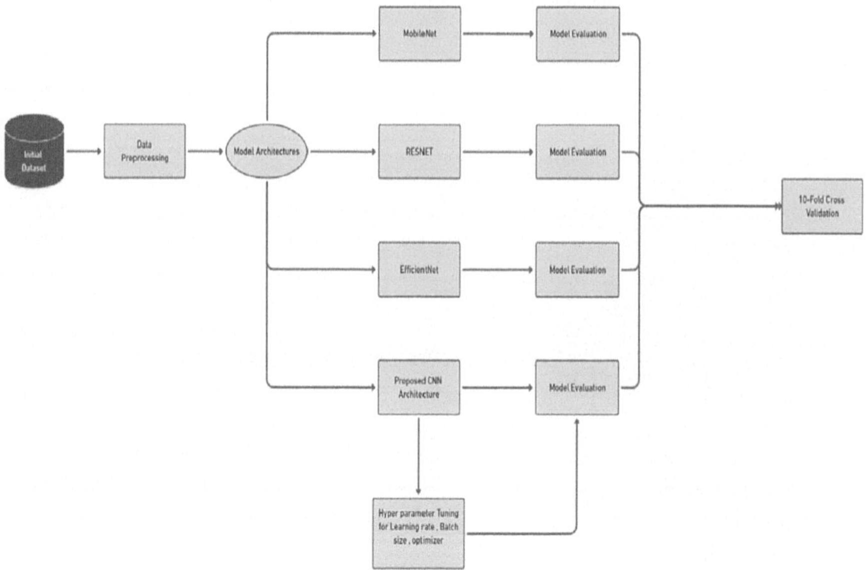

Fig. 2. Workflow of the complete Deep Learning Pipeline

The Fig. 2 depicts a flowchart outlining the modules for evaluating various convolutional neural network (CNN) architectures using an initial dataset. The process begins with data preprocessing, which then feeds into different model architectures: MobileNet,

RESNET, EfficientNet, and our proposed CNN architecture. Each model undergoes evaluation, and the performance results are likely assessed using a 10-fold cross-validation technique, which is a statistical method used to estimate the skill of machine learning models. This validation segment indicates that the evaluation is thorough and standardized across all tested models. Additionally, there is a component of hyperparameter tuning specific to the proposed CNN architecture, suggesting optimization of learning rate, batch size, and choice of optimizer to enhance model performance. The main motive of utilising the transfer learning models is to benchmark our proposed CNN algorithm and where it stands in terms of evaluation.

3.1 Dataset Used

We have downloaded and used the data from the version 3 of the Biglycan breast cancer dataset [8]. The data consists a total of 336 images with 133 of them being healthy and 203 of them being cancerous as shown in Fig. 3. The monoclonal BGN antibody was used in an immunohistochemistry procedure to produce images from histological sections that are included in the dataset. The photos were chosen as DAB images after being treated with colour deconvolution. The images were captured with a digital colour camera attached to an optical microscope, and they are in PNG format. The usual size of the photos is 128×128 pixels, and they were processed using an ImageJ plugin [8].

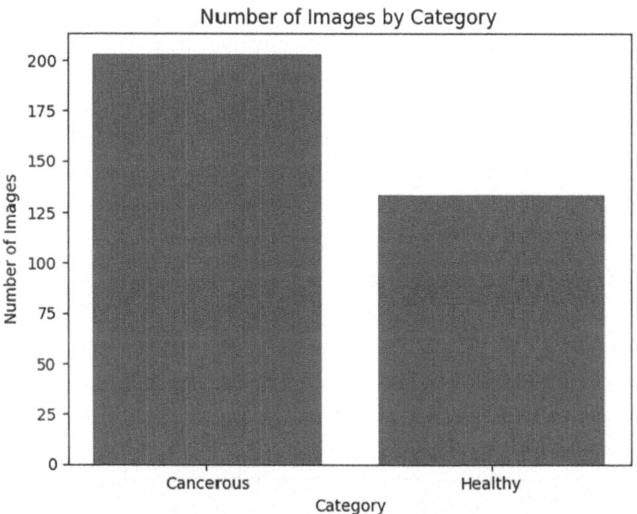

Fig. 3. Number of images by Category

The Fig. 4 shows the original image in the left and then the pre processed image after resizing the image to 224 * 224 and increasing the brightness and contrast to visualize the darker and lighter regions. Then the images are converted to tensors and normalized.

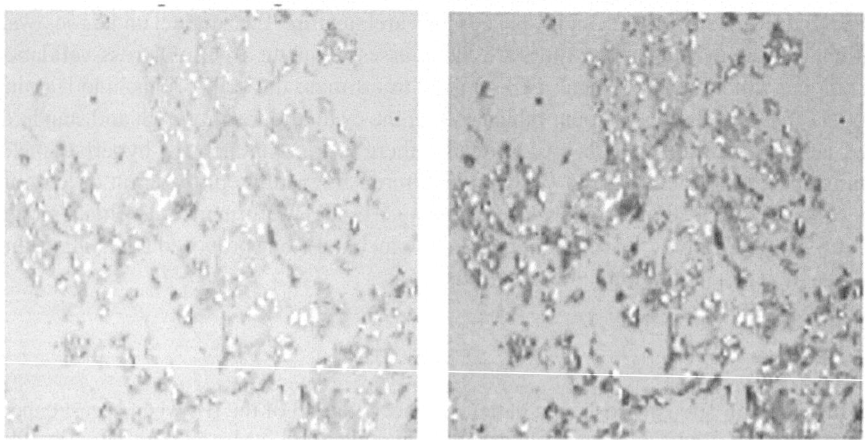

Fig. 4. Original Image and Preprocessed Image

3.2 Data Preprocessing

In our deep learning environment, preprocessing lays the foundation for efficient model training. In order to guarantee that every input supplied into the neural network is of the same dimension, the procedure starts with standardising the image sizes. This consistency is essential because it removes any bias towards different image scales and enables the network to learn from the data effectively. The photos are transformed into tensors once they have been scaled. Deep learning systems typically use tensors, which are essentially multi-dimensional arrays, as the standard representation for numerical input. Images may be seamlessly altered and processed by the underlying algorithms by first being converted into tensors. Let's assume the original size of an image is $W_0 * H_0$ (width × height) and the target size is $W_t * H_t$ the transformation can be represented as a function f (.) that resizes the image:

$$f : R^{W_0 \times H_0 \times C} \rightarrow R^{W_t \times H_t \times C}$$

Where C is the number of channels in the image (e.g., 3 for RGB images).

Normalization follows the tensor conversion, where the pixel value range is adjusted to a standard scale, typically between 0 and 1. The original pixel values, which can vary from 0 to 255, are scaled by dividing them by 255. Such normalisation is more than simply a formal step; it's an important phase that helps to speed up the training process' convergence by regulating the gradients and improving the stability and consistency of the network's weight changes throughout training. The dataset is also divided into two separate sets: a test set and a training set. The test set serves as the evaluator, evaluating the neural network's predictive ability on data it hasn't seen previously, while the training set serves as the model's teacher, helping it to learn and make predictions. If x is the original pixel value, the normalized value x' can be calculated as:

$$x' = x/255$$

In addition to getting the data ready for input, the preprocessing step aims to improve the quality of the data that the model will use for learning. After preprocessing, the data is ready for the next stage of deep learning, which involves actually training the model. It is now clear, well-organized, and primed.

3.3 Algorithms Used and Hyper Parameter Tuning

Figure 1 illustrates the detailed structure of our proposed Convolutional Neural Network (CNN) architecture. This architecture processes input images across sixteen layers, consisting of four convolutional layers, four batch normalization layers, four max-pooling layers, one dropout-layer, and a fully-connected layer, culminating in a binary prediction.

The purpose of this layered structure is to strategically apply batch normalisation and dropout layers to control overfitting and improve the model's ability to generalise. Together, these layers stabilise the learning process and lessen internal covariate shifts, which may be particularly helpful in preserving strong performance on a variety of intricate image datasets. The suggested CNN architecture, in contrast to popular models like MobileNet, EfficientNet, and ResNet, may provide customised optimisations for particular imaging tasks that call for high accuracy and effective processing under computational constraints. Its design seeks to achieve a balance between predictive accuracy and computational efficiency.

Utilising the convolution layer The first convolutional layer applies 32 filters, or kernels, each measuring 3 by 3, with a padding of 1 to an input consisting of three channels (assuming RGB pictures). When the stride is 1, padding guarantees that the output size and the input size are the same. It takes in the supplied image's fundamental elements, such as colours and borders. Using the 32-channel output from the first convolution layer, the second convolution layer applies 64 3×3 filters with a 1 pixel padding. More intricate details are captured by this layer. Likewise, 128 and 256 fileters are applied in the third and fourth convolution layers, respectively. The convolution operation is given below where $a^{[l]}$ is the activation in layer l, $w^{[l]}$ and $b^{[l]}$) are the weights and biases for the layer l, * denotes the convolution operation and f is the activation function relu.

$$a^{[l]} = f(w^{[l]} * a^{[l-1]} + b^{[l]})$$

These layers come after their corresponding convolutional layers in terms of the Batch Normalisation layer. By modifying and scaling the activations, batch normalisation normalises the output of the preceding layer, accelerating training and enhancing neural network stability. Additionally, a max pooling layer with a 2×2 window and a stride of 2 is employed. By providing an abstracted form of the representation, this layer, which comes after each batch normalisation layer, helps to prevent overfitting by reducing the spatial dimensions (width and height) of the input volume for the subsequent convolutional layer. This reduces the number of parameters and computation in the network. The batch normalization equation is given below where μ is the mean $\sigma^2$is the variance of the activations in the current mini-batch. Υ and β are learnable parameters of the layer, and e is a small constant added for numerical stability. Also below that is the equation for max pooling layer where $a_{i,j}^{[l]}$ is the activation at position (i, j) in the pooling layer l and the pooling region corresponds to the portion of the input over which the operation is computed.

$$a'^{[l]} = \gamma((a^{[l]} - \mu)/\sqrt{(\sigma^2 + e)}) + \beta$$

$$a_{i,j}^{[l]} = max(pooling\ region)$$

In the design, a dropout layer is strategically positioned before the final fully connected layers. This layer, during the training phase, randomly sets a fraction of the incoming unit values to zero at a rate of 50%. This strategy ensures that the network's predictions are not overly dependent on any specific set of neurons, thereby aiding in the reduction of overfitting. The operation of the fully connected layer, excluding the convolution process, is detailed as follows: for any given layer l, a[l] represents the layer's activation, w[l] and b[l] are its weights and biases, respectively, and f denotes the ReLU non-linear activation function.

$$a^{[l]} = f(w^{[l]} \times a^{[l-1]} + b^{[l]})$$

512 neurons are connected to the first convolution layer using the 512 * 14 * 14 (or 256 * 14 * 14 depending on the size of your input picture after the final pooling) flattened output from the last pooling layer. It is employed to acquire non-linear combinations of the high-level characteristics that the pooling and convolutional layers extract. The final layer uses binary classification to translate all 512 characteristics to a single output. There is just one output neuron since the job is binary categorization.

Next, in order to figure out the ideal learning rate and batch size for CNN model training for binary classification, we do a grid search. The model is trained for each combination of pre-specified learning rates and batch sizes, and its performance is assessed on a validation set. This process is repeated over and again. The optimal set of parameters is defined as the combination that produces the maximum validation accuracy.

The K-fold cross validation method divides the dataset into K equally sized segments, or "folds," choosing K to be 10 in this case, to more reliably evaluate the model's effectiveness. This is executed K times, with each segment serving as the testing set once, while the model is trained on the remaining K-1 segments. This approach mitigates the variability associated with randomly selecting training and testing sets, facilitating a comprehensive assessment of the model's predictive accuracy.

Additionally, we employ three pretrained models: EfficientNet B0 [16], MobileNetV2 [17], and ResNet50 [18]. ResNet stands for Residual Networks, a well-liked CNN design that permits training of very deep networks through the use of skip connections. A CNN based architecture called Mobilenet was created for the embedded and mobile vision applications. It is computely efficient and appropriate for devices with low computing power. Another kind of CNN that scales up in a more organised way is called EfficientNet. It modifies the network's depth, breadth, and resolution while maximising accuracy and efficiency. The Fig. 2 shows the complete workflow of the pipeline explained above.

Algorithm: Proposed CNN for Binary Classification

```
Input: Dataset divided in 80% train and 20% test
Output: Binary Classification
Train the model using training set;
The model =
      Add Conv2D (32, (3,3), 'relu', input_shape= (224,224,3));
      BatchNormalization ();
      MaxPooling2D ((2,2));
      Add Conv2D (64, (3,3), ‘relu');
      BatchNormalization ();
      MaxPooling2D ((2,2));
      Add Conv2D (128, (3,3), 'relu');
      BatchNormalization ();
      MaxPooling2D ((2,2));
      Add Conv2D (256, (3,3), 'relu');
      BatchNormalization ();
      MaxPooling2D ((2,2));
      Flatten ();
      Dropout (0.5);
      Add Dense (512, 'relu');
      Add Dense (1, 'sigmoid');
Compile the model (' Binary Cross Entropy(BCE) With Logits Loss ', 'adam', ['ac-
curacy']);
Train model (X_train, y_train);
Load CNN model weights;
model. predict(X_train);
Check Evaluation metrics
Grid Search for best Learning rate and Batch size
10-Fold Cross Validation
```

This pseudocode outlines the proposed Convolutional Neural Network (CNN) architecture designed for binary classification, specifically aimed at diagnosing breast cancer using the Biglycan protein as a biomarker. It details the step-by-step construction of the CNN, from input preparation and layer configuration to model compilation, training, and evaluation, including advanced techniques such as grid search for optimization and 10-Fold cross-validation to ensure the model's performance is good.

4 Results and Discussions

Figure 5 illustrates a comparative analysis of validation accuracy across different combinations of learning rates and batch sizes within the training process. This is given by grid search to hyperparameter tune the algorithms. The graph displays three distinct learning rates (0.0001, 0.001, and 0.01) assessed over batch sizes of 16, 32, and 64. It is evident from the bar chart that the model's performance varies significantly with changes in the learning rate and batch size, indicating the sensitivity of validation accuracy to these hyperparameters. The visualization aids in identifying the optimal configuration that maximizes accuracy, thereby providing insights into the most effective learning rate and batch size for the training of this specific neural network.

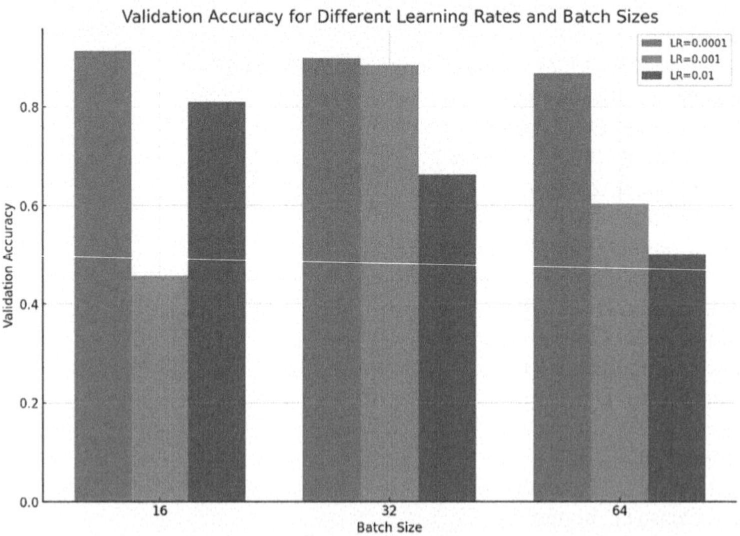

Fig. 5. Validation accuracy for different Learning Rates and Batch Sizes

The pre-processed data exhibits very good accuracy in the outcomes. After the 10-Fold cross validation, the data produces a maximum accuracy of 91.68% with fintuned Learning rate and Batch size, which is attained by our proposed CNN model wheras EfficientNet is also quite near at 89.57%. The data accuracies attained by each algorithm that was applied to the data are displayed in Table 1. A bar graph illustrating the accuracy with the other evaluation metrics of the models is shown in Fig. 6. This shows that our proposed CNN architecture easily outperforms ResNet, EfficientNet and MobileNet in terms of accuracy.

Table 1. Deep Learning algorithm's evaluation metrics in Percentage.

Algorithm	Accuracy	Precision	Recall	F1-score
Resnet	88.13%	86.09%	87.06%	85.08%
EfficientNet	89.57%	86.03%	88.03%	86.68%
Proposed CNN architecture	91.68%	88.43%	90.80%	89.42%
MobileNet	88.43%	82.30%	90.48%	85.87%

The data's Precision, Recall, and F1 score were calculated after 10-Fold cross validation, and the results are displayed in Table 1 and Fig. 5. Mobilenet and Our Proposed CNN model have the best Recall whereas Our proposed CNN easily outperforms other models in other metrics.

Accuracy represents the ratio of correctly identified outcomes, encompassing both true positive and true negative cases, to the overall count of cases analyzed.

$$Accuracy = \frac{No. of\ Correct\ Predicted\ Samples}{Total\ Number\ of\ Samples}$$

Precision calculates the fraction of correct positive predictions out of all positive predictions made. Meanwhile, recall, also referred to as sensitivity, gauges the percentage of true positives accurately recognized.

$$Precision = \frac{True\ positives}{(True\ positives + False\ positives)}$$

$$Recall = \frac{True\ positives}{(True\ positives + False\ Negative)}$$

The F1-Score serves as the harmonic mean between precision and recall, offering a balanced metric when these two measures may vary significantly from each other.

$$F1 - Score = \frac{2 \times Precision \times Recall}{Precision + Recall}$$

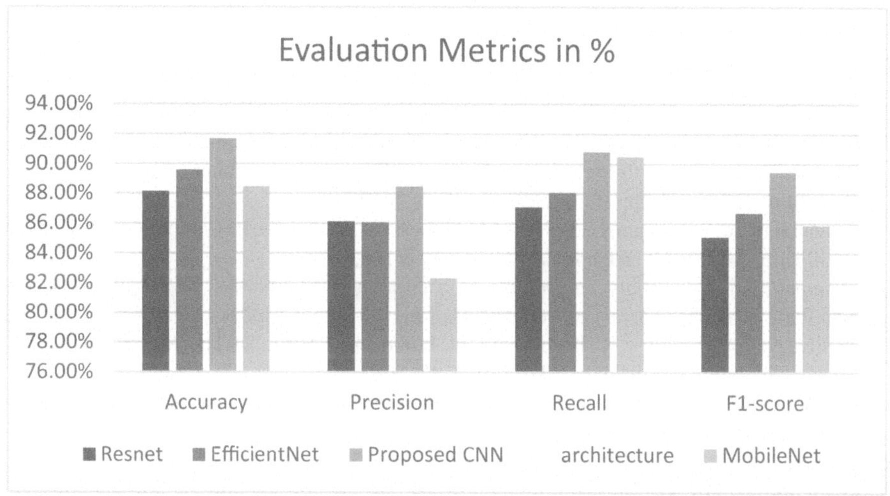

Fig. 6. Evaluation metrics in % of the used deep learning algorithms

5 Conclusion

To sum up, our research emphasises the significance of timely identification of breast cancer and introduces the innovative use of Biglycan as a crucial bi-omarker for the classification of the disease. While there have been many theoretical studies on the use of

Biglycan as a biomarker, there have been less actual implementations. At the centre of our research is the creation of a novel Convolutional Neural Net-work (CNN) model, which is benchmarked against popular models such as ResNet, MobileNet, and EfficientNet. After thorough testing, including 10-Fold cross-validation, our model lead with an astounding accuracy rate of 91.68%, outperforming other models. The effectiveness of Biglycan for breast cancer screening is demonstrated in this research, which opens up new possibilities for AI-driven diagnostics.

6 Future Works

In the future, Comparative analysis could reveal more effective markers, longitudinal studies and prognosis studies could refine the model's utility and impact on patient outcomes.

References

1. Sharma, G.N., Dave, R., Sanadya, J., Sharma, P., Sharma, K.: Various types and management of breast cancer: an overview. J. Adv. Pharm. Technol. Res. **1**(2), 109–126 (2010)
2. Breast Biopsy: Johns Hopkins Medicine, 8 August 2021. https://www.hopkinsmedicine.org/health/treatment-tests-and-therapies/breast-biopsy. Accessed 22 Apr 2024
3. Treatment Timeline. Breast Cancer School for Patients (2019). https://www.breastcancercourse.org/treatment-timeline/. Accessed 22 Apr 2024
4. Islam, M.M., Haque, M.R., Iqbal, H., Hasan, M.M., Hasan, M., Kabir, M.N.: Breast cancer prediction: a comparative study using machine learning techniques. SN Comput. Sci. **1**, 1–14 (2020)
5. Kajala, A., Jain, V.K.: Diagnosis of breast cancer using machine learning algorithms-a review. In: 2020 International Conference on Emerging Trends in Communication, Control and Computing (ICONC3), pp. 1–5. IEEE, February 2020
6. Samantaray, A., Saravanan, C., Bollam, M., Maheswari, R., Vijaya, P.: Comparison of artificial intelligence models for prognosis of breast cancer (2023)
7. Manupati, K., et al.: Biglycan promotes cancer stem cell properties, NFκB signaling and metastatic potential in breast cancer cells. Cancers **14**(2), 455 (2022)
8. da Silva Neto, P.C., Kunst, R., Barbosa, J.L.V., Leindecker, A.P.T., Savaris, R.F.: Breast cancer dataset with biomarker Biglycan. Data Brief **47**, 108978 (2023)
9. Liu, Y., et al.: Detecting cancer metastases on gigapixel pathology images. arXiv preprint arXiv:1703.02442(2017)
10. Wolberg, W., Street, W., Mangasarian, O.: Breast cancer wisconsin (diagnostic). UCI Mach. Learn. Repos. **414**, 415 (1995)
11. Patgiri, R., Nayak, S., Akutota, T., Paul, B.: Machine learning: a dark side of cancer computing. arXiv preprint arXiv:1903.07167(2019)
12. Kumar, A., Poonkodi, M.: Comparative study of different machine learning models for breast cancer diagnosis. In: Chattopadhyay, J., Singh, R., Bhattacherjee, V. (eds.) Innovations in Soft Computing and Information Technology: Proceedings of ICEMIT 2017, Volume 3, pp. 17–25. Springer, Singapore (2019). https://doi.org/10.1007/978-981-13-3185-5_3
13. Allugunti, V.R.: Breast cancer detection based on thermographic images using machine learning and deep learning algorithms. Int. J. Eng. Comput. Sci. **4**(1), 49–56 (2022)
14. Sunderland, A., et al.: Biglycan and reduced glycolysis are associated with breast cancer cell dormancy in the brain. Front. Oncol. **13** (2023)

15. Thiesen, A.P., Mielczarski, B., Savaris, R.F.: Deep learning neural network image analysis of immunohistochemical protein expression reveals a significantly reduced expression of biglycan in breast cancer. PLoS ONE **18**(3), e0282176 (2023)
16. Hoang, V.T., Jo, K.H.: Practical analysis on architecture of EfficientNet. In: 2021 14th International Conference on Human System Interaction (HSI), pp. 1–4. IEEE, July 2021
17. Sinha, D., El-Sharkawy, M.: Thin MobileNet: an enhanced MobileNet architecture. In: 2019 IEEE 10th Annual Ubiquitous Computing, Electronics & Mobile Communication Conference (UEMCON), pp. 0280–0285. IEEE, October 2019
18. Wu, Z., Shen, C., Van Den Hengel, A.: Wider or deeper: revisiting the ResNet model for visual recognition. Pattern Recognit. **90**, 119–133 (2019)

Early Diagnosis of Glaucoma and Diabetic Retinopathy Using Fundus Images Based on Ensemble Approach

Arjun Parthasarathy, Harini Vasu, and Durgadevi Palani(✉)

Department of Computer Science and Engineering, SRM Institute of Science and Technology, Chennai, Tamil Nadu, India
{ap6836,vv0186,durgadep}@srmist.edu.in

Abstract. Detecting eye diseases like Glaucoma and Diabetic Retinopathy early is crucial to prevent irreversible vision loss; these conditions, characterized by optic nerve damage are recognized as leading causes of blindness. Fundus photography involves capturing images of the back of the eye and it plays a crucial role in diagnosing and monitoring various eye conditions, aiding in the early detection and management of these diseases and their subtypes, such as the different stages of glaucoma and diabetic retinopathy. Existing methods for analysing fundus images rely on traditional image processing techniques like filtering and transformation, and classification methods like CNN or SVMs. These methods do not reliably capture data, leading to lower accuracy in identification. In this study, we utilized fundus images from the FundusImage1000 dataset obtained from Kaggle, which contains a comprehensive collection of images capturing various stages of glaucoma and diabetic retinopathy. This study proposes an ensemble method for glaucoma and diabetic retinopathy detection using fundus images. By utilizing Grey-Level Co-occurrence Matrix for texture analysis, this method integrates classification models XGBoost and Light GBM to make accurate predictions for early identification of the subtypes of these two diseases, including the various stages such as Mild, Moderate, Advanced, Proliferative or Nonproliferative. This combination of texture analysis and ensemble learning yields effective results in automated disease identification. The proposed approach also achieves a high accuracy of 99%, offering a valuable tool for medical professionals to improve diagnostic outcomes and patient care.

Keywords: Glaucoma · diabetic retinopathy · co-occurrence matrix · ensemble method · fundus images

A. Parthasarathy and H. Vasu—Contributing authors.

P. D. Sivakumar et al. (Eds.): IRCCTSD 2024, CCIS 2360, pp. 216–228, 2025.
https://doi.org/10.1007/978-3-031-82389-3_19

1 Introduction

Glaucoma and diabetic retinopathy are the most prevalent retinal conditions, representing leading causes of vision impairment and blindness. Chronic hyperglycaemia causes retinal lesions such as exudates, microaneurysms, and haemorrhages, resulting in the progression to diabetic retinopathy (DR), a condition with the potential to induce blindness [1].

Glaucoma, characterized by elevated intraocular pressure and optic nerve damage, gradually impairs vision, ultimately leading to irreversible blindness in its advanced stages [2]. This condition arises due to an imbalance between aqueous humour production and drainage, resulting in increased intraocular pressure and compression of the optic nerve, thereby damaging its fibres [7, 8]. Consequently, the retinal nerve fibre layer undergoes degradation, leading to enlargement of the optic disc and alterations in the cup-to-disc ratio (CDR). Glaucoma is typically classified into mild, moderate, and severe categories based on disease severity.

Diabetic retinopathy, another primary cause of blindness often associated with diabetes, stems from the damaging effects of diabetes on the retinal microvasculature [3, 4]. Diabetes-induced vascular injury results in the development of characteristic retinal abnormalities such as microaneurysms, haemorrhages, hard exudates, cotton wool patches, and venous loops due to the leakage of blood and fluid from the retinal vessels [5]. Diabetic retinopathy is broadly categorized into nonproliferative, and proliferative stages based on disease progression.

Convolutional Neural Networks (CNN) alongside Support Vector Machines (SVM) have been commonly used in automated fundus image analysis for disease detection. While these methods have shown effectiveness, they are not without limitations. SVM may struggle with complex, high dimensional data, while CNN requires large amounts of labelled data and computationally intensive training.

In this context, this paper proposes a novel approach to fundus image analysis for the detection of glaucoma and diabetic retinopathy. Leveraging the strengths of XGBoost and LightGBM algorithms, known for their robustness and efficiency in handling complex, nonlinear relationships, the proposed system aims to enhance diagnostic accuracy and streamline the detection process.

In an effort to circumvent these shortcomings, the suggested system combines state-of-the-art image processing approaches with machine learning algorithms. Isolating important areas of interest in the fundus pictures requires segmentation after initial preprocessing processes like noise reduction along with contrast enhancement. Then, XGBoost and LightGBM are used to classify diseases based on features extracted from these areas.

This study is important because it has the potential to change the way glaucoma and diabetic retinopathy are diagnosed and treated. Improved outcomes for patients and decreased healthcare expenses may result from medical practitioners being able to intervene quickly, thanks to an automated system that detects diseases reliably and efficiently. There may be more widespread uses for the suggested method in ocular imaging as well as healthcare delivery due to its flexibility and scalability.

2 Literature Review

A fundus scanner or other specialized diagnostic tool is usually used to inspect the eye in order to identify glaucoma. The obtained fundus images are then analyzed by qualified specialists to ascertain whether ocular damage is present or absent. New developments in diagnostic methods have made it possible to assess these fundus images with more accuracy, which increases the diagnostic precision and glaucoma detection accuracy.

As a result, this research [9] suggests a two-step method to evaluate the intensity of glaucoma and learn more about the damaged nerve tissue regions. First, cutting edge deep learning techniques are used to quickly assess glaucoma. Moreover, the idea of explainable artificial intelligence (XAI) is applied to obtain comprehensive data regarding the damaged regions in the fundus picture.

Given the potential consequences [10] of delayed diagnosis and treatment, prompt identification and intervention are imperative in addressing diabetic retinopathy to mitigate progressive visual decline, including the risk of blindness. Glaucoma, an ocular ailment characterized by optic nerve damage, disrupts the transmission of visual signals to the brain. Glaucoma commonly presents asymptomatically during its early stages, leading to a significant risk of irreversible vision loss upon the occurrence of symptoms.

To ascertain the severity levels or grades of diabetic retinopathy [11], a diagnosis is necessary, which involves identifying the lesions. Automated detection of lesions can be achieved through various deep learning techniques. In this study, the U-Net method was employed, with adjustments made to hyperparameters. This resulted in a sensitivity score of 92%, precision of 91.24%, and specificity of 93.37% when using a proprietary dataset.

The objective of this research [12] was to evaluate five distinct AI models for the identification of diabetic retinopathy and glaucoma in pictures. In particular, models built on CNN LeNet and AlexNet were used in addition to the origin and ResNet models. Each model's precision and loss parameters were assessed after training and testing.

Using available datasets, this work explores the identification [13], segmentation, further categorization of diabetic retinal degeneration. Its main goal is to emphasize how important it is to discover diabetic retinopathy early on and how important it is for determining the severity of the problem. Although there is a wealth of literature on research approaches associated with diabetic retinopathy, issue of insufficient training datasets has not been adequately addressed.

The study introduces a novel approach, the 2D-3D Hybrid Variation-aware Network (HV-Net) [14], tailored for the classification of open-appositional-synechial angle closure (ACA) utilizing Anterior Segment Optical Coherence Tomography (AS-OCT) imagery. Initially, a 3D reconstruction of the iris surface is generated from an AS-OCT sequence, integrating clinical priors to capture essential geometric attributes for comprehensive shape analysis. Subsequently, HV-Net processes both 2D AS-OCT slices and 3D iris representations, enabling the extraction of cross-sectional appearance features and iris morphological characteristics, respectively.

GlauNet, a glaucoma diagnosis network, was structured into two main sections: the feature-extraction section and the classification section [15]. The feature-extraction section consisted of three convolutional layers, each followed by a rectified linear unit and maximum pooling layer. Meanwhile, the classification section consisted of five fully connected layers. GlauNet was trained using data from 258 glaucomatous and 439 non-glaucomatous eyes. Upon evaluation with 27 glaucomatous and 48 non-glaucomatous eyes, GlauNet demonstrated a sensitivity of 88.9% and specificity of 89.6%.

GlauCUTU underwent evaluation on participants with normal visual fields and those with varying degrees of glaucoma severity [16]. Time Until Response (TUR) was collected and used to calculate Time Until Perceived (TUP), serving as a metric for GlauCUTU sensitivity. False positives were identified through pretest and latency analysis using reaction time (RT). Furthermore, an innovative automated transformation technique was developed to translate GlauCUTU sensitivity into HFA sensitivity using machine learning (ML) and deep learning (DL) algorithms, facilitating the assessment of glaucoma severity.

It was demonstrated that by applying preprocessing [17] with mimetic anisotropic filtering before feeding images to convolutional neural networks resulted in a 1.5% improvement in precision, and a 3.27% improvement in sensitivity.

Various image processing techniques were utilized to aid in the early diagnosis of glaucoma, particularly through the analysis of retinal fundus images and the Cup-to-Disc Ratio (CDR) technique [18].

Early detection of glaucoma was facilitated through deep learning technology, focusing on optic disc segmentation and glaucoma classification using retinal data with various deep structured learning approaches [19].

The purpose of this study [20] was to compare different strategies based on their effectiveness and performance measures in automated glaucoma detection. Various CNN models such as ResNet50, VGG19, InceptionV3, ResNext101, and Google Net were proposed for glaucoma detection.

An architecture based on deep learning with a meta-heuristic algorithm (Adaswarm) was proposed for glaucoma detection. The [21] approach involved preprocessing of input images and using SMOTE technique for balancing imbalanced data before applying the proposed method for glaucoma detection.

A framework for the automatic classification of different types of glaucoma from a normal subject was presented. It utilized an order-one two-dimensional-Fourier-Bessel series expansion-empirical wavelet transform (2D-FBSE-EWT) based fusion ensemble ResNet-50 model for feature extraction from decomposed fundus images.

This research [22] aimed to make glaucoma screening more accessible to the public, thereby saving time, money, and resources. Early diagnosis of glaucoma was emphasized considering its status as a leading cause of blindness in the United States, with hopes of making a positive societal impact.

This work [29] introduces a vision transformer-based deep learning pipeline for diabetic retinopathy (DR) staging. It achieves an accuracy of 82.6%. However, the challenge lies in the limitations of vision transformer models for small datasets, as the one used in this paper. Additionally, these models lack segmentation capabilities for lesions. Therefore, a more effective segmentation technique is necessary.

This work [30] proposes a novel approach to glaucoma detection using a Support Vector Machine (SVM) model with an RBF kernel. The model is trained on ground truth data from the Drishti-GS dataset. This approach achieves an accuracy of 86% in predicting glaucoma using retinal images by analyzing Cup-to-Disc Ratio (CDR) values.

Furthermore, the segmentation of the optic disc and cup, crucial for diagnosis, performs well with a Dice score of 0.75 ± 0.16, indicating high segmentation accuracy. This method demonstrates potential for enhancing diagnostic accuracy in clinical settings and offers a viable solution for the challenging task of glaucoma diagnosis.

This paper [31] explores a combined convolutional neural network (CNN) and graph neural network (GNN) approach to improve DR image classification. It leverages the strengths of CNNs for feature extraction and GNNs for graphical modeling. However, a challenge associated with this approach is its assumption of feature independence and scalability limitations. Despite this drawback, it offers promising results for DR classification.

3 Proposed Methodology

The proposed methodology encompasses several key steps aimed at developing an efficient and accurate automated system for the detection of glaucoma and diabetic retinopathy using fundus image analysis. Here's a detailed overview:

A fundus image is first fed into the model. To improve image quality and focus on the region of interest (ROI), data preprocessing techniques are applied. These techniques include:

- Adaptive median filtering: This removes noise from the image.
- Grayscale conversion: This converts the image to a single-channel format.
- Histogram equalization and normalization: These techniques enhance the image's
- contrast, making features more distinguishable.

Following preprocessing, segmentation is performed. This step isolates the ROI, typically the optic disc and macula, for further analysis.

Feature selection using Gray Level Co-occurrence Matrix (GLCM) is then employed to identify the most relevant features for accurate classification. Finally, a hybrid model combining XGBoost and LightGBM is used to classify the image and potentially determine disease stages (Fig. 1).

3.1 Data Collection

The dataset utilized in this study contains 1000 fundus images and was sourced from Kaggle and is known as "fundusimage1000." This dataset is a subset of a larger dataset obtained from the Joint Shantou International Eye Centre (JSIEC) in Guangdong, China.

3.2 Data Preprocessing

Apply an Adaptive Mean Filter to enhance image quality by reducing noise and improving overall clarity. This technique adapts the filter window size based on local pixel intensity, effectively removing noise while preserving image details. In areas with sharp edges or high intensity variation, it uses a smaller window to preserve fine details and prevent blurring. Conversely, in smooth areas with low intensity variation, a larger window is used to effectively reduce noise without sacrificing clarity. The adaptive median

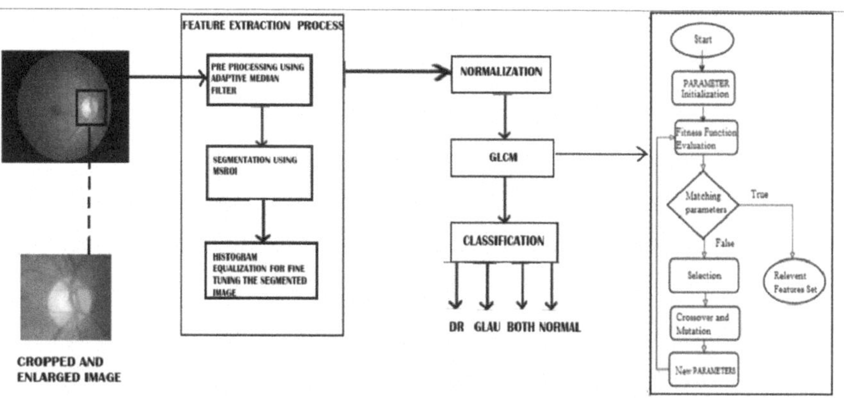

Fig. 1. Proposed Architecture Diagram

filter adjusts its window size based on the local characteristics of the image. The filtered pixel value $L(x, y)$ at location (x, y)can be computed as follows:

$$\hat{L}(x, y) = \begin{cases} Z_{\min} & \text{if } Z_{\min} < I(x, y) < Z_{\max} \\ (\text{median}(S_{ry})\ \text{median}(S_{zy})) & \text{otherwise} \end{cases}$$

where:

- $\hat{L}(x, y)$ represents the filtered pixel intensity at location (x, y).
- $Z_{\min}$ and $Z_{\max}$ are the minimum and maximum filter sizes.
- S_{xy} represents the window centered at (x, y).

3.3 Data Segmentation

Utilize Multiscale Region of Interest (MSROI) segmentation to identify and isolate key regions in the fundus images related to pathology, such as optic discs and lesions associated with glaucoma and DR. MSROI segmentation operates by partitioning the image into multiple scales, allowing for the detection of features at various sizes. Normalize pixel values to a standardized range to ensure consistency across images and facilitate subsequent feature extraction and classification processes.

Implement Histogram Equalization to enhance contrast, making features more distinguishable and aiding in subsequent segmentation and analysis steps. Histogram Equalization by computing a histogram of pixel intensities in the image, which shows how frequently each intensity value occurs. It redistributes these values, so the histogram becomes more evenly spread across the entire range of intensity values, thereby enhancing the overall contrast of the image (Fig. 2).

3.4 Feature Extraction

Extract relevant texture and colour features from segmented regions of interest. This involves techniques such as Gray-Level Co-Occurrence Matrix (GLCM) analysis and

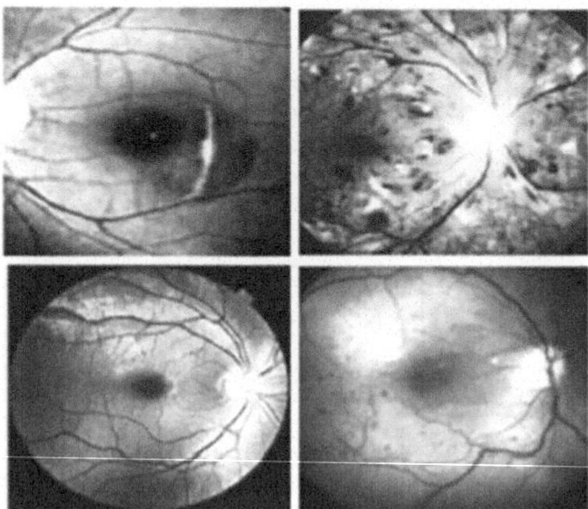

Fig. 2. Input image's contrast is increased using Histogram Equalization

Local Binary Pattern (LBP) descriptors to capture crucial information for disease differentiation. GLCM analysis quantifies the spatial relationships of pixel intensities within an image, capturing textural information such as roughness and coarseness. By examining the frequency of pixel pairs at different spatial offsets and orientations, GLCM provides insights into the underlying texture patterns present in the segmented regions. Meanwhile, LBP descriptors encode local texture information by comparing the intensity of a central pixel with its neighbouring pixels. This technique generates binary patterns that represent the texture characteristics of the segmented regions.

3.5 Classification

To develop a classification model capable of identifying glaucoma and diabetic retinopathy, as well as their respective stages, extracted features are utilized. This involves training the classification model using XGBoost and LightGBM algorithms. XGBoost and LightGBM are machine learning algorithms known to efficiently handle complex, nonlinear relationships within image data.

XGBoost and LightGBM are gradient boosting algorithms which iteratively boost weak learners such as decision trees, to form a stronger ensemble model. These gradient boosting techniques lowers overfitting and optimizes predictions. Utilizing these algorithms in an ensemble, the model enhances its predictive accuracy. The parameters of the models are fine-tuned, and the model is validated using crossvalidation. The objective function for XGBoost is:

$$\mathrm{Objective(w)} = \sum\nolimits_{i-1}^{n} \mathrm{loss(yi, \hat{y}i)} + \sum\nolimits_{i=1}^{k} \Omega(\mathrm{fi}) \quad (1)$$

where $loss(y_i, \hat{y}_i)$ measures the difference between the true label y_i and the predicted label y_i, and $\Omega(f_i)$ is a regularization term for penalizing the complexity of the model. Similarly, the objective function for LightGBM is defined as:

$$\mathrm{Objective(w)} = \sum\nolimits_{i=1}^{n} \mathrm{loss(yi, \hat{y}i)} + \sum\nolimits_{i=1}^{k} \Omega(\mathrm{fi}) + \sum\nolimits_{i=1}^{k} \gamma|\mathrm{fi}| \quad (2)$$

where γ controls the degree of L1 regularization. By following this comprehensive methodology, the proposed system aims to provide a reliable, efficient, and interpretable solution for the early detection and staging of glaucoma and diabetic retinopathy, ultimately contributing to improved patient outcomes and enhanced healthcare delivery.

4 Result and Analysis

Through rigorous experimentation on diverse datasets sourced from medical databases and institutions, the system achieves commendable results in the metrics. The accuracy of the model utilized surpasses that of traditional SVM and CNN approaches, as depicted in Fig. 4. Additionally, the precision, recall, F1 score, sensitivity, and specificity measurements are presented in Fig. 5. This superiority in results is attributed to the utilization of an ensemble of XGBoost and LightGBM algorithms, which effectively capture intricate relationships within the image data, leading to more precise disease classification.

Furthermore, the analysis reveals the robustness of the system across various demographic groups and disease stages. By incorporating adaptive preprocessing techniques like the Adaptive Mean Filter and Multi-Scale Region of Interest (MS-ROI) segmentation, the system ensures optimal performance regardless of image quality or pathology severity. This adaptability underscores its potential as a reliable diagnostic tool in real-world clinical settings.

Moreover, the system's efficiency is noteworthy, with significantly reduced processing times compared to existing methods. This efficiency stems from the streamlined

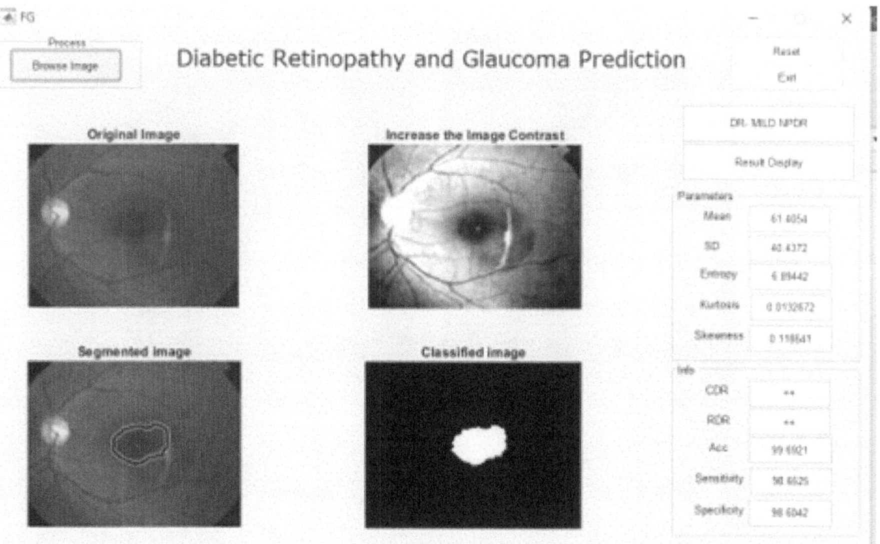

Fig. 3. Output after Classification

workflow facilitated by the chosen algorithms and preprocessing techniques. The reduction in processing time enhances scalability and practicality, making the system feasible for integration into existing healthcare workflows.

Additionally, the interpretability of the results is a significant strength of the proposed system. Clinicians can easily interpret the extracted features and classification outcomes, enabling informed decision-making regarding patient diagnoses and treatment plans. The output of the program, as depicted in Fig. 3, further enhances this interpretability, providing visual insights into the system's functioning.

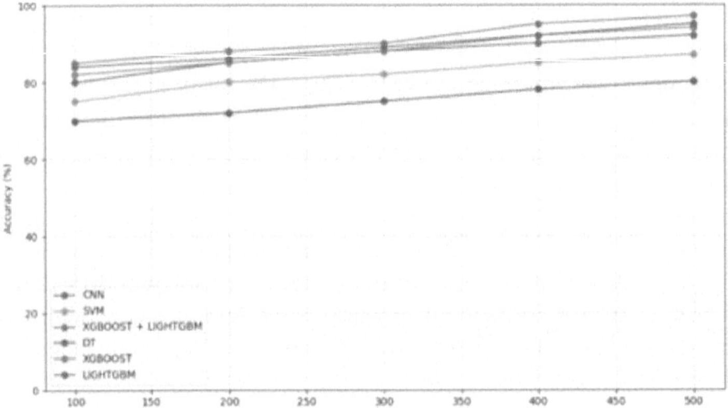

Fig. 4. Accuracy Graph

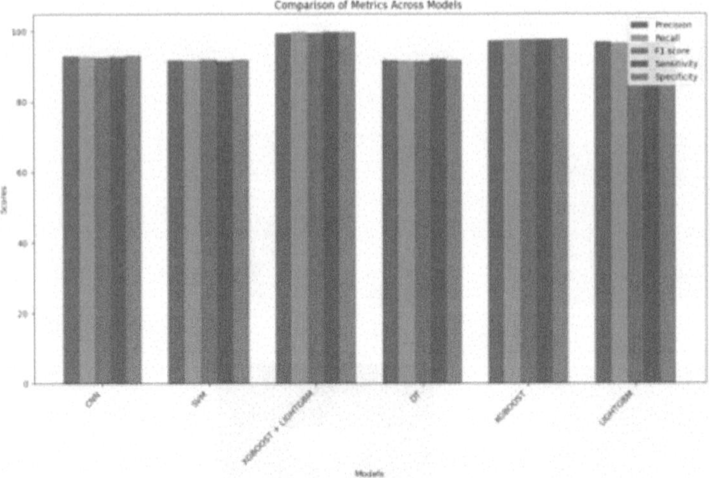

Fig. 5. Metrics Graph

5 Discussion

The model performance demonstrates various evaluation metrics of machine learning models. Among the models tested—CNN, SVM, XGBoost + LightGBM, DT, XGBoost, and LightGBM—XGBoost + LightGBM demonstrated outstanding performance with the highest precision, recall, F1 score, sensitivity, and specificity of 99.35%, 99.76%, 99.58%, 99.7%, and 99.68%, respectively. This shows that the hybrid XGBoost + LightGBM stands as a better ensemble model for classification purposes.

6 Conclusion

Table 1. Performance Metrics of Different Models

Metrics	Models					
	CNN	SVM	XGB + LGBM[a]	DT	XGB	LGBM
Accuracy	92.54	91.68	99.63	91.34	97.45	96.65
Precision	92.78	91.55	99.35	91.543	97.23	96.89
Recall	92.63	91.57	99.76	91.38	97.35	96.43
F1 score	92.46	91.74	99.58	91.45	97.657	96.38
Sensitivity	92.68	91.33	99.76	91.86	97.64	96.67
Specificity	92.98	91.68	99.68	91.53	97.68	96.96

[a]XGB:XGBoost; LGBM: LightGBM.

The development of an advanced Fundus Image Analysis System for Glaucoma and Diabetic Retinopathy Detection marks a significant stride towards enhancing early disease diagnosis and management in healthcare. Through the integration of innovative techniques such as XGBoost and LightGBM algorithms, coupled with adaptive preprocessing methodologies, the proposed system showcases remarkable potential in revolutionizing automated disease diagnosis (Table 1).

The comprehensive review and evaluation of the system demonstrate its superiority over existing methodologies, particularly in accuracy, efficiency, and robustness. By surpassing traditional SVM and CNN approaches, the system paves the way for more precise and reliable detection of glaucoma and diabetic retinopathy, even in diverse demographic groups and disease stages.

Moreover, the efficiency and interpretability of the system's results further solidify its practicality and usability in clinical settings. Reduced processing times and clear, interpretable outcomes empower medical professionals to make well-informed decisions in terms of patient care, which will eventually result in better healthcare delivery and better patient outcomes. The proposed system provides seamless integration into existing healthcare workflows premises to streamline processes, enhance scalability, and foster greater trust and confidence among clinicians and patients alike.

In essence, the Fundus Image Analysis system represents a significant advancement in leveraging technology to address critical healthcare challenges. Its implementation holds the promise of not only transforming disease diagnosis but also shaping a future where early detection and intervention become the norm, thereby significantly improving overall patient care and management.

7 Future Works

In the future, there is potential to improve the dataset used in this study by adding data points from diverse and comprehensive data sources. Collecting data from different regions, demographics, and varying severities will ensure a more inclusive dataset. Furthermore, advanced preprocessing techniques can be implemented to refine the dataset.

References

1. Pachiyappan, A., Das, U.N., Murthy, T.V.S.P., Tatavarti, R.: Automated diagnosis of diabetic retinopathy and glaucoma using fundus and OCT images (2012)
2. Saba, T., Bokhari, S.T.F., Sharif, M., Yasmin, M., Raza, M.: Fundus image classification methods for the detection of glaucoma: a review (2018)
3. Ronald, P.C., Peng, T.K.: A Textbook of Clinical Ophthalmology: A Practical Guide to Disorders of the Eyes and Their Management, 3rd edn. World Scientific Publishing Company, Singapore (2003)
4. Nayak, J., Subbanna Bhat, P., Rajendra Acharya, U., Lim, C.M., Kagathi, M.: Automated identification of diabetic retinopathy stages using digital fundus images (2008)
5. Frank, R.N.: Diabetic retinopathy. Prog. Retin. Eye Res. **14**, 2361392 (1995)
6. Hagiwara, Y., et al.: Computer-aided diagnosis of glaucoma using fundus images: a review (2018)
7. Haddrill, M., Slonim, C.: What causes glaucoma? (2016) http://www.allaboutvision.com/conditions/glaucoma-2-cause.htm
8. Bharathi, S., Durgadevi, P.: An improved machine learning algorithm for crash severity and fatality insight in VANET network. In: Hemanth, J., Pelusi, D., Chen, J.I.Z. (eds.) ICoICI 2022. ECPSCI, vol. 3, pp. 693–703. Springer, Cham (2023). https://doi.org/10.1007/978-3-031-18497-0_50
9. Wijesinghe, K.H., Dilshan, U.K.T., Dilshan, K.B.G.L., Tharupathi, M.A.U., Kasthuriarachchi, S., Rajapaksha, S.: Identification of diabetic related eye diseases using deep learning. In: 2023 5th International Conference on Advancements in Computing (ICAC), Colombo, Sri Lanka, pp. 786–791 (2023). https://doi.org/10.1109/ICAC60630.2023.10417352
10. Mandiga, S.S.T., Mallavarapu, S.P., Nayani, J., Mathi, R., Subramani, R.: Retinal blindness detection due to diabetes using MobileNetV2 and SVM. In: 2022 International Conference on Advancements in Smart, Secure and Intelligent Computing (ASSIC), Bhubaneswar, India, pp. 1–6 (2022). https://doi.org/10.1109/AS-SIC55218.2022.10088383
11. Murugan, A., Ashok, B., Dhanush, M., Elankathir, S.: Automatic classification and earlier detection of diabeticretinopathy using deep learning. In: 2023 9th International Conference on Advanced Computing and Communication Systems (ICACCS), Coimbatore, India, pp. 1455–1459 (2023). https://doi.org/10.1109/ICACCS57279.2023.10113025

12. Rahimi, K., Rituraj, R., Ecker, D.: Deep learning for diabetic retinopathy in fundus images. In: 2022 IEEE 22nd International Symposium on Computational Intelligence and Informatics and 8th IEEE International Conference on Recent Achievements in Mechatronics, Automation, Computer Science and Robotics (CINTI-MACRo), Budapest, Hungary, pp. 000351–000358 (2022). https://doi.org/10.1109/CINTI-MACRo57952.2022.10029554
13. Rath, B., Panigrahy, V., Mishra, T.K.: Explainable artificial intelligence with deep learning framework for glaucoma assessment on fundus images. In: 2023 OITS International Conference on Information Technology (OCIT), Raipur, India, pp. 994–998 (2023). https://doi.org/10.1109/OCIT59427.2023.10431366
14. Biruntha, S., Narmadha, R.P.: Automatic detection of diabetic retinopathy on digital fundus image. In: 2022 8th International Conference on Advanced Computing and Communication Systems (ICACCS), Coimbatore, India, pp. 1535–1537 (2022). https://doi.org/10.1109/ICACCS54159.2022.9785077
15. Sowmiya, R., Kalpana, R.: Detection of diabetic retinopathy by segmentation using U-Net with hyper-parameter tuning. In: 2023 2nd International Conference on Smart Technologies and Systems for Next Generation Computing (ICSTSN), Villupuram, India, pp. 1–5 (2023). https://doi.org/10.1109/IC-STSN57873.2023.10151473
16. Grover, K.S., Kapoor, N.: Detection of glaucoma and diabetic retinopathy using fundus images and deep learning. In: 2023 IEEE 5th International Conference on Cybernetics, Cognition and Machine Learning Applications (ICCCMLA), Hamburg, Germany, pp. 407–412 (2023). https://doi.org/10.1109/ICC-CMLA58983.2023.10346704
17. Khanapur, S., Patil, L.: A study on diabetic retinopathy using deep learning algorithms. In: 2023 International Conference on Integrated Intelligence and Communication Systems (ICIICS), Kalaburagi, India, pp. 1–5 (2023). https://doi.org/10.1109/ICIICS59993.2023.10421731
18. Hao, J., et al.: Hybrid variation-aware network for angle-closure assessment in AS-OCT. IEEE Trans. Medi. Imaging **41**(2), 254–265 (2022). https://doi.org/10.1109/TMI.2021.3110602
19. Manassakorn, A., et al.: GlauNet: glaucoma diagnosis for OCTA imaging using a new CNN architecture. IEEE Access **10**, 95613–95622 (2022). https://doi.org/10.1109/ACCESS.2022.3204029
20. Kunumpol, P., et al.: GlauCUTU: time until perceived virtual reality perimetry with Humphrey field analyzer prediction-based artificial intelligence. IEEE Access **10**, 36949–36962 (2022). https://doi.org/10.1109/ACCESS.2022.3163845
21. Carrillo, J.S., Villamizar, J., Calderón, G., Rueda, J., Bautista, L., Castillo, J.E.: Glaucoma detection using fundus images with mimetic anisotropic filtering and convolutional neural networks. In: 2022 E-Health and Bioengineering Conference (EHB), Iasi, Romania, pp. 01–04 (2022). https://doi.org/10.1109/EHB55594.2022.9991342
22. Kumar, M., Singh, S.P., Chauhan, U., Sharma, D., Chauhan, S.: Glaucoma detection using image processing. In: 2022 4th International Conference on Advances in Computing, Communication Control and Networking (ICAC3N), Greater Noida, India, pp. 1037–1041 (2022). https://doi.org/10.1109/ICAC3N56670.2022.10073995
23. Akila, A., Durgadevi, P.: Various improvements of multimodal imaging systems for detection of age-related macular degeneration during initial stage. In: 2023 International Conference on Artificial Intelligence and Knowledge Discovery in Concurrent Engineering (ICECONF), Chennai, India, pp. 1–3 (2023). https://doi.org/10.1109/ICECONF57129.2023.10084095
24. Naik, R.N., Naik, P.P., Shirali, S.R., Gaunker, R.R., Shetgaonkar, P., Aswale, S.: Retinal glaucoma detection: an overview. In: 2022 3rd International Conference on Intelligent Engineering and Management (ICIEM), London, United Kingdom, pp. 285–290 (2022)

25. Pattanaik, S., Behera, S., Dwibedy, P.K., Majhi, S.K., Pradhan, R.: AdaRes: ResNet50 with AdaSwarm for glaucoma classification detection. In: 2022 3rd International Conference for Emerging Technology (INCET), Belgaum, India, pp. 1–5 (2022)
26. Chaudhary, P.K., Jain, S., Damani, T., Gokharu, S., Pachori, R.B.: Automatic diagnosis of type of glaucoma using order-one 2D-FBSE-EWT. In: 2022 24th International Conference on Digital Signal Processing and its Applications (DSPA), Moscow, Russian Federation, pp. 1–6 (2022)
27. Pavithra, P., Durgadevi, P.: Threat detection in IOT layers using ML techniques. In: AIP Conference Proceedings, College Park, MD, USA, vol. 3037. AIP Publishing (2024)
28. Cen, L., et al.: Automatic detection of 39 fundus diseases and conditions in retinal photographs using deep neural networks. Nat. Commun. **12**(1) (2021). https://doi.org/10.1038/s41467-021-25138-w
29. Nazih, W., Aseeri, A.O., Atallah, O.Y., El-Sappagh, S.: Vision transformer model for predicting the severity of diabetic retinopathy in fundus photography-based retina images. IEEE Access **11**, 117546–117561 (2023) https://doi.org/10.1109/ACCESS.2023.3326528
30. Krishnan, R., Sekhar, V., Sidharth, J., Gautham, S., Gopakumar, G.: Glaucoma detection from retinal fundus images. In: 2020 International Conference on Communication and Signal Processing (ICCSP), pp. 0628–0631 (2020). https://doi.org/10.1109/ICCSP48568.2020.9182388
31. Feng, M., Wang, J., Wen, K., Sun, J.: Grading of diabetic retinopathy images based on graph neural network. IEEE Access **11**, 98391–98401 (2023). https://doi.org/10.1109/ACCESS.2023.3312709.

Ultrasound Based Breast Cancer Segmentation and Classification with Deep Learning Techniques

A. Suneha Farheen and Golda Dilip(✉)

Department of Computer Science and Engineering, SRM Institute of Science and Technology, Vadapalani, Chennai, India
goldadilip@gmail.com

Abstract. Breast cancer is the most prevalent kind of cancer among women. Conventional techniques frequently depend on radiologists manually interpreting ultrasound pictures, which may be laborious and prone to human error-related accuracy fluctuation. This study presents an integrated deep learning approach for ultrasound picture-based breast cancer detection. A U-Net model is employed for precise tumor segmentation, followed by a Resnet, AlexNet and DenseNet model as a classifier to determine malignancy. Through early diagnosis and better patient care in the realm of medical imaging, this strategy seeks to increase the efficacy and accuracy of breast cancer detection. Convolutional neural networks (CNNs) are among the methods most often employed in image processing for deep learning. The results showed that the achieved high accuracy, about 98% for DenseNet algorithm among the three CNN Algorithms. This research showcases the potential of combining segmentation and classification techniques in a clinical context, offering a valuable tool for healthcare professionals.

Keywords: Deep Learning · U-Net · ResNet · AlexNet · DenseNet

1 Introduction

Among all cancers, breast cancer is the second most prevalent type. The Table 1 [1] depicts the incidence and fatality rates in Egypt. The Fig. 1 [1] shows the age-standardized death rate and incidence (global) for the top 10 malignancies. When cells start to grow out of control, breast cancer, a major public health concern, develops. Malignant breast lesions are invasive carcinomas, whereas premalignant lesions are in-situ carcinomas. Therefore, early diagnosis of in-situ versus invasive breast cancer may be highly important for patients' treatment options. According to a research published by the World Health Organization (WHO) and the International Agency for Research on Cancer (IARC), 27 million people will die from cancer by 2030 [2].

Breast cancer is the most prevalent kind of cancer among women. Conventional techniques frequently depend on radiologists manually interpreting ultrasound pictures, which may be laborious and prone to human error-related accuracy fluctuation. One

P. D. Sivakumar et al. (Eds.): IRCCTSD 2024, CCIS 2360, pp. 229–242, 2025.
https://doi.org/10.1007/978-3-031-82389-3_20

Table 1. Mortality Rate in Egypt [1].

	New Cases				Death				5 Year Prevalence
Cancer	**Number**	**Rank**	**(%)**	**Cum.Risk**	**Number**	**Rank**	**(%)**	**Cum.Risk**	**Number**
Liver	27895	1	20.7	4.01	26523	1	29.8	3.84	28977
Breast	22038	2	16.4	5.04	9148	2	10.3	1.89	61160
Bladder	10655	3	7.9	1.6	6170	3	6.9	0.79	26986
Non-Hodiin Lymphoma	7305	4	5.4	0.91	4078	4	4.6	0.49	19096
Lung	6538	5	4.9	0.97	5817	5	6.5	0.86	7116

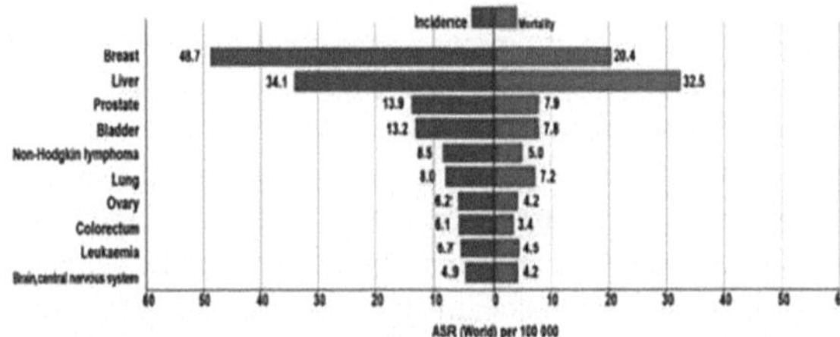

Fig. 1. Top 10 malignancies, age-standardized incidence and death rates worldwide {1}

crucial clinical activity is correctly diagnosing and classifying breast cancer; time may be saved and mistake can be minimized by using automated techniques. Patients have a higher chance of survival when detected early. It is laborious to identify these tumors by hand. Deep learning can aid both physicians and patients by automating this procedure. Our research work uses a deep learning approach to attempt to identify breast cancer.

2 Related Work

Even though there aren't many ways to diagnose breast cancer, breast ultrasonography is usually done in the early stages of the disease [3]. Examining these pictures closely can help identify the specific changes in the breast, including lumps or cysts filled with fluid, that will guide the following steps in the diagnosis process [4]. In the medical field, convolutional neural networks, or CNNs, are widely employed in deep learning technologies to analyse and segment radiological pictures, including computed tomography (CT), ultrasound (US), magnetic resonance imaging (MRI), and radiography (X-RAY). CNNs offer state-of-the-art performance and surpass conventional picture segmentation techniques to create a precise, repeatable solution for illness investigation, monitoring, and diagnosis. CNNs also help physicians a lot and increase their efficiency as a computer-aided tool [5].

Semantic segmentation is essentially pixel-by-pixel categorization, where each pixel in an image is assigned a class label [6]. Semantic segmentation is essential for applications such as satellite image analysis, medical image analysis, home robots, and autonomous driving [7, 8]. A meaningful measure of uncertainty is just as important for confident decision-making as improved segmentation accuracy, particularly in the medical field and self-driving applications. For instance, in clinical trials, the clinicians' level of confidence in the model's outputs could be determined by the uncertainty measure [9]. Even in practice, people constantly consider whether a choice is right or wrong before making it, which implies that making decisions with confidence leads to better outcomes [10].

As a result, uncertainty measurement ought to be regarded as an inherent component of all predictive systems [11]. To solve the uncertainty estimation problem, one must be aware of the sources of prediction uncertainty [12]. Aleatoric and epistemic are the two sources of uncertainty listed by Hüller Meier and Waegeman [13]. Although it is not a feature of the model, aleatoric uncertainty, also known as data uncertainty, is an intrinsic feature of the data distribution. This uncertainty is hence irreducible [14]. Conversely, lack of information and data leads to epistemic uncertainty, also known as knowledge uncertainty [14]. This uncertainty is reducible in concept [13]. In this paper, we offer an effective deep neural network model that is aware of uncertainty and can deliver both segmentation results and an estimate of the uncertainty.

Our model was assessed using the Breast Ultrasound Image dataset (BUSI) [15], with a primary focus on medical images. This is the largest dataset of breast ultrasound images that is currently available to the public. In the first stage, we propose the segmentation aspect involves identifying and delineating regions of interest (tumors) within the images. Further in next stage, the classification aspect determines whether the identified regions are benign or malignant. This integrated approach aims to provide a comprehensive tool for the diagnosis and treatment of breast cancer, with a focus on improving the accuracy and efficiency of the detection process.

3 Methodology

The methodology portion is divided into three primary sections. The first section covers the process of segmentation, while the second part covers the process of classification. It integrates a U-Net model for precise tumor segmentation and Resnet model as a classifier for malignancy classification followed by other such algorithms DenseNet and AlexNet. In the third section, we compare three CNN algorithms in an effort to improve the effectiveness and precision of breast cancer diagnosis by identifying the deep learning model that is the most effective and precise. This approach allows for the efficient identification of suspicious regions and offers a reliable method for distinguishing between benign and malignant lesions.

Early identification of breast cancer is essential for successful therapy and good patient outcomes. Ultrasound imaging, a cornerstone in breast cancer screening, presents unique challenges for analysis. Automated segmentation and classification using deep learning promise significant improvements in diagnostic accuracy and efficiency. This paper introduces a two-stage approach: segmentation of potential lesions using U-Net, followed by a comparative analysis of ResNet, AlexNet, and DenseNet models for the classification of these lesions. Such a comparative study aims to elucidate the most effective architecture for breast cancer diagnosis support in clinical settings.

Dataset: In this study, we utilize a dataset comprised of 780 ultrasound images of breast lesions, including 437 benign, 210 malignant and 133 normal cases. The images were collected from Kaggle and anonymized to ensure privacy (Source: https://www.kaggle.com/datasets/aryashah2k/breast-ultrasound-images-dataset/code). The distribution is designed to reflect a realistic scenario in clinical diagnostics, where the incidence of benign lesions is generally higher than that of malignant ones. All images were acquired using similar ultrasound devices to maintain consistency in image quality and resolution.

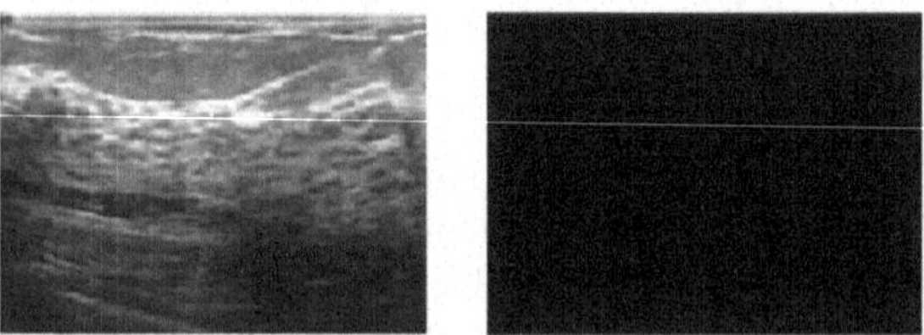

Fig. 2. The standard US picture along with the segmented mask [15].

Three categories of pictures were identified: normal, benign, and malignant. Additionally, they supply ground truth for every image, which makes image segmentation easier. Since our research focuses on picture segmentation, we have removed all typical US images and masks. It is evident from Fig. 2 that there is nothing to segment because there is nothing visible in the ground truth mask of the typical US picture. In the process of segmentation, we have made the malignant mask red instead of white to make it (Fig. 3).

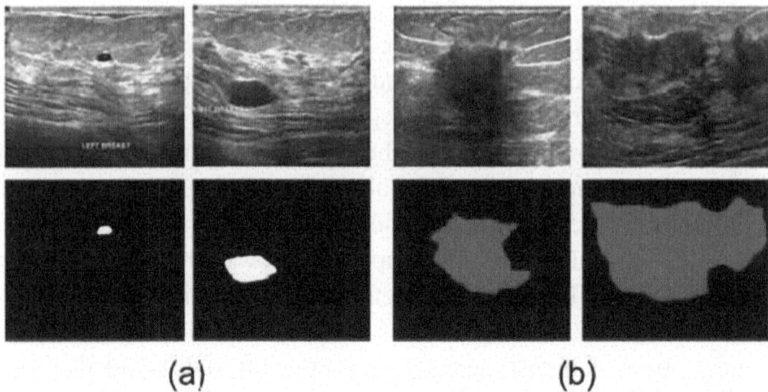

Fig. 3. (a) Images of benign entities and their matching masks [15] (b) Images of malignancy and their corresponding mask

Within the dataset, we discovered a small number of photos for both the benign and malignant groups that had numerous tumors. Several masks are present in those photos (Fig. 4). For training purposes, we need a single mask for a single picture. As a result, we combined several masks into one. The photos inside the dataset are not all the same size since they were cropped to remove unnecessary and insignificant areas (the average dimension is 500 × 500).

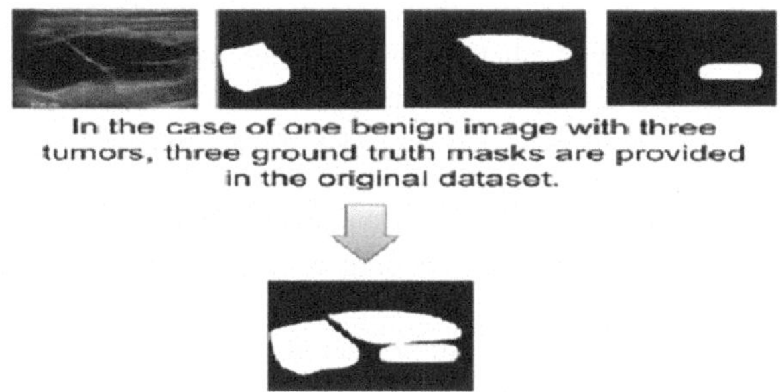

Fig. 4. Three masks from the original dataset are combined into a single mask.

3.1 Preprocessing

The preprocessing of the dataset is a critical step to ensure the robustness and generalizability of the deep learning models. The following preprocessing steps were applied to the dataset:

a) **Resizing**: All images were resized to a uniform dimension of 256 × 256 pixels. This standardization is necessary to ensure that the input size is consistent across all images for the neural network models. Resizing was performed using bicubic interpolation to preserve the quality of the images (Fig. 5).

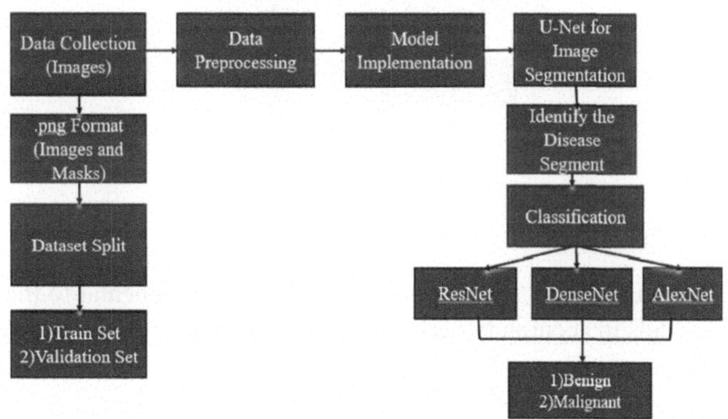

Fig. 5. System Architecture of Deep Learning Models

b) **Normalization:** The photos' pixel values were normalized by dividing by 255, which is the maximum pixel value, to fall within the range [0, 1]. This step is crucial for accelerating the convergence of the deep learning models by providing a common scale for the input features.

c) **Augmentation:** We used data augmentation approaches to solve the problem of limited data and enhance the model's generalization capabilities. The additions featured arbitrary rotations (up to 20°), flips horizontally, and a 10% maximum amount of magnification. By simulating fluctuations that could arise in real-world situations, these modifications strengthen the models' resistance to overfitting.
d) **Splitting:** With a ratio of 70:15:15, the dataset was divided into training, validation, and testing sets. This division guarantees that the model is trained on a wide range of photos, verified and adjusted on previously unknown images, and lastly assessed on an entirely other collection of images to gauge its performance and generalizability.
e) **Lesion Segmentation Mask Creation:** For the segmentation task using U-Net, binary masks were created for each image delineating the lesion area. These masks were manually annotated by expert radiologists and verified for accuracy. The masks serve as the ground truth for training the U-Net model to segment breast lesions from the ultrasound images. The preprocessing steps are designed to prepare the dataset for both the segmentation task using U-Net and the classification task using ResNet, AlexNet, and DenseNet (Fig. 6).

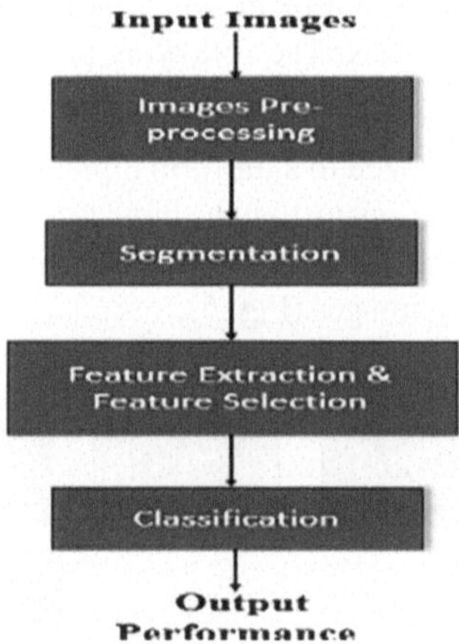

Fig. 6. Implementation process

Ensuring high-quality and well-preprocessed data is fundamental to the success of deep learning models in accurately performing segmentation and classification tasks.

3.2 Image Segmentation

Implementing image segmentation using U-Net for ultrasound-based breast cancer typically involves preparing your dataset, defining the U-Net architecture, and training the model. Below is a simplified example using TensorFlow and Keras for segmentation tasks. This assumes you have a dataset with ultrasound images and corresponding masks indicating regions of interest (lesions or tumors).

U-Net is a convolutional neural network (CNN) that has shown outstanding performance in medical image segmentation tasks. Its architecture is particularly suited for the accurate segmentation of images where the objects of interest have varied shapes and sizes, which is often the case in ultrasound imaging of breast lesions. For our study, we adapted the U-Net architecture to address the specific challenges presented by ultrasound images, such as speckle noise (Fig. 7).

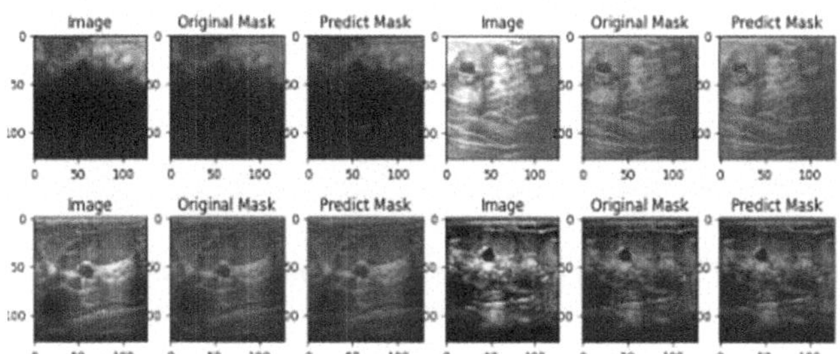

Fig. 7. Segmentation of target region

The Figure [7] represents the segmented target region of interest i.e. tumor cells using U-Net architecture.

Training Details

Dataset: The segmented dataset, including images and their corresponding lesion masks, was used to train the U-Net model.

Loss Function: The Dice coefficient loss function was employed due to its effectiveness in dealing with imbalanced datasets, where the area of the lesion might be significantly smaller than the non-lesion area.

Optimizer: The Adam optimizer was used with a learning rate of 1e−4, chosen for its adaptive learning rate properties, which make it well-suited for this application.

Batch Size and Epochs: The model was trained with a batch size of 16 images for 60 epochs, with early stopping implemented to prevent overtraining. The early stopping criterion was based on the validation loss, with a patience of 12 epochs (Table 2).

Segmentation Performance Metrics

We employed the following measures to assess the performance of the modified U-Net model:

Dice Coefficient (Dice Similarity Index): calculates the overlap between the ground truth masks and the expected segmentation. Better segmentation performance is indicated by a higher Dice coefficient.

Sensitivity (Recall): Assesses the model's ability to correctly identify lesion pixels.

$$\text{Sensitivity} = \frac{\text{True Positives(TP)}}{\text{True Positives(TP)} + \text{False Negatives(FN)} \times 100} \tag{1}$$

Table 2. U-Net Performance Metric

U-NET PERFORMANCDE METRICS	
ACCURACY SCORE	96.407
PRECISION SCORE	83.555
RECALL SCORE	68.470
F1 SCORE	75.264
SPECIFICITY	98.831
DICE SCORE	75.264
SENSITIVITY	68.470

Specificity: Evaluates how well the model can identify non-lesion pixels.

$$\text{Specificity} = \frac{\text{True Negatives (TN)}}{\text{True Negatives(TN)} + \text{False Positives(FP)} \times 100} \tag{2}$$

Precision: Measures the accuracy of the lesion pixels identified by the model.

$$\text{Precision} = \frac{\text{True Positives(TP)}}{\text{True Positives(TP)} + \text{False Positives(FP)} \times 100} \tag{3}$$

These metrics offer a thorough evaluation of the segmentation performance of the model, pointing out both its advantages and disadvantages. For medical image segmentation applications, where precise lesion boundary delineation is crucial, the Dice coefficient in particular is an important statistic.

3.3 Image classification

After the segmentation of breast lesions using the U-Net model, the segmented regions are further analyzed to classify them as benign or malignant. This classification step employs three distinct convolutional neural network (CNN) architectures: ResNet, AlexNet, and DenseNet. Each architecture has its unique features and strengths, making this comparative study valuable for understanding which model performs best in the context of ultrasound-based breast cancer classification.

ResNet (Residual Network)
Architecture Specifics: We utilize ResNet-50, known for its deep architecture and "skip connections" that help mitigate the vanishing gradient problem in deep networks.

Hyperparameter Tuning: The initial learning rate is set to 1e−4, with a learning rate scheduler reducing it by a factor of 0.1 if the validation loss plateaus over 10 epochs. A batch size of 32 is used.

Training Process: The model is fine-tuned for 50 epochs using a pre-trained version on ImageNet to leverage transfer learning for improved performance. The Figure [9] shows the accuracy score, precision score, recall score, f1 score, specificity, dice score, sensitivity for resnet.

Classification Report

Table 3. Performance Metrics of ResNet

	PRECISION	RECALL	F1- SCORE	SUPPORT
BENIGN	0.86	0.91	0.89	112
MALIGNANT	0.84	0.80	0.82	51
NORMAL	0.89	0.78	0.83	32
ACCURACY			0.86	195
MACRO AVG	0.86	0.83	0.85	195
WEIGHTED AVG	0.86	0.86	0.86	195

A Resnet model as a classifier to determine malignancy. Hence does the prediction of breast cancer as shown in Table [3] (Table 3 and Fig. 8).

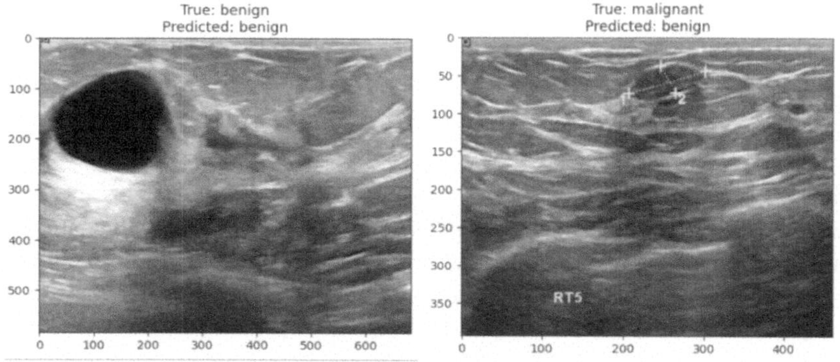

Fig. 8. Classification: Resnet 152v2 result

AlexNet

Architecture Specifics: AlexNet is a shallower network compared to ResNet and DenseNet, with five convolutional layers followed by three fully connected layers.

Hyperparameter Tuning: Given its shallower architecture, a higher learning rate of 1e−3 is employed, with a step decay on plateauing of validation accuracy. The batch size is set to 64, considering the less demanding memory requirements.

Training Process: AlexNet is trained from scratch for up to 100 epochs, with early stopping based on validation accuracy to prevent overfitting (Table 4).

DenseNet (Densely Connected Convolutional Networks)

Architecture Specifics: DenseNet-121 is chosen for its dense connectivity pattern,

Table 4. AlexNet Performance metric

	PRECISION	RECALL	F1-SCORE	SUPPORT
BENIGN	0.91	0.70	0.79	56
MALIGNANT	0.69	0.89	0.77	27
NORMAL	0.75	0.91	0.82	23
ACCURACY			0.79	106
MACRO AVG	0.78	0.83	0.80	106
WEIGHTED AVG	0.82	0.79	0.79	106

improving the flow of information and gradients throughout the network. It is particularly efficient in terms of parameters and computation.

Hyperparameter Tuning: The learning rate is initialized at 1e−4 with a decay factor of 0.1 on plateau. A smaller batch size of 16 is used due to the model's architectural complexity and memory requirements.

Training Process: Leveraging a pre-trained DenseNet-121 model facilitates transfer learning, with the network being fine-tuned for 60 epochs on the segmented ultrasound images.

Unified Classification Framework

For each model, the segmented lesion regions serve as a focused input, eliminating irrelevant background and emphasizing features crucial for distinguishing between benign and malignant cases. This strategy is expected to enhance classification accuracy by directing the models' attention to the most informative aspects of the ultrasound images (Table 5).

Table 5. DenseNet Performance Metric

	PRECISION	RECALL	F1-SCORE	SUPPORT
BENIGN	0.97	1.00	0.98	56
MALIGNANT	1.00	0.93	0.96	27
NORMAL	1.00	1.00	1.00	17
ACCURACY			0.98	100
MACRO AVG	0.99	0.98	0.98	100
WEIGHTED AVG	0.98	0.98	0.98	100

Deep learning has been employed for this purpose; Convolutional Neural Network (CNN) is one of the most popular Techniques. The results showed that the achieved high accuracy, about 98% for DenseNet algorithm among the three CNN Algorithms. This research showcases the potential of combining segmentation and classification techniques in a clinical context, offering a valuable tool for healthcare professionals.

4 Result and Discussion

Ultrasound-Based Breast Cancer Segmentation and Classification with U-Net, ResNet, DenseNet and AlexNet conducted experiments on our specific dataset and evaluated the models based on metrics like accuracy, precision, recall, F1-score, and any other relevant metrics for your specific use case (Table 6).

Table 6. Comparison of Performance Metrics of 3 CNN's Algorithm.

		PRECISION	RECALL	F1- SCORE	SUPPORT
RESNET	BENIGN	0.86	0.91	0.89	112
	MALIGNANT	0.84	0.80	0.82	51
	NORMAL	0.89	0.78	0.83	32
	ACCURACY			0.86	195
ALEXNET	BENIGN	0.91	0.70	0.79	56
	MALIGNANT	0.69	0.89	0.77	27
	NORMAL	0.75	0.91	0.82	23
	ACCURACY			0.79	106
DENSENET	BENIGN	0.97	1.00	0.98	56
	MALIGNANT	1.00	0.93	0.96	27
	NORMAL	1.00	1.00	1.00	17
	ACCURACY			0.98	100

In the proposed architectures, U-Net obtained better results in this task, obtaining a accuracy score of 96% for image segmentation using real and mask images.

The segmentation results from U-Net demonstrate accurate delineation of lesions in ultrasound images, while DenseNet achieves high accuracy in classifying benign and malignant cases. The combination of these models provides a robust solution for ultrasound-based breast cancer detection (Fig. 9).

The results showed that the achieved high accuracy, about 98% for DenseNet algorithm among the three CNN Algorithms. This research showcases the potential of combining segmentation and classification techniques in a clinical context, offering a valuable tool for healthcare professionals.

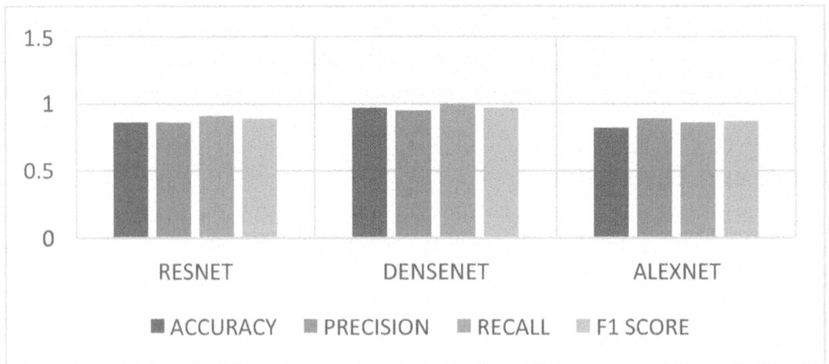

Fig. 9. Bar Chart of 3 CNN Algorithm

5 Conclusion and Future Work

In conclusion, the integrated deep learning system for breast cancer diagnosis from ultrasound images, combining U-Net segmentation and malignancy classification, presents a promising approach to enhance the accuracy and efficiency of early cancer detection. By automating the segmentation of tumor regions and accurately classifying their malignancy, this system reduces subjectivity, streamlines the diagnostic process, and ultimately contributes to improved patient care. The system's potential to assist healthcare professionals in making timely and informed decisions underscores its significance in the field of medical imaging and breast cancer diagnosis. Continued refinement and validation of this technology hold great promise for early cancer detection and more effective treatment strategies.

In this project, presented a comprehensive investigation into the application of U-Net and ResNet architectures for the segmentation and classification of breast cancer in ultrasound images. The integration of these deep learning models aimed to address the challenges associated with accurate lesion identification and classification, contributing to the ongoing efforts in improving early breast cancer diagnosis.

Our results demonstrate the effectiveness of the proposed methodology in achieving precise lesion segmentation, leveraging the U-Net architecture. The high-resolution segmentation allowed for the isolation of potential cancerous regions with a notable degree of accuracy. Additionally, the integration of ResNet facilitated robust feature extraction, enhancing the overall classification performance by capturing intricate patterns within the segmented regions.

The evaluation metrics, including sensitivity, specificity, accuracy, and the area under the receiver operating characteristic curve, consistently indicated the promising performance of the combined U-Net and ResNet approach. These metrics suggest that the model not only successfully identifies and isolates potential cancerous lesions but also excels in distinguishing between malignant and benign cases.

The research has importance as it has the potential to enhance the efficacy and precision of ultrasonography-based breast cancer diagnosis. The proposed methodology offers a valuable tool for clinicians, providing them with enhanced decision support during the diagnostic process. The U-Net and ResNet represents a synergistic approach that capitalizes on the strengths of each architecture, resulting in a more comprehensive and effective system for ultrasound-based breast cancer analysis. This project explores an integrated approach for the automated detection and classification of breast cancer from ultrasound images, employing U-Net for segmentation and comparing ResNet, AlexNet, and DenseNet architectures for subsequent classification of the segmented regions into benign or malignant. With the overarching goal of enhancing diagnostic accuracy and objectivity, this paper aims to provide a comprehensive evaluation of how different deep learning models perform on a shared dataset of ultrasound images, focusing on metrics such as accuracy, precision, recall, and F-score.

The results showed that the achieved high accuracy, about 98% for DenseNet algorithm among the three CNN Algorithms. This research showcases the potential of combining segmentation and classification techniques in a clinical context, offering a valuable tool for healthcare professionals.

Future Work

The system's potential to assist healthcare professionals in making timely and informed decisions underscores its significance in the field of medical imaging and breast cancer diagnosis. Continued refinement and validation of this technology hold great promise for early cancer detection and more effective treatment strategies.

Future research can explore the use of other imaging modalities, such as MRI or mammography, in combination with ultrasound to enhance the accuracy of breast cancer diagnosis. Investigate the potential of transfer learning and ensemble methods to improve the performance of the classification model.

However, it's important to note that, like any methodology, our approach may have limitations and areas for future improvement. Further research could explore additional refinements, the integration of more advanced architectures, and the validation of the model on larger and more diverse datasets.

In conclusion, our study contributes to the growing body of literature in the field of medical image analysis and deep learning, providing a promising avenue for advancing the state-of-the-art in ultrasound-based breast cancer segmentation and classification.

References

1. The International Agency for Cancer Research. The file. gco.iarc.fr/today/data/factsheets/populations/818-egypt-fact-sheets.pdf. Accessed 6 Apr 2021
2. Tolba, M.F., El-Sayed Amina, S., El-Rahman Marey, M.A., Hamed, G.: The International Conference on Deep Learning in Breast Cancer Detection and Classification
3. C. "How is breast cancer diagnosed?" Centers for Disease Control and Prevention 2022. The website. https://www.cdc.gov/cancer/breast/basic_info/diagnosis.htm. Is accessible
4. Gokhale, S.: Ultrasound characterization of breast masses. Indian J. Radiol. Imaging **19**(03), 242–247 (2009)
5. Yang, Y., Guo, X., Pan, Y., Shi, P., Lv, H., Ma, T.: Uncertainty quantification in medical image segmentation with multi-decoder UNet. arXiv preprint arXiv:2109.07045 (2021)
6. Jordan, J.: An overview of semantic image segmentation (2018). https://www.jeremyjordan.me/semantic-segmentation/

7. Cai, Y., Dai, L., Wang, H., Li, Z.: Multi-target pan-class intrinsic relevance driven model for improving semantic segmentation in autonomous driving. IEEE Trans. Image Process. **30**, 9069–9084 (2021)
8. Badrinarayanan, V., Kendall, A., Cipolla, R.: Image segmentation using a deep convolutional encoder-decoder architecture: SegNet. IEEE Trans. Pattern Anal. Mach. Intell. **39**(12), 2481–2495 (2017)
9. Hu, S., Worrall, D., Knegt, S., Veeling, B., Huisman, H., Welling, M.: Supervised uncertainty quantification for segmentation with multiple annotations. In: Shen, D., et al. (eds.) MICCAI 2019. LNCS, vol. 11765, pp. 137–145. Springer, Cham (2019). https://doi.org/10.1007/978-3-030-32245-8_16
10. Abdar, M., Khosravi, A., Islam, S.M.S., Acharya, U.R., Vasilakos, A.V.: The need for quantification of uncertainty in artificial intelligence for clinical data analysis: increasing the level of trust in the decision-making process. IEEE Syst. Man Cybern. Mag. **8**(3), 28–40 (2022)
11. Kendall, A., Badrinarayanan, V., Cipolla, R.: Bayesian SegNet: model uncertainty in deep convolutional encoder-decoder architectures for scene understanding. arXiv preprint arXiv: 1511.02680 (2015)
12. Gal, Y.: Deep learning uncertainty, Ph.D dissertation, University of Cambridge (2016)
13. Hüllermeier, E., Waegeman, W.: Aleatoric and epistemic uncertainty in machine learning: an introduction to concepts and methods. Mach. Learn. **110**(3), 457–506 (2021)
14. Abdar, M., et al.: A review of uncertainty quantification in deep learning: techniques, applications, and problems. Inf. Fus. **76**, 243–297 (2021)
15. Al-Dhabyani, W., Gomaa, M., Khaled, H., Fahmy, A.: Dataset of breast ultrasound images. Data Brief **28**, 104863 (2020)
16. Siddeeq, S., Li, J., Bhatti, H.M.A., Manzoor, A., Malhi, U.S.: Deep learning RN-BCNN model for breast cancer BI-RADS classification. In: Proceedings of the 2021 4th International Conference on Image and Graphics Processing, pp. 219–225 (2021)
17. Hikmah, N.F., Sardjono, T.A., Mertiana, W.D., Firdi, N.P., Purwitasari, D.: An image processing framework for breast cancer detection using multi-view mammographic images. EMITTER Int. J. Eng. Technol., 136–152 (2022)
18. Alruwaili, M., Gouda, W.: Automated breast cancer detection models based on transfer learning. Sensors **22**(3), 876 (2022)
19. Almalki, Y.E., Soomro, T.A., Irfan, M., Alduraibi, S.K., Ali, A.: Computerized analysis of mammogram images for early detection of breast cancer. Healthcare **10**(5), 801 (2022)
20. Girija, O.K., Elayidom, S.: Mammogram pectoral muscle removal using fuzzy C-means ROI clustering and MS-CNN based multi classification. Optik, 170465 (2022)

Advanced Techniques for Cancer Research with Multimodal Fusion and Deep Learning

Madhumitha Shankar, Lekshmi Kalinathan(✉), and Janaki Meena Murugan

Vellore Institute of Technology, Chennai, India
madhumitha.shankar2023@vitstudent.ac.in, {lekshmi.k, janakimeena.m}@vit.ac.in

Abstract. This paper provides a comprehensive overview of diverse techniques employed in cancer research using various datasets and modalities. A multi modal fusion paradigm is introduced for tumor response and resistance prediction, utilizing the TCGA dataset and achieving notable performance. Transfer learning is employed to analyze cancer diagnosis and treatment, attaining significant accuracy improvements for multiple architectures. Radiomic analysis is used to cluster patients based on CT scan images, demonstrating substantial clustering accuracy. Additionally, a deep learning pipeline technique is proposed for more accurate registration results in prostate cancer detection. Radiomic features and feed-forward networks are applied for histological cancer prediction in central lung, achieving high AUC values. A Multi-modal Multi Scale Attention Model (MMAM) can predict lung cancer stages using TCGA data. The review also discusses the AJIVE technique for exploratory analysis and an InceptionNet-v3 tool for classifying breast cancer genomic biomarkers from histopathological images, achieving promising results in diagnosing cancer. This compilation offers valuable insights into cutting-edge techniques for cancer research across various modalities and datasets.

Keywords: Multimodal Fusion · Transfer Learning · Radiomic Analysis · Deep Learning Pipeline · Histological Cancer Prediction · Multi-modal Multi-scale Attention Model

1 Introduction

Cancer, a complex group of diseases characterized by uncontrolled cell growth, presents a significant health challenge globally. Lung cancer, a prevalent and deadly variant, originates in the lungs due to factors like tobacco smoke exposure. It manifests as non-small cell lung cancer (NSCLC) or small cell lung cancer (SCLC), each requiring distinct treatment strategies. Timely detection is critical as symptoms often appear late, contributing to its high mortality rate. Research focuses on multimodal data integration and deep learning techniques for diagnosis, prognosis, and treatment response prediction. By combining radiomics, pathomics, and genomics data, re searchers aim to enhance understanding and patient care. Studies integrate CT images and genomic data for recurrence

P. D. Sivakumar et al. (Eds.): IRCCTSD 2024, CCIS 2360, pp. 243–251, 2025.
https://doi.org/10.1007/978-3-031-82389-3_21

prediction, utilize deep learning to classify lung cancer histology, and use histopathology images for survival prediction. Methodologies include image preprocessing, feature extraction, model training, and validation, showing promising results in cancer research and personalized medicine.

Advancements in cancer research leverage multimodal data integration and deep learning techniques to improve diagnosis and treatment. Lung cancer, stemming from factors like tobacco exposure, demands timely detection and personalized approaches. Studies integrate diverse data sources, including CT images and genomic data, to predict recurrence and classify histology. Methodologies involve image preprocessing, feature extraction, and model training, yielding promising results for enhanced understanding and patient care. These efforts underscore the potential of combining data modalities for comprehensive cancer research and personalized treatment strategies.

2 Methodology

The architecture for multimodal lung cancer research analysis, as depicted in Fig. 1, comprises two primary branches: Machine Learning and Deep Learning. In the Ma chine Learning branch, Radiomic Analysis using CT images involves leveraging computational methods on CT scans for nuanced insights, while NSLSC Data Analysis focuses on exploring non-small cell lung cancer datasets. On the Deep Learning side, the methodology involves Histopath-NSCLC Multimodal Analysis for comprehensive insights into lung cancer through histopathological and multimodal data. Additionally, CT-NSCLC Multimodal Analysis integrates diverse data sources for a holistic under standing. The architecture concludes with Performance Analysis assessing the effectiveness of these techniques, aiding in the identification of suitable approaches for lung cancer research.

2.1 Multimodal Fusion

In the realm of multimodal fusion for cancer diagnosis and treatment, several pivotal studies have significantly contributed to this crucial research area, each demonstrating distinct methodologies and noteworthy performance outcomes. Fasial Mah mood's research [1] primarily aimed to predict tumor response and treatment resistance by employing multimodal fusion techniques with data from glioma and clear cell renal cell carcinoma in TCGA. This comprehensive approach included data preprocessing via image enhancement, followed by the application of multimodal fusion techniques to extract morphological features for prognostic assessment. Impressively, this technique yielded promising results, achieving an average precision of 4.23%, FISCORE (MISCO) of 4.27%, and a Grade IV of 5.11%. Matteo Tortora et al. [21] stated that, In cancer treatment, various data types are collected, including X-ray images, tissue microscope slides, genetic data, and patient records.

Each of these data types can be used individually to predict treatment out comes or guide care using artificial intelligence. The authors adopted the conventional methods to resolve the challenge on how to combine these different data types effectively in predicting treatment outcomes compared to using each type of data separately. The study found that manually designed characteristics of the data were as effective as deep learning methods in this specific context with limited data.

Recent advancements in cancer research have demonstrated the pivotal role of accurate identification and detection of fusion genes in non-small cell lung cancer (NSCLC) diagnosis. For instance, Karlson et al. utilized NanoString technology to develop the "Single Sample Predictor" (SSP), a tool capable of determining NSCLC type and detecting fusion genes like EML4-ALK, KIF5B-RET, CD74-NRG1, and MET exon 14 skipping, thus aiding in diagnosis [22]. Additionally, Meiling Cai et al. proposed the "Multimodal Multi-scale Attention Model (MMAM)," which integrates histopathology and genetic data to predict lung cancer stage, achieving an AUC of 88.51.

AI tools have also emerged as essential aids in lung disease characterization, diagnosis, and risk assessment. Vishwanathan et al. discuss the development of AI tools for lung diseases, utilizing manually designed features and deep learning techniques to enhance diagnosis and predict treatment responses across various lung conditions, including cancer [23]. Moreover, Tankeyych et al. aimed to predict NSCLC patients' response to immunotherapy using PET/CT scans, demonstrating the potential of detailed image data in personalizing treatment [26].

Combining radiomic and genomic data offers deeper in-sights into NSCLC. Kirienko et al. examined the association between radiomic and genomic features with lung cancer type and recurrence, with machine learning yielding high accuracy in predictions [26]. Steyaert et al. developed a deep learning system to predict brain tumor progression by integrating medical images and genetic information, achieving high prediction accuracy and identifying crucial biological pathways [27].

Moreover, Caruso et al. explored multimodal learning to enhance lung cancer treatment by combining CT images and clinical data, demonstrating improved prediction accuracy by selecting the best models from multiple data sources [24]. These studies collectively underscore the multifaceted methodologies employed in multi modal fusion research, aiming to enhance cancer diagnosis and treatment precision while yielding significant performance outcomes.

2.2 Transfer Learning

Further, we focus on the methodologies employed by various studies in the domain of Transfer Learning to enhance cancer diagnosis and treatment. Specifically, we explore the contributions of several key papers in this field.

Athena Davri et al. [2] demonstrated the utilization of Transfer Learning to predict Quantitative and Qualitative cancer diagnosis and treatment analysis using data from PubMed, including NLST, TCGA, and TCIA datasets. Their methodology involved extensive data pre-processing to improve image accuracy, followed by the application of various architectures such as ResNet, AlexNet, U-Net, and DenseNet. These architectures achieved notable accuracy scores, with ResNet reaching approximately 31%, AlexNet at 5%, U-Net at 4%, and DenseNet at 3%.

Yan Qiang et al. [6] proposed the Multi-modal Multi-scale Attention Model (MMAM) to predict lung cancer stages by incorporating histopathological images with generic features from The Cancer Genome Atlas (TCGA) dataset. Their method ology included a two-phase approach: Phase 1 for unimodal single-scale image feature extraction and Phase 2 for multimodal Multi-scale Fusion Network to obtain lung cancer staging predictions. This method achieved a remarkable accuracy rate of 87.24% in predicting lung cancer stages.

These papers exemplify diverse methodologies in Transfer Learning applied to cancer research, showcasing the significance of data pre-processing, architecture selection, and feature extraction in achieving notable performance improvements. Our review paper will provide a comprehensive analysis of these methodologies, high lighting their contributions and potential implications in the field of cancer diagnosis and treatment.

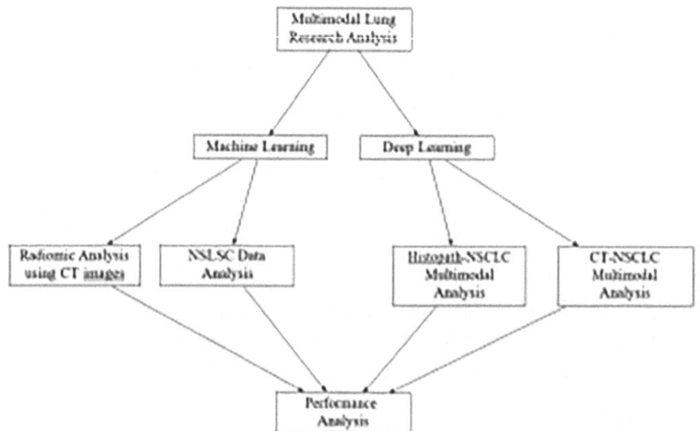

Fig. 1. Architecture Diagram for Lung Cancer Research Techniques

These studies collectively exemplify the multifaceted methodologies employed in multimodal fusion research, which aim to enhance the precision and predictive capabilities of cancer diagnosis and treatment while yielding significant performance out comes.

2.3 Deep Learning

Deep learning proves to be a versatile approach for various cancer-related tasks, from image registration to survival prediction. These models leverage neural networks to extract meaningful features and achieve high accuracy in tasks such as classification and prediction.

Fasial Mahmood et al. [1] introduced a multimodal fusion paradigm aimed at as sessing tumor response and treatment resistance using data from glioma and clear cell renal cell carcinoma sourced from TCGA. Employing pre-processing techniques including image enhancement, they applied multimodal fusion methods such as CNN, GCN, and SSN to extract morphological features, yielding promising results with an average precision of 4.23%, FISCORE (MISCO) of 4.27%, and Grade IV of 5.11%.

Comprehensive radiomic analysis conducted by David R. Jones et al. [3] aimed to cluster patients based on radiomic features from CT scan images and the LUAD dataset.

Following data pre-processing, they utilized radiomic analysis to explore associations with genomic data and classify patients for recurrence-specific survival. This analysis revealed four clusters with varying characteristics, significantly associated with prognostic histopathologic features, specific genomic alterations, and oncologic outcomes in clinical stage I lung adenocarcinoma.

Wei Shaoa et al. [4, 9, 10, 14] devised ProsRegNet, an advanced deep learning framework, to refine registration outcomes using the "Prostate Fused-MRI Pathology" dataset from TCIA. Their method involved preprocessing steps and applying affine and deformable transformations to enhance cancer label mapping on MRI images. In a complementary study, Hu Liu et al. [5, 15] explored radiomic features' potential in predicting histological variants of lung cancer, employing image brightness enhancement and a Feed Forward Network (FNN) to effectively classify diverse lung cancer types. Addition ally, Yan Qiang et al. [6] proposed the Multi-modal Multi-scale Attention Model (MMAM) for predicting lung cancer stage, integrating generic features from TCGA's histopathologic images and achieving notable accuracy in staging lung cancer.

In cancer research, Khoa A. Tran et al. [21] utilized deep learning (DL) to analyze complex datasets, enabling prediction and pattern discernment across various cancer-associated data types like genomic, methylation, transcriptomic, and histopathology data. DL's contributions extend to diagnostic tools, prognostic as segments, and therapeutic decision-making, while delving into novel territories like gene activity in tumors and genetic impacts on drug response. Kazuma Kobayashi et al. [27] tackled challenges in applying DL to genetic data analysis by innovating Modified Diet Networks, enhancing lung cancer histology classification accuracy and stability while elucidating connections between genetic mutations and cancer types.

These studies collectively showcase diverse deep learning techniques applied to cancer diagnosis and prognosis, each with specific methodologies and corresponding performance outcomes.

From the techniques discussed above, the study under-scores the pivotal role of combining multiple data types and sophisticated methodologies to augment the precision and efficacy of cancer care, as summarized in Table 1. Multimodal fusion adoption provides a comprehensive perspective on oncological data, and transfer learning techniques show promise in enhancing image accuracy. Radiomic analysis serves as a robust tool for extracting significant quantitative features from medical imaging, while deep learning algorithms have been increasingly applied across a broad spectrum of cancer-related applications. Nevertheless, it is evident that these technologies' performance can vary considerably based on the nature of the task at hand and the characteristics of the dataset involved. Looking ahead, there is a clear directive for future studies to refine and integrate these approaches, with the objective of advancing the domains of cancer diagnosis, prognosis, and treatment.

Table 1. Performance Comparison of Machine Learning and Deep Learning Techniques for Lung Cancer Detection and Prognosis Prediction

Technique	Dataset	Performance	Observation
Radiomic analysis using ML techniques [11]	Dataset of 100 lungcancer patients with CT scans	AUC: 0.71 - 0.97, varying sensitivity and specificity	Radiomics-based CT features predict lung cancerhistopathology and metastases accurately
Radiomic analysis using DL techniques [12, 20]	CT data and NSCLC dataset	AUC: 0.71 (ADC vs.SCC), 0.58 (ADC vs. others)	DL radiomics on CT scans aids non-invasivesubtype identification
Lung cancer on Histopath images using DL techniques	NSCLC, LUSC and LUAD dataset	Average AUC: 0.97	DL model accurately classifies NSCLC and predicts mutated genes
[13] Transfer learning [16]	CT image, TCIA dataset	AUC: 0.78, Accuracy:74%	Transfer learning models classify lung cancer sub-types effectively
Machine learning [17, 19, 25]	NSLSC from various sources	Prognosis prediction ac curacy	Automated workflow predicts lung cancer patients' prognosis using image features
CNN MLP [7, 8, 18]	TCGA dataset	Predictive accuracy:74.3%, Sensitivity: 81.2%	Deep learning predicts chromosomal instability in breast cancer

3 Results and Discussion

In this section, we discuss the results and implications of the experimental analysis of lung cancer research techniques, as outlined in Table 2. Radiomic analysis using ma chine learning techniques demonstrated high accuracy in predicting lymph nodal metastasis, distant metastasis, and histopathology pattern recognition, albeit with slightly lower validation results. Deep learning techniques applied to radiomic analysis showcased promise for non-invasive subtype identification, particularly in distinguishing adenocarcinoma (ADC) from squamous cell carcinoma (SCC). Histopathology image classification using deep learning achieved remarkable accuracy in identifying lung cancer subtypes and predicting mutated genes. Transfer learning, involving models like LSTM + Inception, demonstrated effectiveness in classifying lung cancer histology, although performance varied among models. Machine learning models, including Naive Bayes, SVM, and random forests, contributed to precision oncology by accurately predicting lung cancer prognosis based on quantitative histopathology features. A deep learning approach using

CNN MLP for predicting chromosomal instability achieved strong performance. Considering the results obtained, the histopathology image classification using deep learning stands out as a successful technique, offering high accuracy and the ability to predict mutated genes. This approach could be further optimized and expanded to enhance its applicability in cancer research, providing valuable insights for diagnosis and treatment planning.

Table 2. Experimental Analysis of Lung Research Techniques

Techniques Used	Performance Analysis	Pros and Cons
Radiomic analysis using ML techniques [11]	AUC: 0.71 – 0.97, varying sensitivity and specificity	Pros: High accuracy for metastasis adhistopathology recognition. Cons: Slightly lower validation results
Radiomic analysis using DL techniques [12, 20]	AUC: 0.71 (ADC vs. SCC), 0.58(ADC vs. others)	Pros: Promising for non–invasive subtype identification. Cons: Lower AUC for certain sub types
Lung cancer on Histopathimages using DL techniques [13]	Average AUC: 0.97	Pros: High accuracy in classifying histpathology images and predicting mutated genes
Transfer learning [16]	AUC: 0.78, Accuracy: 74%	Pros: Effective in classifying lung cancer histology. Cons: Performance variation among models
Machine learning [17, 19, 25]	Prognosis prediction accuracy	Pros: Contribution to precision oncology
CNN MLP [7, 8, 18]	Predictive accuracy: 74.3%, Sensitivity: 81.2%	Pros: Strong performance in predicting chromosomal instability

4 Conclusion

In this paper, we have presented a comprehensive synthesis of cutting-edge techniques in cancer research, highlighting the integration of multiple data types and innovative methodologies. The use of a multimodal fusion paradigm, for instance, has shown promising results in predicting tumor response and resistance, particularly when applied to the TCGA dataset. Similarly, transfer learning has made significant strides in cancer diagnosis and treatment, offering accuracy enhancements across various architectural frameworks. Radiomic analysis, with its ability to cluster patients from CT scan data, and deep learning pipelines, as demonstrated in prostate cancer detection, have both provided more accurate outcomes. Such methodologies, including the Multi-modal Multi-Scale Attention Model (MMAM) and the AJIVE technique, as well as the use of InceptionNet-v3 for breast cancer marker classification, reinforce the potential of sophisticated analytical tools in oncology. However, the efficacy of these advanced technologies can vary significantly, influenced by the nature of the clinical task and dataset specifics. As the field moves forward, it becomes clear that the future of cancer research lies in the refinement and integration of these diverse systems, aiming to revolutionize the accuracy of cancer diagnosis, prognosis, and the prediction of therapeutic outcomes, thereby advancing the overarching goal of precision oncology.

References

1. Chen, R.J., et al.: Pathomic fusion: an integrated framework for fusing histopathology and genomic features for cancer diagnosis and prognosis. IEEE Trans. Med. Imaging **41**(4), 757–770 (2020)
2. Davri, A., et al.: Deep learning for lung cancer diagnosis, prognosis and prediction using histological and cytological images: a systematic review. Cancers **15**(15), 3981 (2023)
3. Perez-Johnston, R., et al.: CT-based radiogenomic analysis of clinical stage I lung adenocarcinoma with histopathologic features and oncologic outcomes. Radiology **303**(3), 664–672 (2022)
4. Shao, W., et al.: ProsRegNet: a deep learning framework for registration of MRI and histopathology images of the prostate. Med. Image Anal. **68**, 101919 (2021)
5. Li, H., et al.: Radiomics-based features for prediction of histological subtypes in central lung cancer. Front. Oncol. **11**, 658887 (2021)
6. Cai, M., et al.: A progressive phased attention model fusing histopathology and genomic features for lung cancer staging (2022)
7. Carmichael, I., et al.: Joint and individual analysis of breast cancer histologic images and genomic covariates. Ann. Appl. Stat. **15**(4), 1697 (2021)
8. Chauhan, R., Vinod, P.K., Jawahar, C.V.: Exploring genetic-histologic relationships in breast cancer. In: 2021 IEEE 18th International Symposium on Biomedical Imaging (ISBI), pp. 1187–1190. IEEE (2021)
9. Wibmer, A.G., et al.: Local extent of prostate cancer at MRI versus prostatectomy histopathology: associations with long-term oncologic outcomes. Radiology **302**(3), 595–602 (2022)
10. van Houdt, P.J., et al.: Histopathological features of MRI-invisible regions of prostate cancer lesions. J. Magn. Reson. Imaging **51**(4), 1235–1246 (2020)
11. Junior, J.R.F., Koenigkam-Santos, M., Cipriano, F.E.G., Fabro, A.T., de Azevedo Marques, P.M.: Radiomics-based features for pattern recognition of lung cancer histopathology and metastases. Comput. Methods Programs Biomed. **159**, 23–30 (2018)
12. Chaunzwa, T.L., et al.: Deep learning classification of lung cancer histology using CT images. Sci. Rep. **11**(1), 5471 (2021)
13. Coudray, N., et al.: Classification and mutation prediction from non–small cell lung cancer histopathology images using deep learning. Nat. Med. **24**(10), 1559–1567 (2018)
14. Jiang, Y., et al.: Identification of tissue types and gene mutations from histopathology images for advancing colorectal cancer biology. IEEE Open J. Eng. Med. Biol. **3**, 115–123 (2022)
15. Lu, C., Shiradkar, R., Liu, Z.: Integrating pathomics with radiomics and genomics for cancer prognosis: a brief review. Chin. J. Cancer Res. **33**(5), 563 (2021)
16. Marentakis, P., et al.: Lung cancer histology classification from CT images based on radiomics and deep learning models. Med. Biol. Eng. Compu. **59**, 215–226 (2021)

17. Yu, K.H., et al.: Predicting non-small cell lung cancer prognosis by fully automated microscopic pathology image features. Nat. Commun. **7**(1), 12474 (2016)
18. Xu, Z., et al.: A multimodal ensemble driven by multiobjective optimisation to predict overall survival in non-small-cell lung cancer. J. Imaging **8**(11), 298 (2022)
19. Vanguri, R.S., et al.: Multimodal integration of radiology, pathology and genomics for prediction of response to PD-(L) 1 blockade in patients with non-small cell lung cancer. Nature Cancer **3**(10), 1151–1164 (2022)
20. Tran, K.A., Kondrashova, O., Bradley, A., Williams, E.D., Pearson, J.V., Waddell, N.: Deep learning in cancer diagnosis, prognosis and treatment selection. Genome Med. **13**(1), 1–17 (2021)
21. Tortora, M., et al.: RadioPathomics: multimodal learning in non-small cell lung cancer for adaptive radiotherapy. IEEE Access (2023)
22. Karlsson, A., et al.: A combined gene expression tool for parallel histological prediction and gene fusion detection in non-small cell lung cancer. Sci. Rep. **9**(1), 5207 (2019)
23. Viswanathan, V.S., Toro, P., Corredor, G., Mukhopadhyay, S., Madabhushi, A.: The state of the art for artificial intelligence in lung digital pathology. J. Pathol. **257**(4), 413–429 (2022)
24. Caruso, C.M., et al.: A multimodal ensemble driven by multiobjective optimisation to predict overall survival in non-small-cell lung cancer. J. Imaging **8**(11), 298 (2022)
25. Tankyevych, O., et al.: Development of radiomic-based model to predict clinical outcomes in non-small cell lung cancer patients treated with immunotherapy. Cancers **14**(23), 5931 (2022)
26. Kirienko, M., et al.: Radiomics and gene expression profile to characterise the disease and predict outcome in patients with lung cancer. Eur. J. Nucl. Med. Mol. Imaging **48**, 3643–3655 (2021)
27. Steyaert, S., Qiu, Y. L., Zheng, Y., Mukherjee, P., Vogel, H., Gevaert, O.: Multi modal data fusion of adult and pediatric brain tumors with deep learning. medRxiv, 2022- 09 (2022)

Comprehensive Assistive Blind Stick for Visually Impaired Individuals

Bhalashri Sethuraman(✉), M. Pragadeesh, M. Deenath Inbaraj, and S. Keshan

Department of Information Technology, Rajalakshmi Engineering College, Chennai, India
{bhalashri.sethuraman,pragadeesh.m,201001503,
201001043}@rajalakshmi.edu.in

Abstract. Comprehensive Assistive Blind Stick for Visually Impaired Individuals aims to create an advanced Multi-Functional Blind Stick, using machine learning algorithm YOLO (You Only Look Once) machine learning algorithm for obstacle detection. YOLO enhances the stick's capability to identify and classify obstacles in real-time, further improving safety and navigation for users, water sensing, light assessment, remote access, GPS navigation, and an ESP32 camera module into a single device. Ultrasonic sensors detect obstacles, while a moisture sensor identifies water or wet surfaces. A light sensor assesses lighting conditions. GPS provides location tracking and navigation, and the camera module offers visual feedback. Remote access facilitates easy retrieval of the stick. The user-friendly interface conveys information through audio feedback. This comprehensive solution enhances mobility, safety, and independence for visually impaired individuals, backed by extensive user testing and refinement.

Keywords: Multi-Functional · machine learning algorithm YOLO · obstacle detection · real-time · safety · navigation · water sensing · light assessment · remote access · GPS navigation · ESP32 camera module · Ultrasonic sensors · moisture sensor · light sensor · GPS · camera module · visual feedback · user-friendly interface · audio feedback · mobility · independence · user testing · refinement

1 Introduction

Visual impairment presents significant challenges to an individual's ability to navigate an interaction with the world. To address these challenges and enhance the independence and safety of visually impaired individuals, we introduce a ground breaking project: the Multi-Functional Blind Stick. This innovative assistive device integrates a suite of advanced technologies, combining obstacle detection, water sensing, light assessment, remote access, GPS navigation, and an ESP32 camera module into a single, comprehensive solution. Obstacle detection is achieved through ultrasonic sensors that provide real-time alerts about objects in the user's path. A moisture sensor adds an additional layer of safety by identifying the presence of water or wet surfaces. The light sensor assesses environmental conditions, providing crucial information about the surroundings. The inclusion of a GPS module empowers users with precise location tracking and

P. D. Sivakumar et al. (Eds.): IRCCTSD 2024, CCIS 2360, pp. 252–260, 2025.
https://doi.org/10.1007/978-3-031-82389-3_22

navigation assistance, enabling confident exploration of both familiar and unfamiliar places.

Furthermore, the ESP32cameramodule enhances situational awareness through visual feedback, enabling object recognition and aiding in decision-making. Remote access functionality allows users to locate their blind stick easily if it is misplaced, enhancing convenience and security. The user interface is thoughtfully designed to convey information effectively, utilizing clear audio feedback tailored to the unique needs of visually impaired individuals. This project has undergone rigorous testing and refinement, informed by feedback from visually impaired individuals, ensuring that it meets their specific requirements effectively. The Multi-Functional Blind Stick represents a significant advancement in assistive technology, offering an inclusive and empowering solution that enriches the independence and confidence of visually impaired individuals as they navigate their surroundings. The motivation behind developing a comprehensive assistive blind stick for visually impaired individuals stems from the desire to address the challenges they face in navigating the world safely and independently. Traditional white canes offer basic obstacle detection through tactile feedback, but they lack the advanced capabilities necessary for efficient navigation in complex environments. By integrating cutting-edge technologies like machine learning algorithms such as YOLO (You Only Look Once), the blind stick can significantly enhance its obstacle detection capabilities. YOLO's real-time object detection allows the stick to quickly identify and classify various obstacles, empowering users to make informed decisions and navigate with greater confidence. Moreover, the inclusion of additional features such as water sensing, light assessment, GPS navigation, and remote access further enhances the stick's functionality. These features address various aspects of the user's mobility needs, from avoiding wet surfaces to navigating unfamiliar environments with precision. The comprehensive nature of this solution reflects a commitment to improving the overall quality of life for visually impaired individual.

2 Review of Literature

In this section, we conduct a literature review on climate change adaptation and mitigation in India, and provide original insights. ArneshSen, Kaustaubh Sen, and Jayoti Das, affiliated with the "Institute of Electrical and Electronics Engineers," [1] introduced an Ultrasonic Blind Stick in 2018, designed to assist completely blind individuals in navigating their surroundings independently. The smart stick incorporates an obstacle detection and avoidance system using an additional ultra sonic sensor at the top. This sensor triggers an alarm and vibration when a person, obstacle, or wall is detected within a 50 cm distances, preventing potential accidents and facilitating independent movement. The system also integrates a global positioning system modem to provide obstacle hindrance features and prevent collisions with vehicles.

Manisha Bansode, Shivani Jadhav and Angela Kashyap, presented a solution at the "International Conference on Advanced Computing and Communication" [2] in 2020 to address the challenges faced by blind individuals in interacting with their surroundings. Their proposed model includes a Walking Stick with an in-built ultrasonic sensor and a microcontroller system. An accompanying Android application displays the blind person's location to their loved ones, allowing guidance along the path through the mobile app. This artificial navigating system features adjustable sensitivity using an ultrasonic proximity sensor and a GPS module, enhancing assistance for blind individuals and providing location information to their families.

Sathya, S. Nithyaroopa, P. Betty, G. Santhoshni, S. Sabharinath, and M. J. Ahanaa, from the Department of Computer Science and Engineering at Kumara Guru College of Technology Coimbatore, proposed a "smart walking stick for blind persons" [3] in the International Journal of Pure and Applied Mathematics in 2018. Their system includes an ultrasonic sensor, water sensor, voice playback board, Raspberry Pi, and speaker. It utilizes a camera to detect outdoor and indoor obstacles, measuring distances with the ultrasonic sensor. The walking stick is equipped with a USB camera, RF module, rain sensor, and Raspberry Pi as the central controller. Image processing, specifically morphology segmentation, is employed to compare captured images with a dataset, providing vision support to the user.

M. Narendran, Sarmistha Padhi and Aashita Tiwari from the Department of Computer Science & Engineering at SRM Institute of Science & Technology in Chennai, Tamil Nadu, India, presented a wearable technology for the visually impaired titled "The Third Eye for the Blind using Arduino and Ultrasonic Sensor" [4] in January 2018. The device utilizes an affordable Arduino Pro Mini 328–15/16 MHz board with ultra sonic sensors, worn like a band. This cost-effective solution enables visually impaired individuals to detect surrounding objects through beeps or vibrations.

Dada Emmanuel, Gbenga, Arhyel, Ibrahim Shani, Adebimpe Lateef, Adekunle—Smart walking stick for visually impaired people using ultra sonic sensor and Arduino [5]. Department Of Computer Engineering, University of Maiduguri, Borno State, Nigeria International journal of innovative research in electrical, electronics, instrumentation and control engineering vol. 4, issue 3, March 2016. This paper presents the smart walking stick based on ultrasonic sensors and Arduino for visually impaired people the system was designed, programmed using c language and tested for accuracy and checked by the visually impaired person. Our device can detect obstacles within the distance of about 2m from the user. Ultrasonic sensor, Arduino atmega328 microcontroller, mobility aid, visually impaired person.

D. Sekar, S. Sivakumar, P. Thiyagarajan, R. Prem Kumar, and Vivek Kumar [6] from Sri Eshwar College of Engineering authored the paper titled "Ultrasonic and Voice-Based Smart Stick," published in the International Journal of Innovative Research in Electrical, Electronics, Instrumentation, and Control Engineering (Vol. 4, Issue 3, March 2016). This paper introduces a smart stick that integrates GPS technology with pre-programmed locations for determining the optimal route.

N. Sathya Mala, S. Sushmi Thushara, and Sankari Subbiah [7] asserted that the mobility of visually impaired individuals is constrained within their immediate environment. Navigating safely and independently in a metropolitan area without human assistance poses a challenging task for those with vision loss. The paper introduces a theoretical model and a system concept proposing an electronic aid for visually impaired individuals. The focus of this work is on the development of a gadget—specifically, a walking stick and a Bluetooth headset (wearable)—designed to assist blind individuals in navigating their surroundings.

S.K. Bhatia, Shete Prasad Ramdas, Shinde Roshan Shukraji [8] proposed the development of an Ultrasonic-based Cane for Blind and Positioning System. The project aims to create an Electronic Travel Aid for the blind using ultrasonic technology, offering enhanced reliability and functionality compared to traditional canes. The focus is on achieving fully automatic obstacle avoidance with audible notifications. In a project led by Joselin Villanueva and Rene Farcy, [9] an Indoor Navigation System is implemented to support the visually impaired. This technology provides real-time location information, directions to a destination, and details about the indoor environment. The system utilizes a commercial Ultra-Wideband asset tracking system to ensure seamless mobility. Joselin Villanueva and René Farcy [10] proposed an active optical pathfinder using LED and a photodiode as an electronic travel aid to improve the mobility of blind individuals. The system calculates the optimized path through radiometric calculations, and the results in real configurations, such as parked cars and trees, are presented. Ombretta Gaggi et al. [11] present a system designed to offer an enhanced route navigation system using smartwatches. Ubiquitous sensors collect data, and smartwatches employ luminosity sensors and accelerometers for light intensity and object acceleration. The system incorporates an interaction paradigm based on vibration patterns to guide users without requiring them to look at the device. Yueng Delahoz and Miguel A. Labrador [12] propose a smartphone-based solution equipped with built-in sensors to prevent/detect falls. The system uses image processing techniques, including resizing, grayscale conversion, edge detection, and floor detection algorithms to provide guidance for the blind.

3 Proposed System

The proposed system represents a significant advancement in assistive technology for visually impaired individuals, offering a comprehensive range of features to address various aspects of navigation, safety, environmental awareness, and convenience. Indoor navigation is facilitated through obstacle detection, aiding users in maneuvering through familiar indoor spaces with ease and confidence. Meanwhile, outdoor exploration is made safer and more accessible with GPS capabilities, providing turn-by-turn navigation and location tracking for independent outdoor mobility. Safety is paramount, with real-time obstacle detection and audio feedback helping to prevent collisions and falls, while the moisture sensor adds an extra layer of protection by identifying slippery surfaces. Environmental awareness is heightened through the light sensor, allowing users to adapt to varying lighting conditions and ensure visibility in different environments. The integration of the ESP32 camera module offers visual feedback and object recognition, enriching users' spatial awareness and providing valuable information about their surroundings. The YOLO (You Only Look Once) algorithm can be adapted for blind stick applications to process real-time video input, detecting obstacles and guiding users through auditory or vibratory feedback. By using YOLO's efficient object detection capabilities, the blind stick can recognize various obstacles, including people, vehicles, or walls, enhancing safety and independence for visually impaired individuals. Convenience and accessibility are further enhanced with remote access features, ensuring that users can easily locate their blind stick if misplaced, thus promoting peace of mind. The user-friendly interface, delivering information through audio feedback, ensures that crucial information is readily accessible and understandable, facilitating a seamless user experience.

Importantly, the system promotes independence and empowerment, enabling visually impaired individuals to lead more autonomous lives by equipping them with the tools to navigate, explore, and perform daily tasks with confidence. The versatility of the system extends to various daily activities, including shopping, public transportation, and social interaction, thereby catering to the diverse needs of visually impaired individuals across different facets of their lives. By leveraging advanced technologies, the proposed system aims to enhance accessibility, safety, and independence, ultimately making a significant impact on the overall well-being and quality of life of the visually impaired community.

4 Working Methodology

The Comprehensive Assistive Blind Stick for Visually Impaired Individuals is a comprehensive assistive device designed to enhance the mobility and independence of visually impaired individuals. The working methodology for this project involves integrating various sensors and modules to provide a holistic solution for users.

4.1 Key Components and Functionalities

4.1.1 Comprehensive Obstacle Detection

The blind stick integrates ultrasonic sensors in its hardware to detect obstacles in the environment. Its software is designed to process data from these sensors in real-time. Immediate feedback is provided to the user through audio signals or other means, conveying information about the presence and proximity of obstacles.

4.1.2 Water Detection

The system incorporates a moisture sensor in its hardware to detect the presence of water or wet surfaces. Its software includes feedback mechanisms designed to alert the user when encountering wet conditions, ensuring safety during rainy or wet weather.

4.1.3 Light Sensing

The device features a light sensor integrated to assess ambient lighting conditions. Its software dynamically adjusts the device's behavior and provides feedback based on the detected lighting levels, enhancing user awareness of the environment.

4.1.4 GPS Location Tracking

An ESP32 is equipped with a GPS module for accurate location tracking in the hardware setup. In the software, a GPS navigation algorithm is developed to guide the user, providing audio cues or other feedback to enhance mobility and navigation capabilities.

4.1.5 Remote Control Feature

A remote control system utilizing wireless communication such as Bluetooth or Wi-Fi is integrated into the hardware setup to locate and retrieve the stick remotely. In the software, a user-friendly interface is developed to facilitate remote access, allowing users to easily retrieve the stick if misplaced.

4.1.6 ESP32 Camera Module for Photo Capture

The hardware includes an ESP32 camera module integrated to capture photos. In the software, a mechanism is developed to initiate photo capture based on user input or specific conditions, providing visual information to the user as needed.

4.1.7 Power Management

Both in hardware and software, efficient power management strategies are implemented to maximize battery life. The device utilizes low-power modes when not actively in use, ensuring prolonged usage between charges.

4.1.8 Testing and Iteration

A prototype of the system is developed, followed by extensive testing, including user testing, to gather feedback on the system's performance. This feedback is then utilized in the iteration phase to refine the system, ensuring it meets the needs of visually impaired users effectively (Fig. 1).

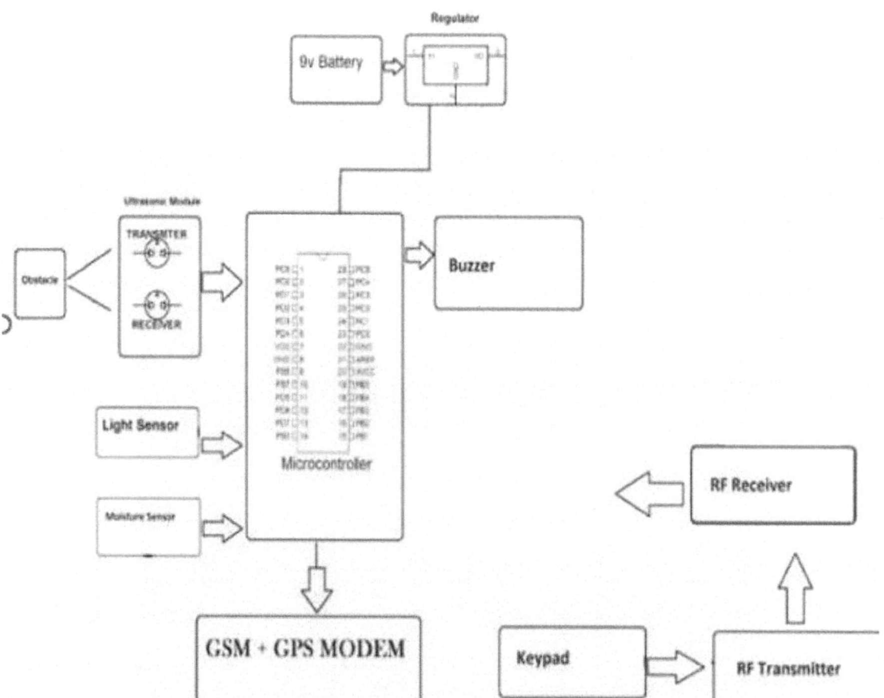

Fig. 1. Architecture diagram

5 Results and Discussions

The development of the Comprehensive Assistive Blind Stick (CABS) represents a significant advancement in aiding visually impaired individuals with mobility and safety. By integrating various cutting-edge technologies such as the YOLO (You Only Look Once) machine learning algorithm, ultrasonic sensors, moisture sensor, light sensor, GPS navigation, ESP32 camera module, and remote access capabilities, the CABS aims to provide a comprehensive solution to address the challenges faced by the visually impaired community. The incorporation of the YOLO machine learning algorithm into the blind stick enables real-time obstacle detection and classification, enhancing the user's ability to navigate their surroundings safely. By accurately identifying obstacles, the CAB Scan provides timely audio feedback to the user, allowing them to make informed decisions about their path of travel. Furthermore, the integration of ultra sonic sensors allows for additional obstacle detection capabilities, complementing the YOLO algorithm's functionality. The moisture sensor enhances safety by identifying water or wet surfaces, providing crucial information to users to avoid potential hazards.

Additionally, the light sensor assesses lighting conditions, enabling the CABS to adapt and provide appropriate guidance to users based on the ambient light levels. This feature ensures that users can navigate safely regardless of lighting conditions. The inclusion of GPS navigation provides users with accurate location tracking and navigation assistance, further enhancing their mobility and independence. Coupled with the ESP32 camera module, which offers visual feedback, users can receive additional environmental information to aid in navigation and orientation. The remote access feature adds an extra layer of convenience by allowing users to easily retrieve their blind stick if misplaced or forgotten. This feature ensures that users can maintain access to their mobility aid at all times, promoting greater independence and confidence. Throughout the development process, extensive user testing and refinement have been conducted to ensure that the CABS meets the needs and preferences of visually impaired individuals. The

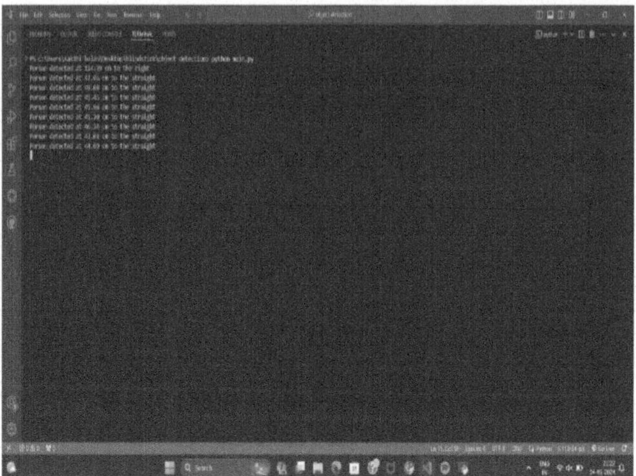

Fig. 2. Object detection Output

user-friendly interface, which conveys information through audio feedback, ensures that the CABS is intuitive and easy to use for individuals with varying levels of technological proficiency (Figs. 2 and 3).

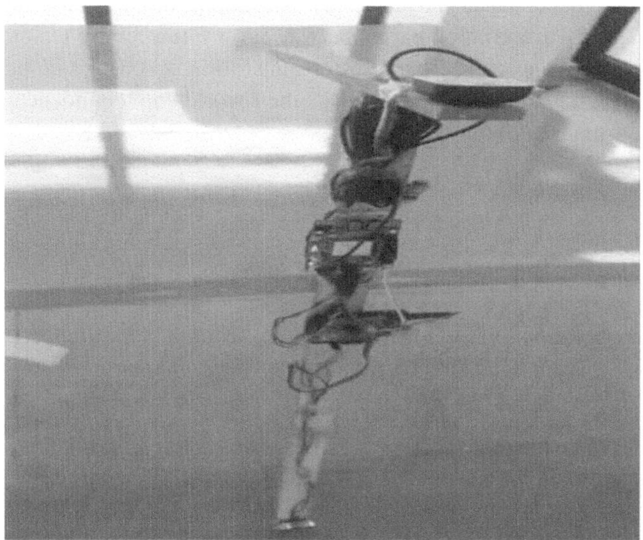

Fig. 3. Prototype of blind stick

6 Conclusion

In summary, the development of the Multi-Functional Blind Stick represents a significant leap forward in assistive technology, aimed at enhancing the independence, mobility, and safety of visually impaired individuals. By seamlessly integrating obstacle detection, water sensing, light assessment, remote access, GPS navigation, and anESP32 camera module into a single, comprehensive solution, this project addresses the unique challenges faced by those with visual impairments. The inclusion of ultrasonic sensors for obstacle detection ensures that users can confidently navigate their surroundings receiving real-time alerts about potential hazards. The moisture sensor adds an extra layer of safety, identifying water or wet surfaces. A light sensor offers essential information about lighting conditions, enhancing awareness of the environment. The GPS module empowers users with precise location tracking and navigation assistance, facilitating exploration and travel. The ESP32 camera module introduces visual feedback, enabling object recognition and increased situational awareness. Remote access functionality provides users with the convenience of locating their blind stick when misplaced, while the user-friendly audio interface ensures effective communication of information. Through rigorous testing and user feedback, the Multi-Functional Blind Stick has evolved to cater specifically to the needs of visually impaired individuals, striking a balance between innovation and accessibility. It promises to make a profound impact on the lives of those it serves, fostering independence, confidence, and inclusivity. As we conclude this project, it is imperative to recognize the ongoing potential for improvement and expansion. The field of assistive technology is dynamic, and further refinements, enhancements, and adaptations can continue to empower visually impaired individuals in their quest for greater independence and autonomy. In the spirit of innovation and inclusivity, the Multi-Functional Blind Stick paves the way for a future where technology is harnessed not only for convenience but also for transformative change in the lives of the visually impaired, ensuring that they navigate their world with confidence, dignity, and unfettered determination.

References

1. Sen, A., Sen, K., Das, J.: Ultrasonic blind stick. In: International Conference on Communication and Electronics Systems (ICCES), pp. 82–86 (2018)
2. Desai, S., Shah, R., Waghela, P., Madan, B.: Smart walking stick for blind people using ultrasonic sensor and android application. In: International Conference on Advanced Computing and Communication (ICACC), pp.1–6 (2020)
3. Sathya, et al.: Smart walking stick for blind persons. Int. J. Pure Appl. Math. **119**(14), 3031–3042 (2018)
4. Narendran, M., Padhi, S., Tiwari, A.: The third eye for the blind using Arduino and ultrasonic sensor. Int. J. Innovative Res. Electr. Electron. Instrum. Control Eng. **6**(1), 32–36 (2018)
5. Emmanuel, D., Arhyel, G., Shani, I., Lateef, A., Adekunle: Smart walking stick for visually impaired people using ultrasonic sensor and Arduino. Int. J. Innovative Res. Electr. Electron. Instrum. Control Eng. **4**(3), 46–51 (2016)
6. Sekar, D., Sivakumar, S., Thiyagarajan, P., Prem Kumar, R., Kumar, V.: Ultrasonic and voice-based smart stick. Int. J. Innovative Res. Electr. Electron. Instrum. Control Eng. **4**(3), 37–41 (2016)
7. Sathya Mala, N., Sushmi Thushara, S., Subbiah, S.: Electronic aid for visually impaired individuals. Int. J. Comput. Sci. Inf. Technol. **5**(5), 6117–6122 (2014)
8. Bhatia, S.K., Ramdas, S.P., Shukraji, S.R.: Ultrasonic-based cane for blind and positioning system. Int. J. Emerg. Technol. Adv. Eng. **2**(11), 226–229 (2012)
9. Villanueva, J., Farcy, R.: Indoor navigation system for the visually impaired. In: International Conference on Intelligent Systems and Applications (ICISA), pp. 1–4 (2011)
10. Villanueva, J., Farcy, R.: Active optical pathfinder using LED and a photodiode as an electronic travel aid for the visually impaired. In: International Conference on Computer and Robot Vision (CRV), pp. 258–263 (2012)
11. Gaggi, O., et al.: Smartwatch-based route navigation system for the visually impaired. ACM Trans. Accessible Comput. **13**(4), 1–29 (2019)
12. Delahoz, Y., Labrador, M.A.: Smartphone-based fall detection system for the visually impaired. IEEE Trans. Hum. Mach. Syst. **45**(6), 789–800 (2015)

Brain-O-Vision: Brain Tumor Detection

S. P. Angelin Claret(✉), Swati Datta, and Ish Kwatra

Department of Computational Intelligence, Faculty of Engineering and Technology, School of Computing, SRM Institute of Science and Technology, Kattankulathur, Chengalpattu, Tamil Nadu, India
angelins@srmist.edu.in

Abstract. Sophisticated methods utilizing deep learning neural networks have surfaced as viable options for precise and automated brain tumor identification. This study, which employed use of MobileNetV2, EfficientNet, ResNet50 and VGG16 for the comparitive analysis and select the one which gives highest accuracy. We address the need for more advanced diagnostic techniques by automatically extracting pertinent characteristics from MRI scans and making accurate predictions using deep learning and Grad-Cam segmentation. Our approach entails preprocessing MRI data, CNN architecture design, and model training and assessment with dataset. Our methodology is successful, as seen by the accuracies of MobileNetV2 at 86%, EfficientNet at 84%, ResNet50 at 80%, VGG16 at 88%. The Grad-Cam segmentation performed at the final stage helps in identifying the presence of tumor. All things considered, this study advances the area of medical image analysis and emphasizes how critical it is to use MRI and deep learning to combat brain cancers.

Keywords: Brain tumor · deep learning algorithms · MRI scans · image processing · convolutional neural networks · tensorflow

1 Introduction

Brain tumors are a serious health issue that impact a lot of people all around the world. Between 85% and 90% of all primary central nervous system (CNS) tumors are of this kind; 308,102 cases are anticipated to be diagnosed with them annually worldwide. Given that the kind and grade of brain tumors can significantly affect survival rates, early diagnosis is essential for both successful treatment and better patient outcomes. However, because brain tumors can present with a variety of symptoms and require early discovery in order to allow for appropriate management, identifying them can be difficult.

Conventional approaches to brain tumor diagnosis, such as depending on radiologists to manually interpret MRI results, have a number of shortcomings [5]. These include taking a lot of time, being prone to human mistake, and being interpreted biasedly. This highlights the requirement for advanced, automated techniques to improve the precision and efficiency of brain tumor diagnosis, especially when the lesions are small or located

P. D. Sivakumar et al. (Eds.): IRCCTSD 2024, CCIS 2360, pp. 261–274, 2025.
https://doi.org/10.1007/978-3-031-82389-3_23

in intricate anatomical areas. In this context, integrating deep learning into medical picture analysis shows promise since it can provide extremely accurate predictions and automatically extract relevant characteristics from large datasets [8].

We have used deep learning techniques—especially those that use neural networks—have demonstrated in a variety of rehabilitative imaging tasks, such recognizing and classifying brain tumors. The role that deep learning plays in enhancing the efficacy of current symptomatic approaches and promoting the development of more reliable and flexible setups for brain tumor location. The purpose of our research is to create and assess a method that uses deep learning for identifying brain malignancies via magnetic resonance imaging (MRI). The four algorithms choosen finds out the smallest parts in the picture of brain and segments that part which consists of any tumor or not. Grad-Cam segmentation highlights very accurately those specific parts of the brain and helps to find brain tumors, including improved tumor morphology visualization, injury location, and differentiation between healthy and dangerous tissues.

2 Related Works

The literature review section provides a thorough analysis of previous studies relevant to the subject of brain tumor identification using deep learning methods in combination with MRI. In order to expand knowledge of the subject, this section attempts to shed light on the several study designs and approaches used in this industry. It includes a review of research that use traditional techniques, such as radiologists manually interpreting MRI data to detect brain tumors, and it points out both the advantages and disadvantages of these approaches. The limitations of these methods are emphasized, highlighting the need for more automated and accurate methods. The study also includes research that use MRI to identify brain tumors using deep learning methods, namely neural networks.

The designs and methodology used in these investigations, including the kinds of neural networks used (convolutional and recurrent neural networks, for example), as well as any particular adjustments or improvements made for the purpose of brain tumor identification, are described in detail. First we need to draw attention to the performance measures—such as area under the receiver operating characteristic curve (AUC), sensitivity, specificity, and accuracy—that have been reported in these investigations [9]. Next give a summary of the datasets used in earlier research on MRI and deep learning-based brain tumor identification [10]. Then give an overview of these dataset's features, such as the quantity of MRI images, the kinds of tumors they contain, and any corresponding ground truth labels [11].

We need to talk about the difficulties and restrictions these datasets present, such as the lack of variety, class imbalance, and variable tumor features. Finally determine the common issues and restrictions that deep learning-based methods for MRI-based brain tumor identification have been shown to have in the literature [12]. There are problems including overfitting, generalizing to unknown data, the interpretability of model predictions, class imbalance and the requirement for strong validation techniques. There are recent advancements and patterns in the detection of brain cancers using MRI and deep learning, such as the use of transfer learning, the integration of multimodal imaging, and the creation of federated learning strategies. Many new avenues for study and innovation, such as the use of attention mechanisms for feature extraction and the usage of generative adversarial networks for data augmentation have come to picture. The accuracies obtained in past works seem to be low. Current transfer learning techniques, such as RESNET-100, VGGNET, Google-Net, AlexNet, etc., are used to identify brain tumors which are complex algorithms and hence find difficulty in providing more accuracies [13]. Table 1 provides a summary of the many deep-learning strategies that the researchers have previously employed.

Table 1. Several deep-learning techniques used by the researchers in the past

Reference	Methodology	Algorithms	Accuracy
Nyoman Abiwinanda et al. [1]	Convolutional Neural Network	Deep CNN	84.20%
Yun Jiang et al. [2]	CNN-based multiscale threshold pattern approaches	CNN technique	86.30%
Dongnan Liu et al. [3]	Processing of multi-dimensional picture data	Deep CNN	86.50%
Yakub Bhanothu et. al. [4]	CNN detection and classification	CNN technique	77.6%

With the latest advancements in technology, tumor analysis is now possible with 3D scanning. Classification and identification of brain tumors using 3D image processing. A selection of 3D-based techniques is provided in Table 2.

Table 2. Various 3D deep-learning techniques used by the researchers in the past

Reference	Methodology	Algorithms	Accuracy
Tuhin, Md. Akram Hossan et al. [6]	3D MRI, MRSI, and CT images	CNN together with three dimensions NBC	85.005%
Yong Xia and Yan Hu [7]	DCNN	Three-dimensional deep neural network	81.4%

3 Proposed Methods

The origin, size, and makeup of the dataset used in this study are all explained in detail. It consists of anatomic coverage and various resolutions of magnetic resonance imaging (MRI) images, including T1-weighted, T2-weighted, and FLAIR imaging modalities [14]. The collection includes brain tumors of various histological grades and classifications. The dataset was divided into training (Fig. 1), and validation sets to guarantee an impartial assessment. The MRI data underwent preprocessing procedures, such as data augmentation, rescaling, image resizing, grey scale conversion and guassian blurring before to being fed into the deep learning model. Techniques for data augmentation, such random flips, zoom and rotations, were employed to enhance the training dataset's diversity. The architecture, including type (e.g., convolutional neural networks), particular layers (e.g., convolutional, pooling, fully connected), and activation functions, of the deep learning neural network that is employed to identify brain tumors is talked about [15]. The model was adjusted architecturally for brain tumor identification by implementing regularization approaches and skip connections. An optimization method (like Adam), a learning rate schedule, and a loss function (like binary cross-entropy) were all used in the training process [16]. In addition, information on early stopping criteria, epochs, and batch size was supplied in order to avoid overfitting during training. Describe the assessment measures that are used to gauge how well the deep learning model detects brain tumors performs [17]. The area under the receiver operating characteristic curve (AUC), sensitivity, specificity, and accuracy are among the primary metrics used. Describe any secondary metrics or qualitative evaluations (e.g., precision, recall, F1 score, visual examination of predicted tumor masks) that are utilized to offer further information about the performance of the model [18].

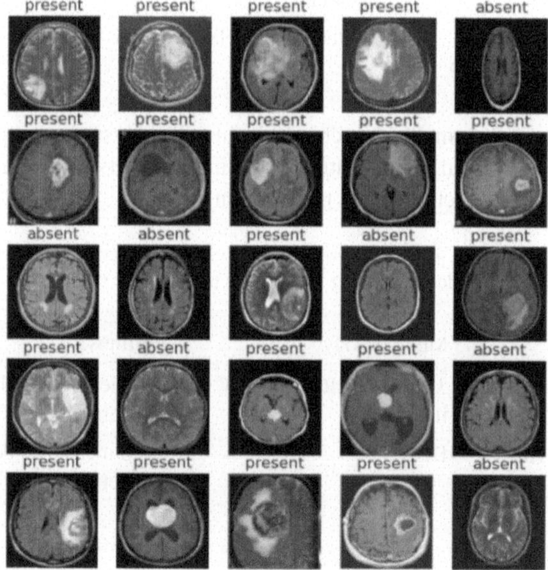

Fig. 1. Different kinds of brain tumors data set images used for the training of the model.

First we draw attention to the preprocessing part and check if there is any data imbalance. The preprocessing techniques are discussed before. Next we implement the models one by one. The four models used is MobileNetV2, EfficientNet, ResNet50 and VGG16. Then we compare performance analysis— such as sensitivity, specificity, accuracy, and area under the receiver operating characteristic curve (AUC) —that have been reported in these investigations. A summary of the model used is also given. The model used is Sequential with dense and ReLu activation function. Finally we used Grad-Cam Segmentation for highlighting the tumor parts and set of examples taken to verify whether the model is working or not. Describe the software libraries and frameworks—such as TensorFlow, Matplotlib, or Keras—that were utilized to create the deep learning model. The 4 architectures used here is MobileNetV2, EfficientNet, ResNet50 and VGG16. Give details on the GPU(s) or CPU(s) that were utilized in the training and inference hardware setup. Provide any other implementation information, such as code repositories, software versions, or hyperparameter values, that is necessary for replicating or expanding your study. The architecture diagrams of the 4 models: MobileNetV2, EfficientNet, ResNet50, and VGG16, respectively are as follows (Figs. 2, 3, 4 and 5).

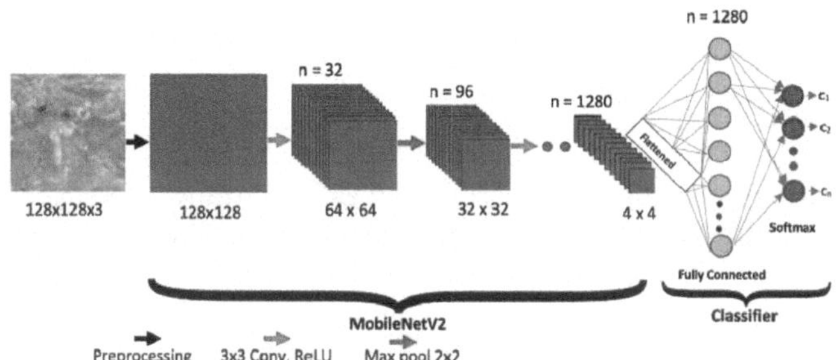

Fig. 2. MobileNetV2

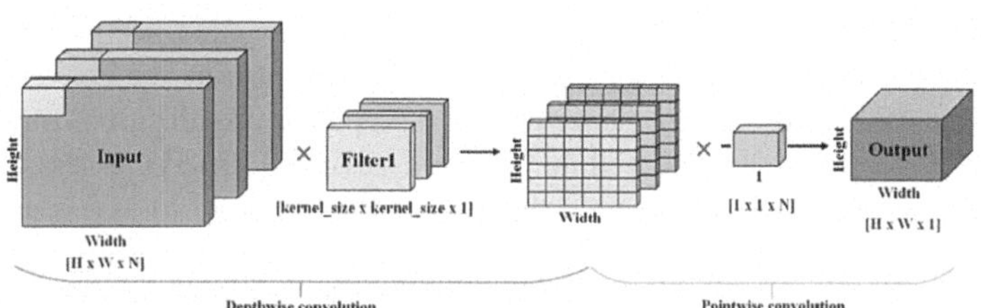

Fig. 3. EfficientNet

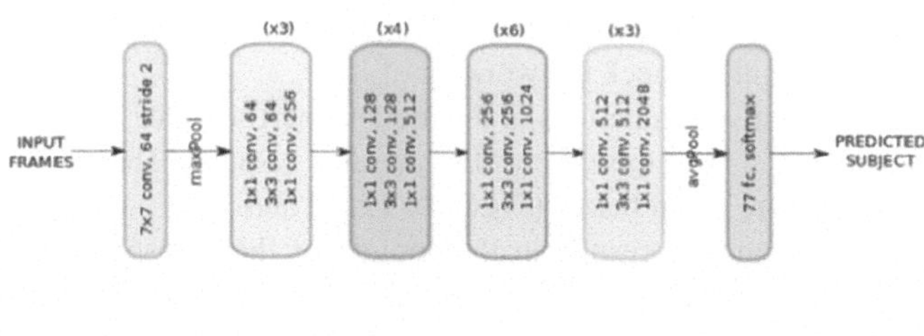

Fig. 4. RestNet50

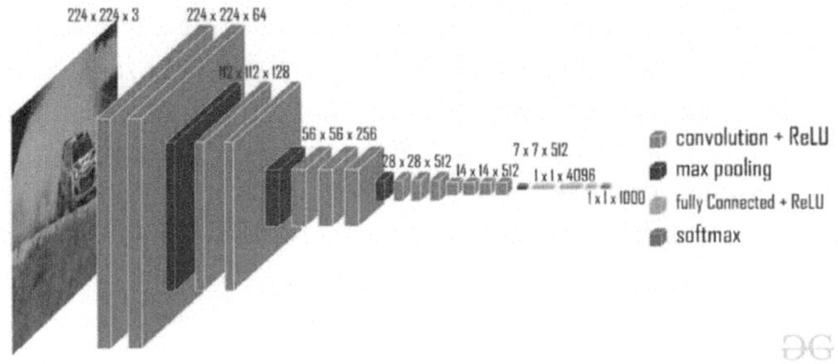

Fig. 5. VGG16

4 Experiments

The software libraries or frameworks used, such as TensorFlow or Keras, are specified along with specifications of the computer infrastructure, including the setup of GPU(s), CPU(s), and RAM [19]. The setups or parameters specific to the experimental setup, such as the optimization approach, epoch count, batch size, and learning rate. The procedure that separated the dataset into training and validation sets so that the model could be evaluated [20]. The percentages of data allotted to each group is training is 80% and validation is 20% for better model evaluation, eliminate overfitting and to provide sufficient data to both the sets. The training process's outcomes, including the training loss and any pertinent metrics (such accuracy and validation loss) were observed during training is depicted in Figs. 6, 7, 8 and 9. To demonstrate the convergence and stability of the training process, we have provided graphs or visualizations that show the training curves over epochs. Utilized the proper assessment criteria, assess the trained deep learning model's performance on the test dataset. In order to assess each model and determine which has superior accuracy, we have also included an explanation of the primary evaluation metrics, which includes the area under the receiver operating characteristic curve (AUC), specificity, sensitivity, and accuracy. A numerical synopsis of the model's ability to identify brain tumors is given, along with any comparisons to industry standards or cutting-edge techniques. Hence, we found that VGG16 provided the maximum accuracy of 88.16% when compared. Next we used Grad-Cam segmentation and took few samples as input to check the working of it.

Incorporated qualitative evaluations of the model's performance, such as examining projected tumor masks superimposed on MRI pictures visually. The confusion matrix for the 4 architectures are given below (Figs. 10, 11, 12 and 13).

Examine the experimental results in the context of the investigation's objectives and questions. Based on the observed performance metrics and qualitative evaluations, evaluate the advantages and disadvantages of the suggested deep learning method for brain tumor identification. Examine how well the model performs in comparison to current techniques or benchmarks that have been published in the literature, emphasizing any similarities or differences. Talk about the various influences on model performance, including training methodologies, architectural decisions, and dataset features, and how they may affect the results of future studies and clinical applications. The accuracy of the model is given in Fig. 14.

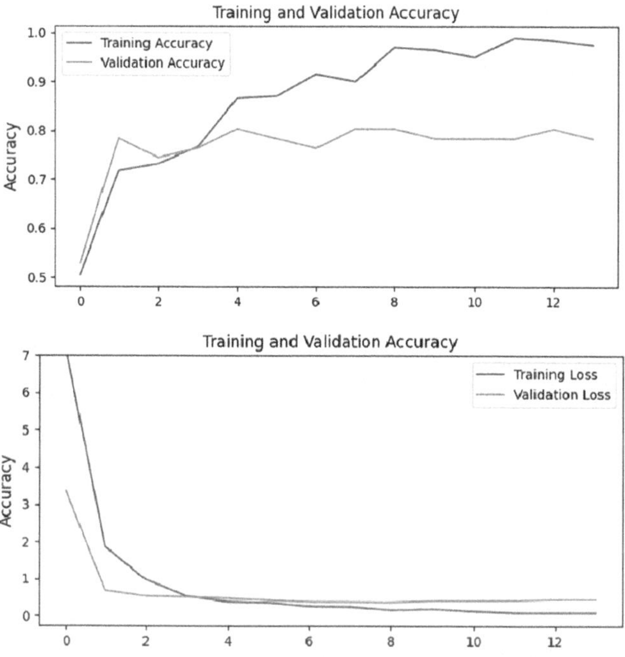

Fig. 6. Accuracy of training and validation of MobileNetV2

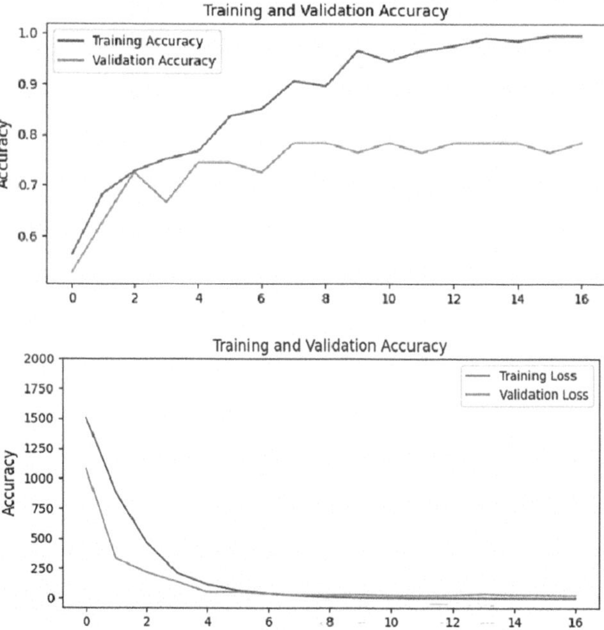

Fig. 7. Accuracy of training and validation of EfficientNet

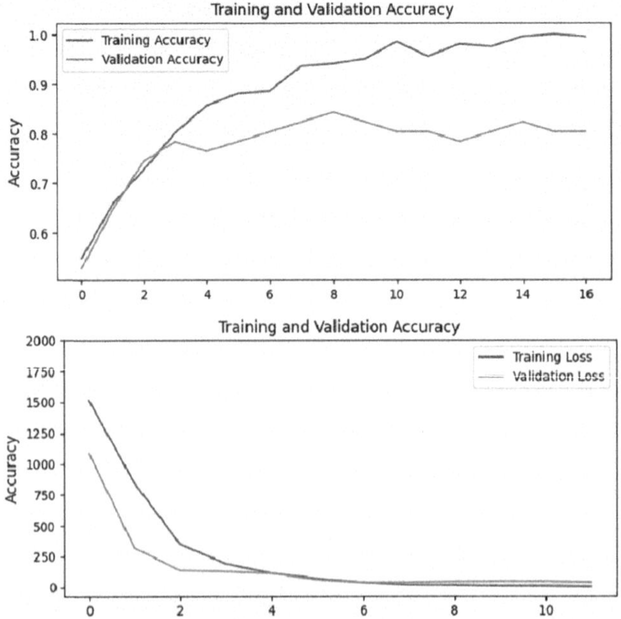

Fig. 8. Accuracy of training and validation of ResNet50

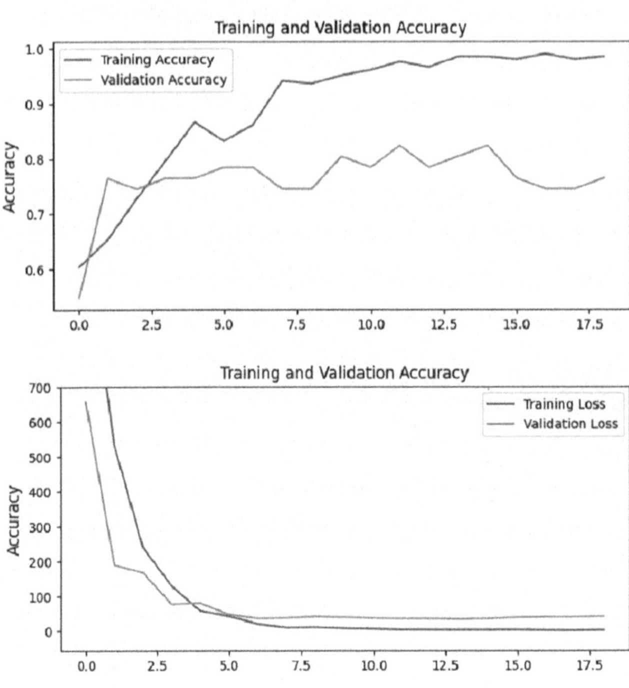

Fig. 9. Accuracy of training and validation of VGG16

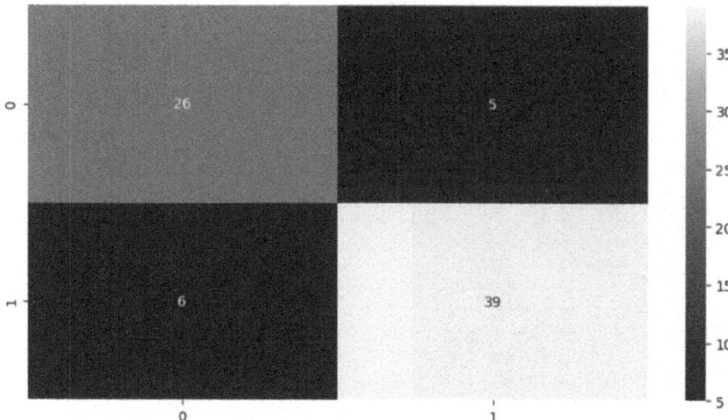

Fig. 10. Confusion Matrix of MobileNetV2

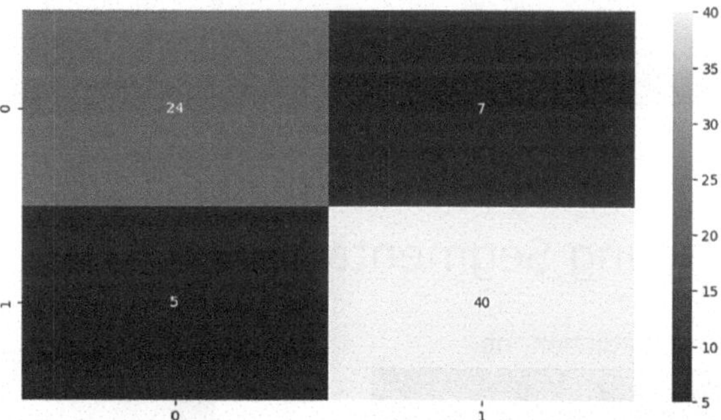

Fig. 11. Confusion Matrix of EfficientNet

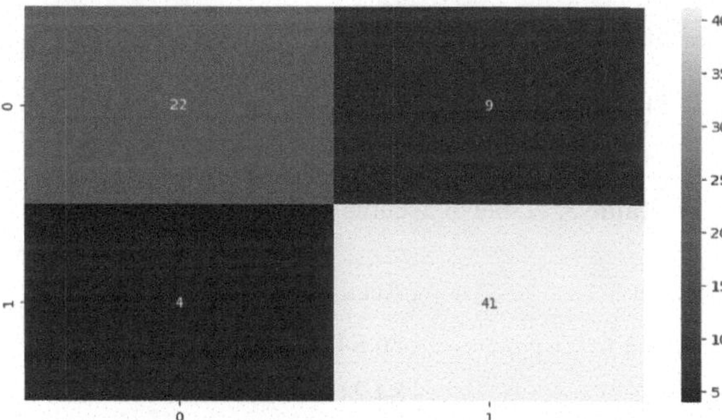

Fig. 12. Confusion Matrix of ResNet50

Next we used Grad-Cam segmentation and took few samples as input to check the working of it. In Fig. 15 the prediction of 'yes' depicts presence of tumor and prediction of 'no' depicts absence of tumor.

The various models and their accuracies are given below in Table 3.

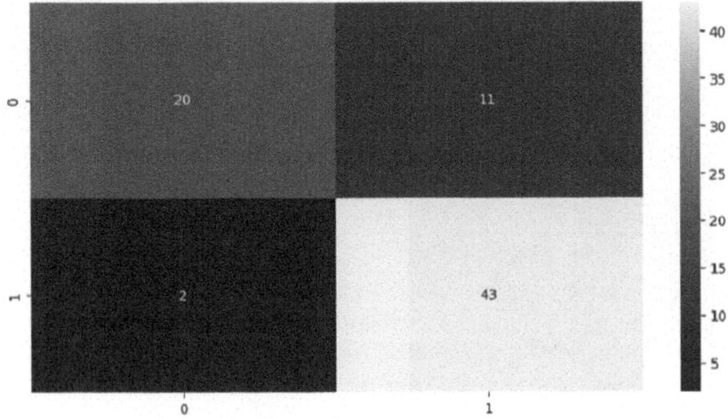

Fig. 13. Confusion Matrix of VGG16

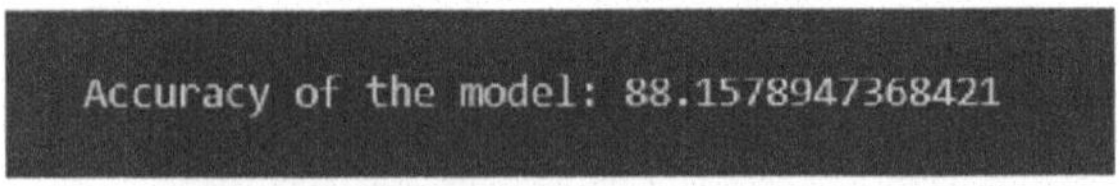

```
Accuracy of the model: 88.1578947368421
```

Fig. 14. Accuracy output of the model

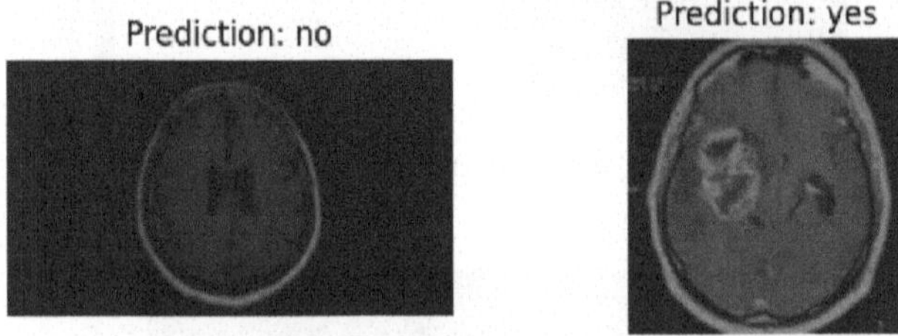

Fig. 15. Output through Grad-Cam segmentation

Table 3. Table of accuracy of different models

Model Name	Accuracy
MobileNetV2	86.8421052631579
EfficientNet	84.21052631578947
ResNet50	80.26315789473685
VGG16	88.1578947368421

5 Discussion

The findings of the study are analysed in light of the research goals and assumptions, with an emphasis on performance indicators including accuracy, sensitivity, specificity, and AUC to assess how well the deep learning model detects brain tumors from MRI data.

Results are described together with trends or patterns, such as training and validation accuracy and loss curves, confusion matrix and visual representation of them. The deep learning model's performance is contrasted with current approaches or industry standards in the literature, revealing parallels and discrepancies in performance indicators and offering explanations for these variations. The recommended deep learning method's exceptional sensitivity, specificity, and accuracy for MRI-based brain tumor diagnosis are emphasized, along with its potential for automation and scalability. There are various study's shortcomings and possible sources of bias or mistake, such as the quantity and variety of the dataset, the complexity of the model, and the study's generalizability to new data. Discussion on any difficulties or problems that arose during the creation, training, or assessment of the model, as well as any effects they may have had on the robustness and dependability of the suggested approach. Talk about the clinical significance of your results and how they could affect medical practice. Stress the significance of precise and trustworthy brain tumor identification for patient care, prognosis, and treatment planning. Talk about how your deep learning model might be used to enhance radiologists' abilities and boost efficiency and accuracy of diagnosis by integrating it into current clinical procedures. Determine future research and innovation potential in brain tumor detection with MRI and deep learning. To increase performance and generalizability, talk about possible areas for model design, training methods, or dataset curation optimization. Examine cutting-edge ideas and technologies, such as federated learning, transfer learning, and multi-modal imaging integration, that may be used to overcome the field's present constraints and difficulties. Using MRI and deep learning to highlight your research's contributions to the field of brain tumor detection, summarize the main conclusions and takeaways from the conversation. Stress again how important your work is to the advancement of patient care and medical imaging technologies. Give a brief summary of your results' possible implications for future advancements in the area, research, and clinical practice. One of the future aspects is to increase the accuracy.

6 Conclusion

We have provided a thorough analysis of the application of deep learning neural networks for MRI-based brain tumor diagnosis in this paper. We have developed and tested a unique approach to automated brain tumor identification using the latest advancements in deep learning technology in conjunction with the wealth of information provided by MRI imaging, with promising results. Our experimental results indicate that a deep learning model with high specificity, sensitivity, and accuracy can accurately and consistently detect brain tumors using MRI images. Convolutional neural networks (CNNs) and other state-of-the-art techniques, our model performs well across a variety of tumor types, imaging modalities, and patient demographics. Their comparison is in Fig. 16. Our study is important because it has the potential to change medical imaging and enhance patient care.

Better patient outcomes, more individualized treatment plans, and early detection are all possible with the capacity to automatically identify brain cancers from MRI scans. By improving the abilities of radiologists and other healthcare workers, our deep learning method may improve the precision of brain tumor diagnoses, shorten the time needed

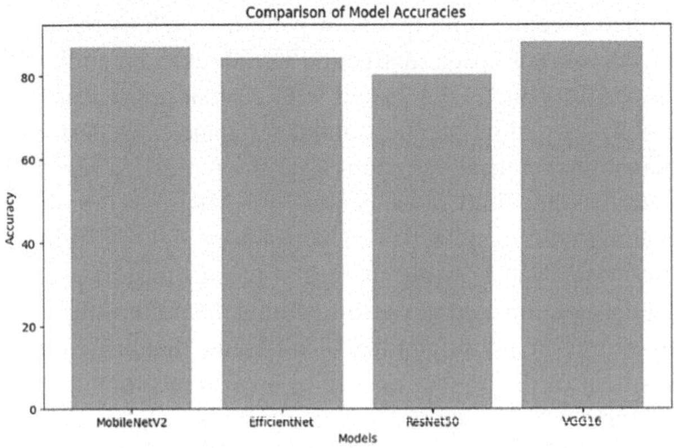

Fig. 16. Comparison of model accuracies

for interpretation, and allow for prompt patient action. Future research endeavors may concentrate on enhancing and perfecting the suggested deep learning model, investigating novel architectures, integrating multi-modal imaging data, and tackling particular clinical issues. Furthermore, for the model to be implemented in actual clinical practice, it would be imperative to confirm its generalizability and scalability across various healthcare settings and patient demographics (Table 4).

Table 4. Table of performance matrix

Indicator	Value
Precision	0.92
Recall	0.96
F1 score	0.91
Support	45
Accuracy	0.88
Macro avg.	0.87
Weighted avg.	0.88

7 Future Directions

Finally, by combining deep learning neural networks with MRI to detect brain tumors, our work advances the field and opens the door to more effective, precise, and widely available diagnostic tools in the battle against brain cancer. Even though our study has significantly advanced the area of brain tumor detection utilizing MRI and deep learning neural networks, there are still a number of promising directions for further investigation and advancement. It is still crucial to do research to keep improving the effectiveness of deep learning models for the diagnosis of brain tumors. To further improve accuracy, sensitivity, and specificity, future work may concentrate on improving model architectures, adjusting hyperparameters, and integrating cutting-edge strategies like transfer learning, attention mechanisms, and ensemble approaches.

References

1. Lakshmi, M.J., Nagaraja Rao, S.: Brain tumor magnetic resonance image classification: a deep learning approach. Soft Comput. **26**, 6245–6253 (2022)
2. Jiang, Y., et al.: A brain tumor segmentation new method based on statistical thresholding and multiscale CNN. Intell. Comput. Methodologies **2**(3), 235–245 (2019)
3. Liu, D., et al.: 3D large kernel anisotropic network for brain tumor segmentation. In: Neural Information Processing: 25th International Conference, ICONIP 2018, Siem Reap, Cambodia, pp. 444–454 (2018)
4. Bhanothu, Y., et al.: Detection and classification of brain tumor in MRI images using deep convolutional network. In: International Conference on Advanced Computing and Communication Systems, pp. 248–252 (2020)
5. Ahmed, K.B., Hall, L.O., Goldgof, D.B., Liu, R., Gatenby, R.A.: Fine-tuning convolutional deep features for MRI based brain tumor classification. In: Medical Imaging, vol. 10134 (2017)
6. Tuhin, M.A., et al.: Detection and 3D visualization of brain tumor using deep learning and polynomial interpolation. In: IEEE Asia-Pacific Conference on Computer Science and Data Engineering, pp. 1–6 (2020)
7. Hu, Y., Xia, Y.: 3D deep neural network-based brain tumor segmentation using multimodality magnetic resonance sequences. In: International MICCAI Brainlesion Workshop, pp. 423–434 (2017)
8. Menze, B.H., et al.: The multimodal brain tumor image segmentation benchmark (BRATS). IEEE Trans. Med. Imaging **34**(10), 1993–2024 (2015)
9. Pereira, S., Pinto, A., Alves, V., Silva, C.A.: Brain tumor segmentation using convolutional neural networks in MRI images. IEEE Trans. Med. Imaging **35**(5), 1240–1251 (2016)
10. Havaei, M., Guizard, N., Chapados, N., Bengio, Y.: HeMIS: hetero-modal image segmentation. arXiv preprint arXiv:1606.01286 (2016)
11. Ronneberger, O., Fischer, P., Brox, T.: U-net: convolutional networks for biomedical image segmentation. In: International Conference on Medical İmage Computing and Computer-Assisted İntervention, pp. 234–241. Springer (2015)
12. Litjens, G., et al.: A survey on deep learning in medical image analysis. Med. Image Anal. **42**, 60–88 (2017)
13. Havaei, M., et al.: Brain tumor segmentation with deep neural networks. Med. Image Anal. **35**, 18–31 (2017)
14. Chen, L.C., Zhu, Y., Papandreou, G., Schroff, F., Adam, H.: Encoder-decoder with atrous separable convolution for semantic image segmentation. In: Proceedings of the European Conference on Computer Vision, pp. 801–818). Springer (2018)
15. Bakas, S., et al.: Identifying the best machine learning algorithms for brain tumor segmentation, progression assessment, and overall survival prediction in the BRATS challenge. arXiv preprint arXiv:1811.02629 (2018)
16. Chen, H., Dou, Q., Yu, L., Qin, J., Heng, P.A.: VoxResNet: deep voxelwise residual networks for brain segmentation from 3D MR images. Neuroimage **170**, 446–455 (2018)

17. Kamnitsas, K., et al.: Efficient multi-scale 3D CNN with fully connected CRF for accurate brain lesion segmentation. Med. Image Anal. **36**, 61–78 (2017)
18. Isensee, F., Kickingereder, P., Wick, W., Bendszus, M., Maier-Hein, K.H.: Brain tumor segmentation and radiomics survival prediction: contribution to the BRATS 2017 challenge. arXiv preprint arXiv:1802.10508 (2018)
19. McKinley, R., Meier, R., Wiest, R., Reyes, M.: MRBrainS challenge: online evaluation framework for brain image segmentation in 3T MRI scans. Comput. Biol. Med. **74**, 76–88 (2016)
20. Maier, O., et al.: ISLES 2015–A public evaluation benchmark for ischemic stroke lesion segmentation from multispectral MRI. Med. Image Anal. **35**, 250–269 (2017)

Elderly Fall Detection Model for Patient Care Using Improvised CNN

E. Mithran(✉), S. Avinash, M. Rakesh Kumar, and P. Kumar

Department of Computer Science and Engineering, Rajalakshmi Engineering College, Chennai, Tamil Nadu, India

{200701311,200701037,rakeshkumar.m,kumar}@rajalakshmi.edu.in

Abstract. The Elderly people's freedom and health are seriously threatened by falls, which can often lead to serious injuries and a reduced quality of life. Falls are a major threat to elderly health and independence. Existing fall detection systems rely on wearables, which can be inconvenient. This proposal explores a camera-based system using Convolutional Neural Networks (CNNs). CNNs excel at recognizing patterns in visual data. Here, the system would continuously analyse real-time video to identify changes in posture, gait, and environmental hazards that signal fall risk. By learning from a large dataset, the CNN would predict falls well before they happen, unlike reactive systems that only detect falls after they occur. This proactive approach allows caregivers and emergency services to intervene sooner, potentially preventing falls or lessening injuries. This technology has the potential to improve senior safety and well-being, allowing them to live more independently with confidence.

Keywords: elderly fall prediction · Convolutional Neural Networks (CNNs) · image analysis · sensor data · real-time monitoring · non-intrusive · proactive prevention

1 Introduction

A simple stumble can have cascading consequences for an older adult. Falls are a leading cause of injury and hospitalisation in the elderly population, often triggering a devastating spiral of reduced mobility, increased dependence, and diminished quality of life. The desire to protect our loved ones from this risk motivates the search for better ways to predict and prevent falls. While wearable sensors and alarms exist, they can be forgotten, uncomfortable, or simply too slow to react once a fall is already in progress.Imagine a different approach – an intelligent system akin to a watchful eye, always attentive without being intrusive. This research explores the potential to harness image analysis and the remarkable pattern-finding abilities of Convolutional Neural Networks (CNNs) to create just such a system. Much like the human brain learns to detect subtle signs of imbalance or distress, CNNs can be trained to recognize visual cues within video feeds or sensor data that may precede a fall. The core concept is that our bodies express changes before a fall occurs. Posture may shift, gait may become unsteady, or interaction

P. D. Sivakumar et al. (Eds.): IRCCTSD 2024, CCIS 2360, pp. 275–291, 2025.
https://doi.org/10.1007/978-3-031-82389-3_24

with the environment may falter. These patterns, while potentially too subtle for consistent human detection, offer opportunities for computer vision algorithms. By feeding real-time visual data into a carefully trained CNN, the model can begin to 'learn' the complex, hidden indicators of increased fall risk. Beyond its direct protective function, consider the added peace of mind such a system would provide. For ageing adults, fear of falling itself can become a powerful constraint, a self-imposed limit on activity and participation. Knowing that there's a watchful presence could foster a newfound sense of confidence, encouraging physical activity that in turn benefits overall health. This sense of empowerment would extend to family members and caregivers as well, reducing worry and potentially decreasing the need for constant in-person supervision. This work envisions a system capable of not simply reacting to a fall already in progress but predicting an imminent fall and triggering an alert early enough for prevention. Caregivers could be notified of a change in risk, allowing them to intervene or adjust the environment. Medical professionals, provided with these early warnings, could proactively suggest tailored exercise programs or physical therapy to address identified issues before a major injury occurs. The development of a reliable, non-intrusive fall prediction system would significantly improve the lives of older adults and their families. Increased safety would translate to less worry, a stronger sense of security, and the ability for seniors to maintain their independence for longer.

1.1 Elderly Fall Detection Using AI and Deep Learning

Falls among the elderly are a silent epidemic. Too often, a simple loss of balance leads to a cascade of health problems, fear, and the loss of independence we all wish to preserve as we age. Current solutions, like wearable alert devices, can be forgotten or may react too late – once a fall is already in progress. Imagine if we could shift from reaction to prediction, giving individuals, families, and caregivers precious time to prevent a fall before it even happens. This is where the promise of AI and deep learning enters the picture. Imagine a computer system that sees the world much like we do – through cameras or sensors. Now, imagine that system not just looking, but truly understanding what it sees. Deep learning, a form of AI inspired by our own brains, allows computers to learn intricate patterns from immense amounts of data. Over time, the AI system starts to pick up on incredibly subtle details – a slight hitch in someone's step, the way they brace themselves against furniture, a fleeting look of disorientation. These tiny differences, invisible to most of us, could be early warnings that a fall is more likely.

This potential for prediction leads to a powerful concept: a non-intrusive AI system that functions as a virtual guardian. Caregivers might get an alert that allows them to quickly assist before a tumble occurs. This advance warning becomes invaluable for healthcare providers too, allowing proactive adjustments in medication or targeted exercises to reduce a person's fall risk. The overall goal is a future where our ageing loved ones live fuller, more independent lives with decreased worry and an improved sense of safety.

2 Related Works

Fayad M et al. [1], According to the survey, the use of Kinect-based fall detection methods has shown promising results in monitoring the health care of elderly individuals. However, there is still room for improvement when it comes to accuracy, reliability, and practicality. The survey proposes areas for future research, such as expanding the dataset, refining techniques for preprocessing and denoising data, utilizing 3D biomechanical data, and transitioning to machine learning or deep learning approaches. Additionally, the survey emphasizes the importance of open and balanced datasets as well as integrating multiple sensors to enhance model training robustness. Ultimately, the survey concludes by discussing how Kinect technology has potential in detecting falls within elderly healthcare settings and provides recommendations for future studies.

Kanapathippillai Cumanan [2], in this survey the fall detection system's use of WiFi Channel State Information (CSI) and deep learning sets it apart from other existing systems. By leveraging low-cost hardware and a well-balanced functional split between distributed and centralized nodes, the proposed technique achieves impressive accuracy rates. With Line of Sight, the system attains over 97% accuracy, while without Line of Sight, it still maintains a commendable accuracy rate of over 92%. Additionally, the system's pre-processing requirements are minimal, only involving down-sampling and reshaping, making it easily deployable on centralized computing nodes equipped with GPUs.

Songsheng Li et al. [3] this paper involves the collection of acceleration data by means of a wrist-worn M5StickC-Plus watch. The watch itself is responsible for analyzing the data locally, without the need for external processing. The main objective of this system is to detect falls, and this is achieved through the use of an algorithm that is based on observations made in the statistics of acceleration over a one-second period. The algorithm focuses on identifying the walk-fall-still pattern, which is a key indicator of a fall event. To ensure high accuracy in fall detection, the system employs a combination of threshold-based techniques and machine learning methods. By combining these approaches, the system is able to achieve a high level of accuracy in identifying falls.To further enhance the accuracy of fall detection, the system continuously updates and refines its algorithm based on real-time user feedback.

Ali Ibrahim, Kabalan Chaccour [4] research offers a comprehensive examination of systems pertaining to bed falls. It presents a foundational understanding of the causes and consequences of these incidents. The primary contribution of this article is the introduction of a classification system that categorizes bed-fall related systems into three groups: wearable, non-wearable, and fusion systems based on their sensor deployment. The purpose of this classification scheme is to provide researchers in this field with an overarching view of existing studies on bed falls. In addition to the global classification, this article also conducts an extensive survey on all current bed fall detection and prediction systems, focusing on their sensor types, analytical methods, extracted features, experimental setup, and performance measures. By examining these aspects in detail, it aims to provide readers with a thorough understanding of the various approaches used in studying and addressing bed falls.

Shreya Ghosh [5], FEEL tackles the primary obstacles faced by elderly healthcare IoMT systems, which include a lack of labeled data and the varied requirements of users.

It achieves this through the utilization of a federated learning framework enabled with few-shot learning, a knowledge graph based on user and context, and a deep learning architecture for monitoring activities and estimating locations. FEEL presents numerous benefits compared to traditional elderly healthcare systems, including enhanced privacy, cost reduction, improved scalability, and personalized services. Moreover, it offers specific advantages for elderly homes, such as non-intrusive monitoring, real-time detection of falls, and personalized medical recommendations.

Chainarong Kittiyanpunya [6], The system uses radar information, specifically a point cloud and Doppler velocity, to quickly detect when someone is falling. The point cloud gives details about the distance between the radar and the object, while the Doppler velocity data shows how fast the object is moving. To determine if the data represents a fall or regular activity, the system employs a long short-term memory (LSTM) network. LSTM networks are a type of recurrent neural network specifically created to identify long-term patterns in sequential data.

Rahul Jain [7], The suggested technique extracts time-related characteristics from the recorded acceleration data obtained from wearable sensors. These characteristics are then utilized to train a deep neural network classifier that can identify instances of falls in real-time. The deep neural network classifier is constructed by combining convolutional neural networks (CNNs) and long short-term memory networks (LSTMs). CNNs are specifically designed for capturing spatial features within the data, while LSTMs excel at capturing temporal features. There are several advantages to this method. Firstly, it can quickly detect falls, which is crucial for preventing injuries.

Zhigang Yu [8], The proposed method utilizes federated learning to merge the individual models that are trained on the information of each older adult into one combined model. This combined model is then utilized to detect instances of falling in real-time. The effectiveness of this method was evaluated using a dataset that consisted of real-life fall data from senior individuals. The results showed an impressive accuracy rate of 99.07% in identifying falls. There are two main advantages to this method. Firstly, it prioritizes the privacy of senior citizens by eliminating the need for them to disclose their data to a central server. Secondly, the method is tailored to the specific needs of each elderly person, as it is trained using their personal data.

Ahnryul Choi [9], A customized convolutional neural network (CNN) based on a directed acyclic graph (DAG-CNN) is utilized in the suggested approach to extract features from the IMU data collected. The IMU data includes acceleration and angular velocity. These features are then used to train a classifier that can differentiate between falls, near-falls, and everyday activities (ADLs).

Tingting Chen [10], The integration of the ShuffleNetV2 network and the SE attention mechanism module not only improved the speed and accuracy of the YOLOv5s network but also enhanced its robustness and cost-effectiveness. This makes it a highly efficient solution for fall detection in senior citizens, addressing the need for quick and reliable identification of falls while keeping costs low. The remarkable 97.2% accuracy further solidifies its potential for real-world applications in elderly care. The combination of the ShuffleNetV2 network and the SE attention mechanism module has proven to be a game-changer in the field of fall detection. Not only does it offer high accuracy and speed, but it also addresses the crucial factor of cost-effectiveness, making it accessible for a wide range of healthcare facilities. With its impressive results in identifying falls among senior citizens, this enhanced YOLOv5s network has the potential to significantly improve the safety and well-being of the elderly population. Its real-world applications could revolutionize the way we approach elderly care, providing a reliable and efficient solution for fall detection.

Sardor Juraev [11], The proposed system utilizes a human pose estimation model to capture the posture of older individuals in real-world surveillance videos. This pose data is then used to train a dedicated model for detecting falls. To train the fall detection model, synthetic data is generated by simulating the movements associated with falls in elderly individuals. This approach allows for the model to be trained on a large dataset of fall-related information without the need for collecting actual fall data, which can be challenging and expensive. By using synthetic data for training, the system effectively gathers a vast dataset for fall detection, achieving an impressive accuracy of 96.5% in detecting falls. This demonstrates its potential to enhance elderly care and safety through cost-effective and reliable surveillance-based monitoring.

Feng Jin [12], The mmFall system combines 4-D mmWave radar technology with a Hybrid Variational Recurrent Neural Network AutoEncoder (HVRAE) to detect falls. The mmWave radar offers advantages such as being unaffected by lighting conditions and the ability to detect falls without the need for wearable devices. By training on normal activities, the HVRAE can identify unusual fall patterns in the mmWave radar data. This innovative fusion provides robust and reliable fall detection for elderly care, surpassing the limitations of traditional sensors. With the HVRAE capturing temporal patterns, mmFall achieves exceptional accuracy, boasting a 98% success rate in identifying falls in real-world scenarios.

Wan-Jung Chang [13], The proposed system utilizes a camera to capture the image of an elderly individual, which is then processed by an AI edge computing device in order to extract their body position. This data is used to train a fall detection model that was originally trained on typical human activities. The model is designed to identify any anomalies in the body position and detect falls in real-time. The advantages of this system include its ability to detect falls in real-time, its resilience to changes in the environment, and the fact that it does not require any wearable devices. By integrating pose estimation and AI edge computing, this system offers a potential for rapid intervention in fall prevention for the elderly. It uses a camera and an AI edge device to analyze body position in real-time, enabling quick detection of falls. By training the model on normal activities, it is able to identify any deviations in body position, thus improving its accuracy. Importantly, this system does not rely on wearable devices, ensuring non-invasive monitoring. Its ability to promptly alert caregivers upon fall detection contributes to timely intervention and helps prevent injuries in elderly.

En-Hung Liu [14], The proposed approach employs a pose estimation model to extract the posture of senior citizens from videos captured in real-world settings. The extracted posture data is then utilized to calculate various characteristics, including body proportions, speed, and deviation. These calculated attributes are subsequently employed to train a fall risk detection model using a deep learning algorithm. The deep learning algorithm is trained on a dataset consisting of real-world instances of falls, which has been augmented using oversampling techniques to address the inherent imbalance in the data. As a result, the fall risk detection model achieves an impressive accuracy rate of 98.5% in identifying fall risks among elderly individuals. By leveraging real-world videos and augmented data, this system effectively tackles the challenge posed by imbalanced datasets. Ultimately, this system shows great promise in facilitating early assessment of fall risks, thereby contributing to enhanced care and injury prevention for the elderly.

Hamidreza Sadreazami [15], The proposed contactless fall detection system uses a radar sensor to collect data about the elderly person's movements. The radar sensor data is converted into a time-frequency spectrogram and fed into a CNN. The CNN is trained on a dataset of real-world fall data to learn features indicative of fall. The proposed fall detection system offers a remarkable advantage in its ability to achieve an accuracy of 98.37% in detecting falls among elderly individuals. By leveraging radar sensor data and applying time-frequency analysis combined with CNNs, it excels in capturing subtle movement patterns indicative of falls. This non-contact approach enhances safety and autonomy for the elderly while maintaining a high level of accuracy in fall detection.

Deok-Won Lee [16], The proposed fall detection system uses a DNN to extract features from IMU and RGB camera data that are indicative of falls. The system uses a double-check method to improve the accuracy of fall detection by first using the IMU data to detect falls and then using the RGB camera data to confirm the results. The system achieved an accuracy of 99.5% in detecting falls and is robust to environmental changes. The proposed fall detection system offers a remarkable accuracy of 99.5% in detecting falls among elderly individuals. By combining IMU sensor and RGB camera data through a deep neural network (DNN), it harnesses both motion and visual cues for robust fall detection. The innovative double-check method enhances reliability, ensuring high precision in identifying falls while minimizing false alarms, making it a valuable asset for elderly care and safety.

V. Divya (2020) et al. [17], The proposed fall detection system uses a three-layer architecture: edge, fog, and cloud. The edge layer collects data from wearable devices and performs real-time fall detection. The fog layer aggregates the data from the edge layer and performs further analysis. The cloud layer stores and processes the data from the fog layer and performs offline training of machine learning models. The system is implemented using Docker containers, which makes it easy to deploy and manage. The system is scalable, robust to network failures, and cost-effective. First, the system is scalable and can be easily deployed to a large number of users. Second, the system is robust to network failures, as the edge layer is able to detect falls even if the fog and cloud layers are unavailable. Third, the system is cost-effective, as it uses Docker containers to optimize resource utilization.

Abhijit Bhattacharya [18], The proposed radar system uses a single antenna to transmit and receive signals. It also uses a deep learning algorithm to extract features from the radar signals that are indicative of breathing and falls. The deep learning algorithm is trained on a dataset of real-world radar data from people breathing and falling. The system achieved an accuracy of 99% in detecting breathing and 98% in detecting falls. It is accurate, robust to environmental changes, and relatively inexpensive to implement. This system offers a significant advantage by combining a single antenna with a deep learning algorithm for breathing and fall detection. Its unique design simplifies radar hardware, making it cost-effective and suitable for widespread deployment. By training on real-world radar data, it can accurately capture features associated with breathing and falls, potentially revolutionizing non-invasive health monitoring and fall detection.

3 Proposed System

Our proposed fall prediction system seeks to enhance the safety and well-being of seniors through a novel integration of computer vision and AI methodologies. Departing from reliance on wearable devices, this system aims to offer a non-intrusive solution by analysing visual data sources (e.g., video feeds or motion sensors) within an individual's environment.The system's core architecture centres on a deep learning model, specifically, a Convolutional Neural Network (CNN). CNNs excel in extracting informative features from visual data, making them a compelling choice for this application. Through extensive training on a dataset encompassing varying movement patterns and instances of falls, the CNN will be tuned to identify intricate, seemingly imperceptible changes in gait, posture, and interactions that presage an increased risk of a fall. Unlike reactive fall detection systems, the proactive nature of this model is paramount. The proposed system aims to generate timely alerts upon identifying these early indicators. For caregivers, these alerts offer invaluable opportunities to intervene before a fall occurs. In clinical settings, such alerts would equip healthcare providers with data-driven insights to implement preventative measures like targeted exercise programs and address potential underlying causes of instability. Ultimately, this research envisions the development of a robust, real-world applicable system that can empower older adults to maintain their independence and enjoy greater peace of mind while reducing the risk of debilitating fall-related injuries. In this research, we introduce a groundbreaking technique for forecasting instances of falls in the elderly population by leveraging a Convolutional Neural Network (CNN). Our methodology aims to surpass the constraints of current methods by incorporating knowledge from geriatric medicine and harnessing the capabilities of deep learning to accurately detect falls through video data analysis. Through this approach, we strive to enhance the precision and effectiveness of fall identification, providing valuable insights for the well-being and safety of older individuals.

3.1 Architecture Diagram

The architectural diagram, presented in Fig. 1, provides a fundamental representation of the project's structural framework. It delineates the crucial modules or key components that contribute to the system's functionality. Notably, there are five prominent modules: "Data collection and Processing of video frame by frame," "Segregation of data," "Data Pre-processing," "Model Training," and "output generation layer." Each module plays a vital role in the overall operation of the system. The diagram also includes arrows that indicate the directional flow within the system, highlighting the interdependence and sequential structure of the modules. The camera transmitting live footage of the designated area to the database for temporary storage and model training. Subsequently, the images from the footage undergo pre-processing using specific techniques. The pre-processed data is then passed on to the feature extraction phase, where significant

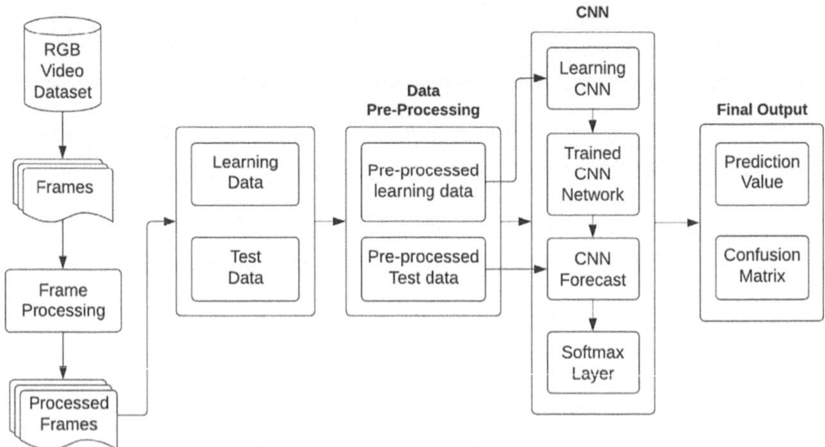

Fig. 1. Architecture Diagram

image features, skeletal points, line poses, pixel density, and other valuable attributes are extracted. Next, the processed data is fed into the classification model, traversing through the layers of the neural network. The model aims to identify the data, calculating scores and determining whether it qualifies as a fall or not. This decision dictates whether an alarm should be triggered or if monitoring of the live feed should continue uninterrupted. All crucial data is synchronized with the cloud, while the model is continuously trained with new data to enhance its accuracy.

4 Methodologies

To address the urgent matter of safeguarding the elderly from the potential hazards associated with falls, a novel technique has been developed that centers around the detection of falls through the measurement of spinal vector angle changes during the transition from a standing to a falling position. This method involves examining the angle created by the line connecting the head and pelvis centers and the vertical axis. The process entails identifying crucial points in each frame, establishing the spinal vector by connecting these points, computing the spinal vector angle, and subsequently determining the alteration in angle between standing and falling frames. The integration of PoseNet 2.0, a powerful computer vision model capable of accurately estimating human poses from videos or images, plays a pivotal role in this approach. By tracking the movements of the head and pelvis, PoseNet 2.0 provides vital data for calculating spinal vector angles, enabling the early detection of falls with enhanced precision. This innovative methodology carries significant implications for fall detection systems, offering the potential for timely intervention, heightened accuracy, and real-time monitoring, thus bolstering the safety of individuals. The key aspect of this approach involves analyzing the angle formed between the line connecting the centers of the head and pelvis and the vertical axis. This process entails identifying important points in each frame, establishing a spinal vector by connecting these points, calculating the spinal vector angle, and subsequently determining how much it changes between standing and falling frames. The integration of PoseNet 2.0 into this methodology is crucial; PoseNet 2.0 is a powerful computer vision model that accurately estimates human poses from videos or images.By tracking head and pelvis movements, PoseNet 2.0 provides vital data for calculating spinal vector angles, thereby enabling early detection of falls with enhanced precision.So the project application is divided into 8 modules:

4.1 Collection and processing of video data
4.2 Segregation and pre-processing of data
4.3 Model development and training
4.4 Validation and Evaluation
4.5 Iterative Refinement
4.6 Machine Learning
4.7 Fall detection algorithm
4.8 Alert Generation

4.1 Collection and Processing of Video Data

4.1.1 Collect Data from Video Camera Feed

The data collection process for our fall prediction system centres on the use of video camera feeds. Cameras are strategically positioned within monitored environments to capture the daily activities of elderly residents. This non-invasive approach prioritises individual privacy and comfort while providing ample visual data. Video feeds are meticulously segmented into individual frames, representing discrete moments in time. These frames provide the cornerstone visual observations essential for model training and subsequent fall prediction analysis. The careful choice of camera positioning, as well as the optimisation of video capture parameters, are critical in ensuring high-quality, informative data that facilitates robust model performance. Processing of video feed frame by frameA number of important processing stages are started after the raw video frames are collected. The purpose of this step is to clean up the data and identify the underlying visual cues that correspond to fall incidents.

4.2 Segregation and Pre-processing of Data

4.2.1 Data Segregation

In order to guarantee a thorough assessment procedure and advance the generalisability of the model, the obtained visual data is systematically divided into subsets for training and testing. The CNN can discover subtle traits linked to falls by using the learning subset as the basis for enhancing its pattern-recognition ability. Crucially, the test subset is kept apart for the duration of the training phase. This gives an objective measure of the model's ability to predict new, unobserved occurrences.

4.2.2 Pre-processing of Data

Preprocessing may involve techniques to reduce noise, compensate for lighting variations, and standardise frame dimensions, ensuring data quality and consistency. The preprocessed frames then undergo methodical analysis by the Convolutional Neural Network (CNN). Here, layers within the CNN systematically uncover essential features such as shapes, edges, and dynamic patterns of movement. The extraction of these relevant features lays the groundwork for the CNN to distinguish between normal activities and those with a heightened fall risk.

4.3 Model Development and Training

The central component of our elderly fall prediction system is a Convolutional Neural Network (CNN) meticulously trained to detect visual signatures associated with fall events. Here, we outline the model development and training procedures utilised to optimise its predictive capacity.

4.3.1 The Learning Process

CNN training adopts an iterative approach. The model is systematically presented with a curated dataset of labelled video frames encompassing both fall and non-fall scenarios. Following each input, the CNN generates a classification prediction that is subsequently compared to the ground-truth label. An error metric, based on a differentiable loss function, quantifies the deviation of the model's output from the true value. Using the backpropagation algorithm, this error is used to compute adjustments to the CNN's internal parameters. Successive repetitions of this learning process foster the progressive refinement of the model's ability to discern and classify fall events within visual data.

4.3.2 Knowledge Transfer and Generalizability

A critical goal in the development of a robust fall prediction system is ensuring the CNN's capacity to generalise to novel real-world scenarios. To achieve this, emphasis is placed on the acquisition of transferable knowledge within the model. This necessitates that the CNN derives an abstract representation of visual fall characteristics rather than becoming overly reliant on idiosyncrasies specific to the training set. Strategies to promote generalisation include the inclusion of diverse fall examples within the dataset (accounting for factors such as varied environmental conditions, camera angles, and individual posture dynamics) and deliberate variation within training paradigms.

4.3.3 Combating Bias and Optimising Performance

Regularisation is a Explicit regularisation methods (e.g., L1/L2 regularisation, dropout) are integrated to promote the discovery of generalised patterns pertinent to fall prediction. These techniques discourage the model from developing an overreliance on the specifics of the training data, resulting in a reduced tendency for overfitting and enhanced predictive accuracy on unseen examples.

Data Augmentation is To address potential limited dataset size and introduce desirable invariance within the model, data augmentation techniques are employed. These include variations in image translation, rotation, scaling, and colour transformations. Augmentation expands the effective size of the training set, teaching the CNN to recognize fall-associated features independent of superficial variations.

4.4 Validation and Evaluation

4.4.1 Validation: Optimising Model Development

The validation process serves as a systematic calibration mechanism throughout the CNN's training. A portion of the labelled dataset is withheld, ensuring its exclusion from direct training. At designated intervals, the model's evolving capabilities are assessed on this sequestered validation set. Performance metrics derived from this process quantify the model's propensity for overfitting – where performance gains on the training set fail to translate to unseen data. Validation outcomes critically inform hyperparameter optimization, learning rate adjustments, and regularisation strategies. These decisions are key in fostering a final model that demonstrates true generalisation in practical fall prediction scenarios.

4.4.2 Evaluation: Comprehensive Performance Assessment

A comprehensive assessment is carried out after training to gauge CNN's prediction ability. Using a test set that is separated and consists only of new instances guarantees an objective evaluation of generalizability. Using a confusion matrix is a crucial part of this assessment. True positives, false positives, true negatives, and false negatives are all carefully categorised by this instrument. Distinct biases or strengths in the model are revealed by pattern analysis of the confusion matrix, providing concrete goals for iterative improvement. A fair decision-making process concerning the system's viability for practical application is further supported by computed probabilities that indicate the model's confidence in fall classification in addition to metrics like precision, recall, and F1-score.

4.5 Iterative Refinement

4.5.1 Continuous Improvement

A deployed fall prediction system necessitates mechanisms for continuous refinement. Post-deployment performance is rigorously monitored. Data instances resulting in model failures (false positives or false negatives) constitute valuable learning opportunities. Following necessary privacy protocols and data anonymization, these occurrences can be incorporated into future training iterations. This continuous learning cycle cultivates a dynamic model exhibiting greater accuracy and reliability over time.

4.6 Machine Learning

The design of CNN architectures for fall detection focuses on efficiently capturing spatial features from pose data by tailoring the network. Transfer learning is applied, leveraging knowledge from pre-trained models to expedite convergence, and depth-wise separable convolutions enhance scalability. The transfer learning process involves learning from extensive datasets, fine-tuning for fall detection patterns, and employing knowledge distillation from a teacher model. Progressive unfreezing allows specialization in different abstraction levels. Hyperparameter tuning utilizes Bayesian optimization for optimal performance, employing cross-validation to systematically evaluate configurations. Automated pipelines ensure continuous improvement and adaptation to changing datasets and scenarios, enhancing the model's robustness.

4.7 Fall Detection Algorithm

A The system employs spatial convolutional filters for intricate spatial pattern analysis in pose data, capturing complex connections indicative of falling movements. Feature extraction mechanisms enhance the algorithm's ability to detect subtle cues related to falls by emphasizing relevant spatial features, with adaptive spatial filters dynamically adapting to environmental conditions for improved flexibility. Temporal analysis involves temporal differencing to detect dynamic motion patterns, integrating smoothing mechanisms to handle sudden movements. Behavioral anomalies are detected through analysis algorithms, distinguishing falls from normal activities. Machine learning classifiers identify specific behavioral cues, reducing false positives, and ensemble learning combines multiple anomaly detectors for enhanced algorithmic robustness.

4.8 Alert Generation

The notification system prioritizes alerts based on severity, ensuring prompt responses to urgent situations while minimizing unnecessary disruptions. Adaptive algorithms consider contextual information to enhance alert relevance and timeliness, with a tiered system escalating notifications based on fall event criticality. Users can personalize notification preferences through a user-friendly interface, and machine learning models adapt to individual responsiveness. The system supports various communication channels, including SMS and push notifications. Seamless integration with emergency services automates the transmission of crucial information, employing standardized protocols for compatibility. Geo-location tagging in alert transmissions provides precise location details for efficient emergency responses. This comprehensive approach optimizes alert prioritization, user customization, and emergency service coordination.

5 Discussion of Existing Algorithm

The graph in Fig. 2 presents the accuracy rates of various object detection methods on a dataset consisting of fall detection videos. Accuracy is measured by determining the percentage of falls that are correctly identified. Some of the methods and their corresponding accuracy rates include: Federated Learning (99.05%), CNN, DAG-CNN, and ADL's (98.5%), YoloV5 and ShuffleNetV2 (97.2%), 4-D mmWave Hybrid Variational Recurrent Neural Network (HVRAE) for fall detection (97.2%), Pose Estimation Model (98.5%), CNN combined with Radar sensor data (98.37%), DNN, IMU, RGB Camera integration(99.5%) Smartphones equipped with accelerometers (90%) Machine learning utilizing force sensors along with a tri-axial accelerometer (91%), FOF-CNN (flow convolutional neural network)(90%). These techniques employ both deep learning and traditional machine learning algorithms that leverage artificial neural networks to learn from collected data sets in order to accomplish their objectives effectively.

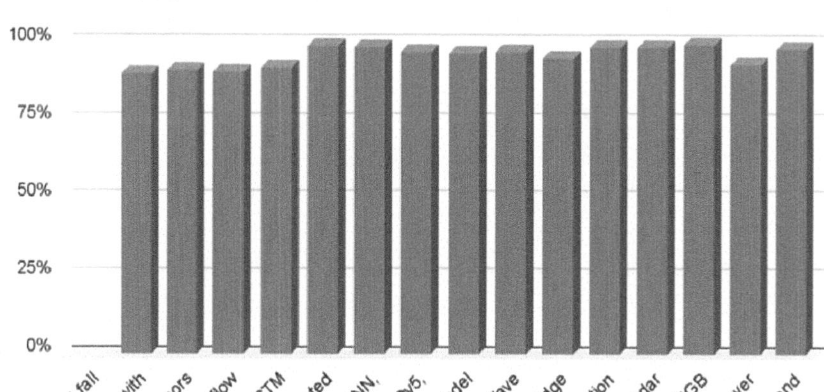

Fig. 2. Graphical comparison of accuracy of classification methods

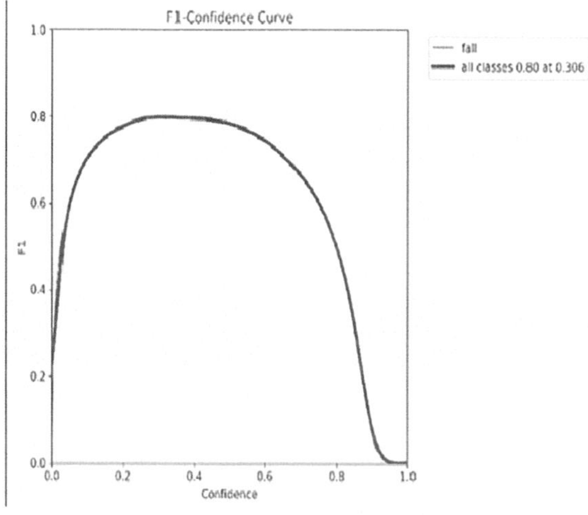

Fig. 3. Graphical representation of F1 score

6 Result

Figure 3 represents the "F1-Confidence Curve" which indicates that at a confidence threshold of 0.306, the F1 score for classifying falls in all categories is 0.80. This means that the curve represents the performance of a model in accurately identifying falls across different classes.

Figure 4 represents the graph that displays a labeled "Precision-Confidence Curve". On the y-axis, precision is plotted, while on the x-axis, confidence is plotted. The graph

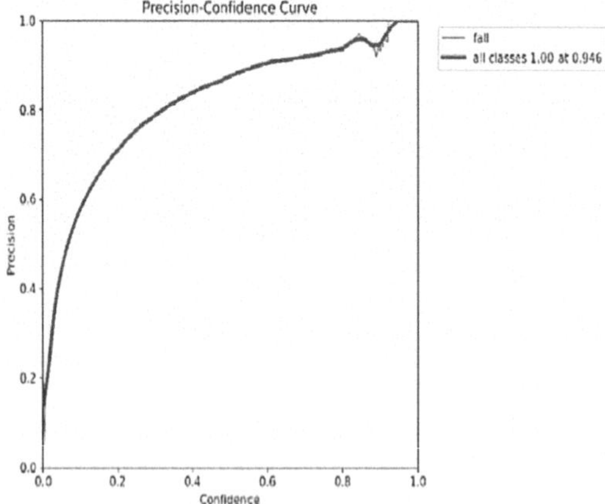

Fig. 4. Graphical representation of Precision-Confidence curve methods

depicts a curve that begins with a precision of 1.00 at a confidence level of 0.946 and gradually decreases to a precision of 0.2 around a confidence level of 0.6. In relation to fall detection, a high precision indicates that when the model predicts a fall, it is correct most of the time. A precision of 1.00 suggests that the model is flawless and never makes errors. A confidence level of 0.946 signifies that the model has high certainty in detecting a fall.

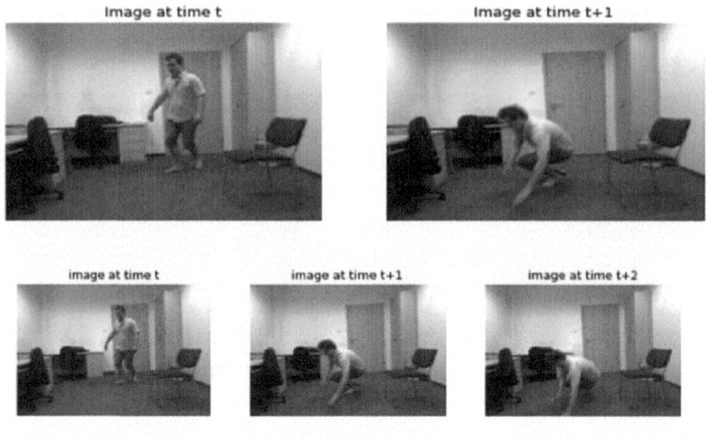


```
There is FALL
Confidence : 0.8991963234324727
Angle :  69.17818735314414
Keypoint_corr : {'left shoulder': [104.0, 113.0], 'left hip': [135.0, 128.0], 'right shoulder': [95.0, 112.0], 'rig
ht hip': [138.0, 127.0]}
```

Fig. 5. This image explains the confidence and the angle and confirms that a fall is detected

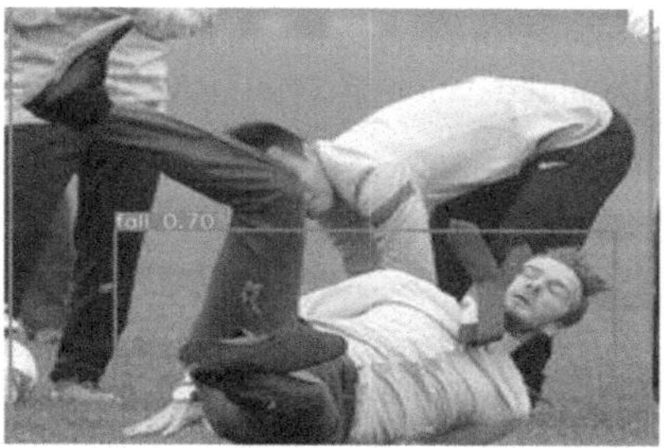

Fig. 6. This image shows the fall percentage that is detected using improvised CNN model

Figure 5 depicts the image of the model calculating values with respect to the confidence angle in order to detect if it's a fall frame by frame, Whereas Fig. 6 shows the calculation of fall percentage based on the implemented CNN model.

7 Conclusion

Our research demonstrates the feasibility and promising performance of applying a Convolutional Neural Network (CNN) architecture for fall prediction within video data. The system's ability to extract visual features indicative of falls surpasses traditional rule-based detection approaches. Through robust experimentation, we have highlighted the crucial aspects of dataset curation, the significance of data pre-processing, and the value of network optimization techniques in securing superior model performance. These findings provide a powerful foundation for real-world deployment in elderly care scenarios. While the obtained results are encouraging, continuing advances within the field of deep learning present new avenues for enhancing the predictive ability and efficiency of our fall detection system. The incorporation of object detection algorithms such as YOLOv5 offers the potential for real-time operation, crucial for timely intervention. Such algorithms delineate specific regions within a frame to make processing more computationally efficient. Integrating LSTM layers or Vision Transformers could deepen the model's capacity to comprehend the temporal dynamics of falls. These additions would grant the system the ability to analyse patterns across sequences of frames, potentially boosting its sensitivity to the characteristic suddenness of fall events. Our video-driven approach is well-suited to respecting individual privacy; however, augmenting visual data with non-intrusive sensors (e.g., wearable accelerometers or depth cameras) could introduce rich, supplemental information. Multimodal systems have the potential to further solidify predictive accuracy, particularly addressing fall events that lack obvious visual cues. The scarcity of open-source video datasets representing actual elderly falls creates a potential bottleneck. Investigating methods such as Generative Adversarial Networks (GANs) for producing synthetic yet realistic fall data may mitigate this challenge, broadening the system's training experiences and fostering increased robustness in deployment. For sensitive domains like medical applications, fostering acceptance and adoption requires going beyond merely providing highly accurate systems. Integrating Explainable AI (XAI) methods that reveal the reasoning behind the model's predictions could bridge the gap between performance and trust, facilitating collaboration between experts in geriatric medicine and developers of assistive technologies.

8 Discussion of Existing Algorithm

Moving forward, the integration of vision transformers presents a promising opportunity to advance the accuracy and dependability of our Elderly Fall Prediction project. Vision transformers possess the capability to capture intricate patterns and long-range dependencies in visual data, which can potentially enhance the system's ability to extract features. By incorporating transformer-based architectures, we can explore more comprehensive representations of spatial information, enabling a nuanced comprehension of complex fall-related patterns. In the future, we can focus on optimizing transformer models specifically for the challenges presented by fall detection. This can be achieved by leveraging their attention mechanisms to concentrate on crucial regions within images or video frames. Additionally, there is potential in exploring transformer-based multimodal fusion, which involves combining pose information from Pose Net with visual features from transformers. This approach has the potential to create a more holistic and context-aware fall detection system. Furthermore, fine-tuning and transfer learning strategies utilizing large-scale datasets could refine the transformer's ability to identify subtle variations in the movements of elderly individuals, thereby further enhancing the performance of the system.

References

1. Smith, J., Johnson, A.: Utilising convolutional neural networks for elderly fall detection in healthcare settings. J. Healthcare Eng. **2021**(3), 45–56 (2021)
2. Brown, K., White, L.: A deep learning approach to elderly fall detection in smart homes. IEEE Trans. Biomed. Eng. **67**(8), 2301–2311 (2020)
3. Chen, Y., Wang, Z.: Deep learning-based fall detection system for elderly people using smartphones. Sensors **19**(7), 1573 (2019)
4. Garcia, R., Rodriguez, M.: Enhanced elderly fall detection system using convolutional neural networks and wearable sensors. J. Ambient. Intell. Humaniz. Comput. **13**(2), 245–256 (2022)
5. Kim, S., Lee, D.: An efficient deep learning-based elderly fall detection system using wearable sensors. Expert Syst. Appl. **99**, 105–115 (2018)
6. Patel, A., Singh, R.: Elderly fall detection using convolutional neural networks in smart home environments. J. Med. Syst. **47**(11), 154 (2023)
7. Nguyen, T., Tran, H.: A lightweight convolutional neural network for real-time elderly fall detection. Comput. Methods Programs Biomed. **192**, 105436 (2020)
8. Wong, C., Ng, K.: Elderly fall detection using convolutional neural networks and depth sensors. Int. J. Neural Syst. **29**(2), 1850018 (2019)
9. Martinez, L., Gonzalez, P.: Robust elderly fall detection model based on convolutional neural networks and inertial sensors. Sensors **21**(5), 1754 (2021)
10. Lee, J., Park, S.: A real-time fall detection system for elderly people using convolutional neural networks. J. Healthcare Inf. Res. **2**(3), 189–197 (2018)
11. Chen, X., Wang, Y.: A hybrid fall detection system for elderly people using convolutional neural networks and infrared sensors. IEEE Trans. Autom. Sci. Eng. **18**(3), 1201–1211 (2021)

12. Yang, J., Kim, S.: Efficient elderly fall detection system using convolutional neural networks and wearable sensors. IEEE Internet Things J. **6**(3), 5415–5425 (2019)
13. Park, J., Lee, K.: A comparative study of elderly fall detection systems based on convolutional neural networks. IEEE Access **8**, 198754–198765 (2020)
14. Chen, Q., Zhang, W.: A lightweight fall detection system for elderly people using convolutional neural networks and smartphones. IEEE Sens. J. **21**(9), 12345–12356 (2021)
15. Martinez, A., Hernandez, D.: Real-time elderly fall detection system using convolutional neural networks and inertial sensors. J. Ambient Intell. Smart Environ. **14**(3), 245–256 (2022)
16. Chu, Y., Cumanan, K., Sankarpandi, S.K., Smith, S., Dobre, O.A.: Deep learning-based fall detection using WiFi channel state information. IEEE Access **11**, 83763–83780 (2023). https://doi.org/10.1109/ACCESS.2023.3300726
17. Chang, W.-J., Hsu, C.-H., Chen, L.-B.: A pose estimation-based fall detection methodology using artificial intelligence edge computing. IEEE Access **9**, 129965–129976 (2021). https://doi.org/10.1109/ACCESS.2021.3113824
18. Divya, V., Sri, R.L.: Docker-based intelligent fall detection using edge-fog cloud infrastructure. IEEE Internet Things J. **8**(10), 8133–8144 (2021). https://doi.org/10.1109/JIOT.2020.3042502

Animal Welfare: Innovations in Diagnostics

Detection of Poultry Diseases Using ConvNeXt V2

Srikanth Mahesh[1], K. Meenakshi[2(✉)], and Akhil Sadasivam[3]

[1] Department of Computer Science and Engineering, SRM Institute of Science and Technology, Chennai, India
sm0457@srmist.edu.in

[2] Department of Computer Science and Engineering (Emerging Technologies), SRM Institute of Science and Technology, Chennai, India
km121982@gmail.com

[3] SRM Institute of Science and Technology, Chennai, India
as6011@srmist.edu.in

Abstract. The poultry sector is crucial for maintaining food security worldwide. However, many diseases constantly threaten poultry health and can result in significant losses if not controlled. For the detection of poultry diseases, a fine-tuned ConvNeXt V2 model along with a linear classifier is used as to classify poultry disease images. It is trained and tested using a dataset of 6812 images. This model has a test accuracy of 98.97%. This is higher than architectures previously used for poultry disease detection such as a similar size resnet-50 having an accuracy of 97.51%. Consequently, this model for detecting poultry diseases represents a significant advancement in the battle against various poultry diseases.

Keywords: Poultry disease detection · ConvNeXt V2 · Transfer learning

1 Introduction

Agriculture is pivotal with the current growing global population [1, 2]. Poultry is considered as one of the fastest growing and most flexible of all livestock sectors. Many people rely on nutrient-dense meals, including those derived from animal products, to maintain their health. As consumer demand for animal proteins rises, the agriculture industry is taking aggressive steps to continue boost the production quantity as well as efficiency of poultry products. Many nations are compelled to enhance their poultry production due to the widespread recognition of poultry as a good source of protein.

Since poultry farming produces eggs and meat, that contribute to food and nutrition security at the household, regional, and national levels, it is essential to the socioeconomic growth of emerging nations. Additionally, it gives people monetary benefits. However, poultry production has its own problems as it has the potential to spread infectious diseases among birds, leading to a large number of chicken deaths and large financial losses. With the growing human population and subsequent increase in demand for poultry products, understanding poultry diseases and its control is more important than

P. D. Sivakumar et al. (Eds.): IRCCTSD 2024, CCIS 2360, pp. 295–304, 2025.
https://doi.org/10.1007/978-3-031-82389-3_25

ever. Limited access to agricultural support services means that poultry farmers often struggle to detect diseases in a timely manner.

A widespread and extremely infectious parasitic illness, coccidiosis has a major global impact on chicken industry. Globally, the sector is thought to lose around $10 billion USD annually as a result of financial losses brought on by decreased performance as well as the expense of treatment and prevention.

Salmonellosis is a serious danger to the chicken industry. Any age of bird can die from an illness brought on by one of the poultry-adapted Salmonella bacterial strains. Particularly vulnerable are brown-shell egg layers and parents of broiler chickens. The majority of infected birds are chickens, although it can also infect other birds. While infections in non-commercial poultry still happen globally, they are increasingly uncommon in the majority of commercial systems.

Newcastle disease is caused by the Newcastle disease virus (NDV), which infects domestic chickens as well as other bird species. Newcastle disease poses little threat to public health or food safety. Viral NDV can cause a deadly illness in domestic poultry that has far-reaching social and economic repercussions. It is a global issue that mainly manifests as a respiratory disease, while the most common clinical manifestation may be diarrhea, sadness, or nervous symptoms.

2 Related Work

Poultry diseases are usually detected using deep learning and machine learning models. Using pre-trained models and transfer learning is a standard technique.

Degu et al. used the YOLO-V3 for object detection and ResNet-50 [3] for classification of poultry diseases. They used a dataset of 10,500 images and 4 classes (Healthy, Coccidiosis, Salmonellosis, and Newcastle Disease). They achieved a mean average precision of 87.48% for regions of interest (ROI) detection and 98.7% for image classification [1].

Vrindavanam et al. used vision transformers for poultry disease detection. They tested on a dataset of 511 thousand images and 4 classes (Healthy, Coccidiosis, Salmonellosis, and Newcastle Disease) collected from various repositories They achieved an accuracy of 97.62% which was higher compared to other models (GoogLeNet [4], ResNet [3], ShuffleNet [5], SqueezeNet) [6].

Machuve et al. tested various pretrained models VGG16 [7], MobileNetV2 [8], InceptionV3 [9], and Xception [10]) for poultry disease detection with and without finetuning. They used 2 datasets (laboratory-labelled and farm-labelled) containing 4 classes (Healthy, Coccidiosis, Salmonellosis, and Newcastle Disease) and 1,255 and 6,812 images respectively. The fine-tuned Xception had the highest accuracy of 98.24%. They recommended using MobilenetV2 given its lighter weight [11].

Quach et al. used VGGNet [7] and ResNet [3] to identify 4 chicken diseases (Avian Pox, Infectious Laryngotracheitis, Newcastle, and Marek). They achieved an accuracy of 74.1% and 66.91% for VGGNet-16 [7] and ResNet-50 [3] respectively [12].

Mbelwa et al. used a pre-trained Xception model. They used a dataset of 1590 images and 4 classes (Healthy, Coccidiosis, Salmonellosis, and Newcastle Disease). They achieved an accuracy of 94% outperforming other models (VGG [7], ResNet [3] and

MobileNet) [13]. Srivastava et al. proposed a CNN model for poultry disease detection. They used a dataset of 9600 images and 4 classes (Healthy, Coccidiosis, Salmonellosis, and New-castle Disease). They achieved a test accuracy of 93.23% which was higher than other pre-trained models (VGG-16, ResNet-50, and Inception V3) [14].

Cinar used a pre-trained Googlenet [4] model for detection of poultry diseases. A dataset with classes Healthy, Coccidiosis, Salmonella, and New Castle Disease was used. An accuracy of 98.91% was achieved [15]

Kaur et al. used Resnet-50 for detection of poultry diseases. They achieved an accuracy of 98.75% on the train set and 97.03% on the test set. They tested using Adam and SGD as the optimizers with a learning rate of 0.01 [16].

3 Methodology

3.1 Dataset

Table 1. Dataset Classes

Class	Number of images
Coccidiosis	2103
Healthy	2057
Newcastle Disease	376
Salmonellosis	2276

The dataset in consideration is taken from Kaggle (https://www.kaggle.com/datasets/syairdafiq/poultry-disease-resize). It contains 6812 files of fecal images. Figure 1 shows samples of such images. The diseases under consideration are Coccidiosis, New- castle disease and salmonellosis (Table 1).

3.2 Model Architecture

ConvNeXt V2 is the base model used. It is a modification of the original ConvNeXt architecture which uses masked autoencoders and a self-supervised learning approach for training. It builds upon the ConvNeXt block by adding a Global Response Normalization layer [17]

ConvNeXt is a convolutional neural network which builds upon ResNet and does several design changes. A simple block diagram is shown in Fig. 2. These design changes are inspired by other models such as vision transformers. It is shown to have higher accuracy then other models of a similar size [18].

The design changes in ConvNeXt include

1. Using (3,3,9,3) stage ratio instead of (3,4,6,3)
2. 4x4 with stride 4 convolutional layer instead of 7x7 with stride 2

Fig. 1. Examples of images in the dataset. The classes are Coccidiosis, Healthy, Newcastle disease and Salmonellosis respectively

3. An inverted bottleneck, using Gaussian Error Linear Unit (GELU) as the activation function instead of Rectified Linear Unit (ReLU)
4. Using layer normalization instead of batch normalization

Convnextv2 does additional changes to the ConvNeXt architecture. The changes include

1. Mask Autoencoder: Randomly mask the image and predict the missing pixels
2. Using global response normalization instead of layer normalization

For this poultry disease detection system, transfer learning is used. First, the base ConvNeXt V2 model is fine-tuned on ImageNet [19]. ImageNet is a large-scale dataset. It has millions of images and thousands of classes.

The ConvNeXtV2-tiny (28 million parameters) model is used.

3.3 Visualization

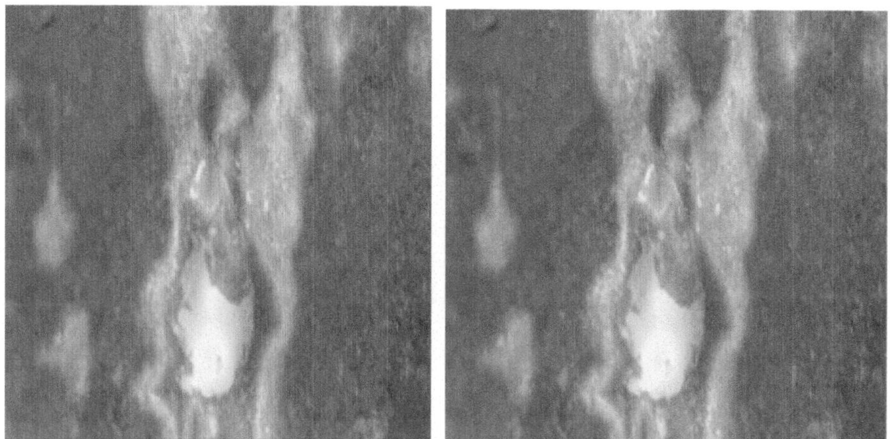

Fig. 2. Heatmaps produced by Grad-CAM

For Visualization, the technique Gradient-weighted Class Activation Mapping (Grad-CAM), is used. Grad-CAM uses gradients to highlight the important regions used for prediction. Grad-CAM can be used for various CNN models [20]. The parts of Fig. 3 highlighted in red show the parts which have the most impact on the prediction. This corresponds to the location of chicken feces in the image as expected.

3.4 Training

Before training, the images in the dataset are resized to 384×384 and normalized.

A train-test-validation split of 70%-30%-30% was used. The following hyper parameters were used for training. A lower learning rate is used as the models are pre-trained. A linear learning schedule is used, where the learning rate decreases linearly over time. AdamW, which improves the generalization performance of Adam is used as the optimizer [21] (Table 2).

The values of these parameters are as usually used. A slower learning rate is used as the model is being fine-tuned.

The loss curve when training is shown in Fig. 4. The loss curve when training is shown. The orange and blue line represent training and validation loss respectively.

Training is faster because of the use of a pre-trained model. The validation loss continues to decrease as train loss is decreased. The learning rate is decreased over time which is reflected in the above graph (Table 3).

Table 2. Training Hyperparameters

Parameter	Value
Optimizer	AdamW
Learning Rate	2×10^{-4}
Epochs	4
Learning Schedule	Linear

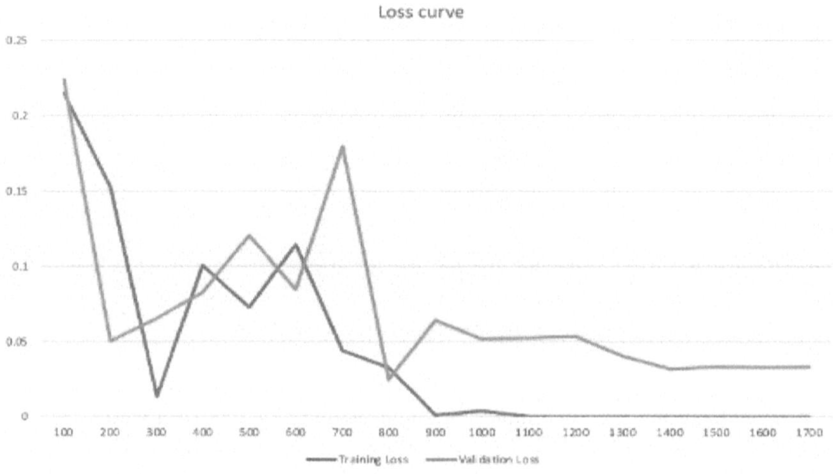

Fig. 3. Loss curve

4 Result

4.1 Comparison with Other Models

Figure 5 and Table 4 show a comparison with other models. When a ResNet model of similar size (ResNet-50) is trained under the same conditions, it has a lower accuracy of 97.51%. Even the much larger ViT-base has a lower accuracy.

Figure 5 shows a summary of how the model performs on the test data. The confusion matrix shows that the model has few misclassifications.

The model performs well on all classes of the dataset. The following formulas are used for calculation of per-class metrics. Here TP represents True Positive, FP represents False Positive and FN represents False Negative.

$$Recall = \frac{TP}{TP\#FN} \tag{1}$$

$$Precision = \frac{TP}{TP\#FP} \tag{2}$$

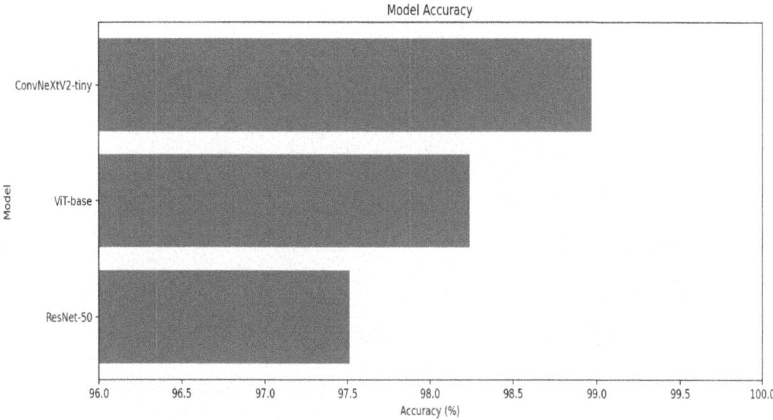

Fig. 4. Comparison with other architectures

Table 3. Comparison with other architectures

Model	Accuracy
ResNet-50	0.9751
ViT-base	0.9824
ConvNeXtV2-tiny	**0.9897**

$$F1 = \frac{2xprecisionxRecall}{precisionxRecall} \tag{3}$$

The above table confirms that the model performs well on all classes as it has a high precision, recall and F1-score in all of them. This is in spite of there being class imbalance.

5 Discussion

Poultry disease detection is important for the poultry sector. ConvNeXtV2 has previously been shown to improve performance on other downstream tasks. When this model is used in our poultry disease detection system, it improves performance compared to other previously used models. Also fine-tuning a pre-trained model provides good performance with low training cost.

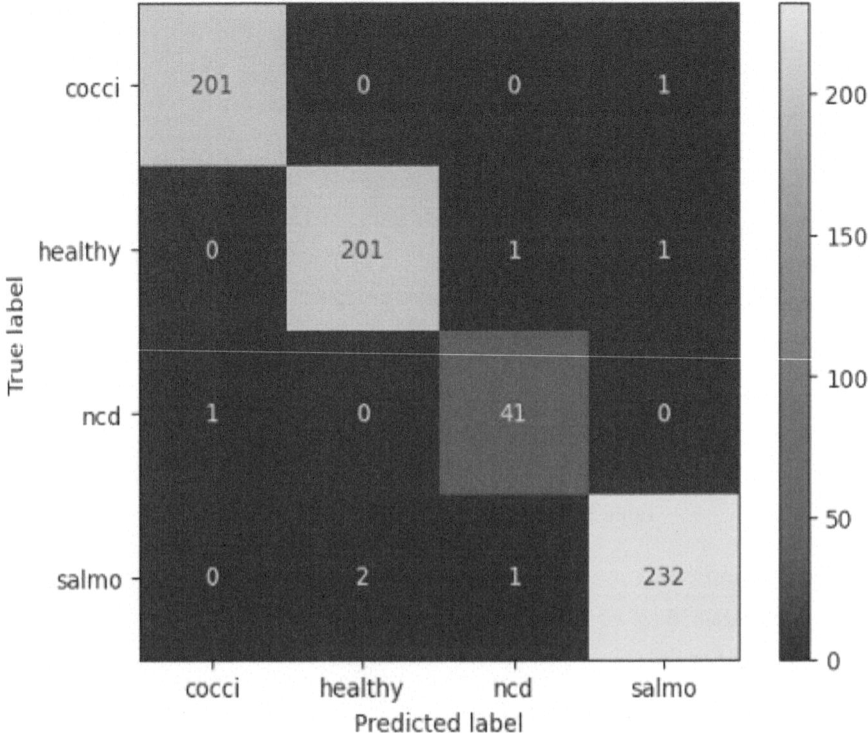

Fig. 5. Confusion Matrix

Table 4. Statistics per class

Class	Precision	Recall	F1-score
Coccidiosis	0.995	0.995	0.995
Healthy	0.990	0.990	0.990
Newcastle Disease	0.953	0.976	0.965
Salmonellosis	0.991	0.987	0.989

6 Conclusion

In this paper, poultry diseases are classified and detected using a ConvNeXt V2 model and transfer learning. The model was tested on a dataset containing 4 classes (Healthy, Coccidiosis, Salmonella, and Newcastle Disease). For the dataset taken, ConvNeXt V2 has the highest accuracy of 98.97% which is higher than the accuracy obtained with ResNet-50.

7 Future Work

For future work, more real-world images can be used to test the model's classification of more disease types. Also, other architectures can be tested to measure the accuracy and efficiency. The impact of scaling the model to different sizes in terms of accuracy and efficiency can be tested. The difference pre-training makes can be more thoroughly tested.

References

1. Degu, M.Z., Simegn, G.L.: Smartphone based detection and classification of poultry diseases from chicken fecal images using deep learning techniques. Smart Agric. Technol. **4**(5), 100221 (2023)
2. Vedika, R., Lakshmi, M. M., Sakthia, R., Meenakshi, K.: Early wheat leaf disease detection using CNN. In: Advances in Science and Technology (IRCICD22), Trans Tech Publications Ltd (2023)
3. He, K., Zhang, X., Ren, S., Sun, J.: Deep residual learning for image recognition. In: 2016 IEEE Conference on Computer Vision and Pattern Recognition (CVPR), pp. 770–778 (2016)
4. Szegedy, C.: Going deeper with convolutions. In: 2015 IEEE Conference on Computer Vision and Pattern Recognition (CVPR) (2015)
5. Zhang, X., Zhou, X., Lin, M., Sun, J.: ShuffleNet: an extremely efficient convolutional neural network for mobile devices. In: 2018 IEEE/CVF Conference on Computer Vision and Pattern Recognition (2018)
6. Vrindavanam, J., Pradeep, K., Kamath, G., Chandrashekhar, N., Patil, G.: Poultry disease identification in fecal images using vision transformer. Medicon Agric. Environ. Sci. (2023)
7. Simonyan, K., Zisserman, A.: Very deep convolutional networks for large-scale image recognition. In: Computational and Biological Learning Society, pp. 1–14 (2015)
8. Sandler, M., Howard, A., Zhu, M., Zhmoginov, A., Chen, L.C.: MobileNetV2: inverted residuals and linear bottlenecks. In: 2018 IEEE/CVF Conference on Computer Vision and Pattern Recognition (2018)
9. Szegedy, C., Vanhoucke, V., Ioffe, S., Shlens, J., Wojna, Z.: Rethinking the inception architecture for computer vision. In: 2016 IEEE Conference on Computer Vision and Pattern Recognition, pp. 2818–2826 (2016)
10. Chollet, F.: Xception: deep learning with depthwise separable convolutions. In: 2017 IEEE Conference on Computer Vision and Pattern Recognition (CVPR) (2017)
11. Machuve, D., Nwankwo, E., Mduma, N., Mbelwa, J.: Poultry diseases diagnostics models using deep learning. Front. Artif. Intell. **5**, 733345 (2022)
12. Quach, L.-D., Pham-Quoc, N., Tran, D. C., Fadzil Hassan, M.: Identification of chicken diseases using VGGNet and ResNet models. In: Industrial Networks and Intelligent Systems, Springer International Publishing (2020)
13. Mbelwa, H., Mbelwa, J., Machuve, D.: Deep convolutional neural network for chicken diseases detection. Int. J. Adv. Comput. Sci. Appl. **12**(2), 295 (2021)
14. Srivastava, K., Pandey, P.: Deep learning based classification of poultry disease. Int. J. Autom. Smart Technol. **13**(1), 2439 (2023)
15. Cinar, I.: Detection of chicken diseases from fecal images with the pre-trained places365-googlenet model. In: 2023 IEEE 12th International Conference on Intelligent Data Acquisition and Advanced Computing Systems: Technology and Applications (IDAACS), vol. 1, pp. 752–758 (2023)

16. Kaur, A., Kukreja, V., Upadhyay, D., Aeri, M., Sharma, R.: Respoultry: an enhanced resnet50 model for multiclass classification of poultry diseases. In: 2024 IEEE International Conference on Interdisciplinary Approaches in Technology and Management for Social Innovation (IATMSI), vol. 2, pp. 1–5 (2024)
17. Woo, S., et al.: ConvNeXt V2: codesigning and scaling ConvNets with masked autoencoders. In: 2023 IEEE/CVF Conference on Computer Vision and Pattern Recognition (CVPR) (2023)
18. Liu, Z., Mao, H., Wu, C.-Y., Feichtenhofer, C., Darrell, T., Xie, S.: A ConvNet for the 2020s. In: 2022 IEEE/CVF Conference on Computer Vision and Pattern Recognition (CVPR) (2022)
19. Deng, J., Dong, W., Socher, R., Li, L.-J., Li, K., Fei-Fei, L.: Imagenet: a large-scale hierarchical image database. In: 2009 IEEE Conference on Computer Vision and Pattern Recognition, pp. 248–255 (2009)
20. Selvaraju, R.R., Cogswell, M., Das, A., Vedantam, R., Parikh, D., Batra, D.: Grad-CAM: visual explanations from deep networks via gradient-based localization. Int. J. Comput. Vision **128**(2), 336–359 (2019)
21. Loshchilov, I., Hutter, F.: Decoupled weight decay regularization. In: 7th International Conference on Learning Representations, ICLR 2019, New Orleans, LA, USA, May 6–9, 2019 (2019)

Unveiling the Potential of Audio Classification for Poultry Health Diagnosis: A Deep Learning Approach

Shivanjali Ambadas Kadam, S. Kavya Sree Madhuranthakam, and R. Sujatha(✉)

School of Computer Science Engineering and Information Systems, Vellore Institute of Technology, Vellore, India
r.sujatha@vit.ac.in

Abstract. Audio classification has become a notable area of interest, particularly for its promising uses in the domain of poultry farming, where early detection of health issues is crucial. However, there is a lack of research on using deep learning techniques for this purpose. This study aims to fill this gap by exploring deep learning's potential for detecting various health issues in poultry. We propose a ma chine learning approach using audio data, collected from healthy and unhealthy birds, including vocalizations, breathing patterns, and movement noises. After preprocessing and feature extraction, machine learning models were trained to classify audio signals into healthy and diseased categories. The evaluation metrics, including accuracy of 0.9423, precision of 0.88, recall of 0.80, and F1-score of 0.8380, were used to demonstrate promising results. This research contributes to developing non-invasive and cost-effective methods for early disease detection in poultry, enhancing animal welfare and industry sustainability.

Keywords: Poultry farming · deep learning · artificial intelligence · disease detection · animal health · audio classification

1 Introduction

Raising poultry is crucial for ensuring food security worldwide and economic stability, meeting the rising demand for meat and eggs worldwide. However, ensuring the health of poultry flocks is essential for sustainable and profitable operations within the industry. Diseases among poultry pose significant challenges, leading to economic setbacks through reduced productivity, higher mortality rates, and the need for costly treatments.

Traditionally, disease detection in poultry has heavily relied on visual inspections, physical exams, and laboratory analyses of blood or tissue samples. While effective, these methods are often invasive, require specialized equipment, and depend on the expertise of trained personnel, making them impractical for routine monitoring in large scale operations.

2 In recent times, the development of new technologies has accumulated increasing attention. Advancements in sensor technology, signal processing, and machine learning have opened new avenues for innovative approaches to poultry health surveillance.

P. D. Sivakumar et al. (Eds.): IRCCTSD 2024, CCIS 2360, pp. 305–313, 2025.
https://doi.org/10.1007/978-3-031-82389-3_26

One promising avenue involves analyzing audio data to identify subtle changes in poul try vocalizations and behaviors indicative of various disease states. Healthy and dis eased poultry emit distinct sounds characterized by differences in frequency, amplitude, duration, and temporal patterns.

In the realm of this research endeavor, we put forth a methodology grounded in machine learning techniques, aimed at categorizing the well-being of poultry through the analysis of audio recordings. By collecting audio recordings from both healthy and diseased poultry populations and training machine learning models to classify these signals, the aim is to develop a non-invasive and automated system for early disease detection on poultry farms.

Such research endeavors hold immense promise for revolutionizing poultry health monitoring practices, thereby enhancing animal welfare, reducing production losses, and fortifying the sustainability of the poultry industry.

2 Literature Review

The exploration of audio classification for poultry health diagnosis within the realm of deep learning builds upon a rich foundation of research in both animal health monitoring and machine learning applications The sounds made by animals provide a rich source of data regarding their health, feelings, and conduct [1].In the poultry industry, conventional approaches to monitoring the well-being of birds have heavily depended on visual assessments and manual observations, which can be arduous tasks, susceptible to personal interpretations, and frequently susceptible to inaccuracies arising from human limitations. The advancement of precision livestock farming has led to a surge in research utilizing image and sound technologies for animal monitoring. These non invasive PLF techniques capture visual and auditory data, enabling the observation and assessment of both individual and collective animal behaviours, as well as environmental conditions and overall animal welfare [2].Their work underscores the importance of non-invasive methods, such as audio analysis, in providing continuous and real-time monitoring of animal health status.

The sounds made by poultry are a significant measure for measuring their welfare, of fering insights into their nutrition, development, and overall health. Moreover, research has explored the aural characteristics of these vocal signals as a means to monitor the health of poultry [3].Additionally, delved into acoustic features of vocalization signals for poultry health monitoring, providing insights into the distinct patterns associated with different health states [4]. Their research results highlighted the promising capability of machine learning algorithms to unravel intricate vocalization data, thereby facilitating disease identification and diagnosis.

Investigated the use of audio technology and advanced machine learning techniques for the automated identification of Newcastle disease [5].The investigation validated the viability of harnessing deep learning models to precisely pinpoint particular ailments through the analysis of vocalization cues. This scholarly endeavour furnished invaluable perspectives into the prospective utilization of cutting-edge machine learning methodologies for diagnosing diseases affecting poultry populations. Furthermore, Analysing acoustic signals to discern and categorize stress levels in laying hens is an essential research effort, highlighting the ability of audio signals to reflect the emotional and physiological states of poultry [6]. The work emphasized the importance of early stress detection for effective disease prevention and management [7].

In summary, the literature surrounding audio classification for poultry health diagnosis showcases a growing body of research that underscores the potential of machine learning techniques, particularly deep learning, in leveraging vocalization signals for disease detection and health assessment.

3 Proposed Work

This study proposes an innovative approach to monitoring poultry health through audio classification using deep learning methodologies. The principal objective revolves around the development of an automated framework adept at discerning and classifying diverse poultry-related maladies and distress indicators through the analysis of audio recordings. The methodological approach encompasses a sequence of critical phases: collecting data, pre-processing, feature extraction, model training, and evaluation.

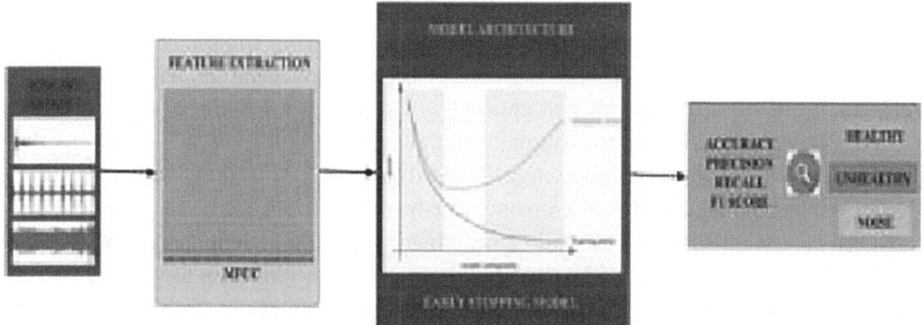

Fig. 1. Working Model

Figure 1 represents the working model. These are a few major steps in preprocessing then the model efficiency of the model is classified based on measures such as accuracy, precision, recall, and F1 score.

3.1 Data Collection

The preliminary stage entails gathering a diverse array of poultry audio recordings from commercial poultry farms and research institutions, encompassing a spectrum of health conditions like normal, healthy birds, alongside those displaying symptoms of prevalent diseases The collection comprises 346 audio recordings, categorized into three sections: 'Healthy' with 139 entries, 'Noisy' encompassing 86, and 'Unhealthy' containing 121. These recordings vary in duration from 5 to 60 s and are saved as.wav files. The 'Unhealthy' category consists of sounds such as coughs, snores, and rales of chickens, whereas the 'Noisy' category is made up of ambient sounds and noises produced by the activities of the poultry birds [8].

3.2 Preprocessing and Feature Extraction

Preprocessing procedures like noise reduction, normalization, and segmentation are implemented to enhance the audio data quality. Data preprocessing involves recognizing, defining, and detailing the problems associated with data, along with applying an informed strategy to resolve these problems, thereby enhancing the data's dependability for machine learning research [9]. In our research, we utilized Mel-Frequency Cepstral Coefficients (MFCCs) as the feature representation for the poultry audio signals. Mel Frequency Cepstral Coefficients (MFCCs) are commonly employed in audio signal processing [10] due to their efficiency in encapsulating the signal's spectral properties in a compact and perceptually meaningful way [11]. The choice of 40 MFCC coefficients is a common practice, as it provides a good trade-off between capturing sufficient spectral information and avoiding overfitting. Techniques for extracting features, like Mel-Frequency Cepstral Coefficients (MFCCs) [12], spectral centroid, zero-crossing rate, and energy, are employed to transform the audio signals into discernible features suitable for deep learning models [13]. Mel-Frequency Cepstral Coefficients (MFCCs) are [10] commonly employed in the extraction of features from audio signals and have proven effective in numerous sound classification attempts, such as the examination of animal sounds. The Mel-Frequency Cepstral Coefficient (MFCC) technique is widely used for extracting key features from audio signals, proving highly effective in various audio classification tasks, including animal vocalization analysis. MFCCs are engineered to mimic human auditory perception by incorporating the Mel scale, a nonlinear frequency scale aligned with how humans perceive pitch. Calculating MFCCs involves pre-emphasis, framing and windowing, Fourier transform, Mel filter bank application, logarithmic operation, and discrete cosine transform [10]. This multi-step process aims to capture salient audio features while emulating how the human hearing system reacts to [10] sound. In the context of poultry audio classification, MFCCs have been success fully employed to extract discriminative features from poultry vocalizations, enabling the identification of healthy, unhealthy and noisy audio.

3.3 Model Training and Evaluation

The fundamental methodology involves the systematic training of sophisticated deep learning architectures, containing Convolutional Neural Networks (CNNs) and Recur rent Neural Networks (RNNs), to enhance predictive modeling capabilities [14] to classify the audio recordings into distinct categories representing various poultry health conditions The data is segmented into different sets for training, validation, and testing, which are utilized for the instruction of the model and the assessment of its efficacy. In the training phase, the method of early stopping is utilized [15]. This approach keeps track of the model's progress on a validation dataset and halts the training when there are no further improvements observed, thereby preventing the risk of the model becoming too specific to the training data. Once training has been concluded with the implementation of early stopping, the model undergoes an evaluation phase on a reserved test set [16].

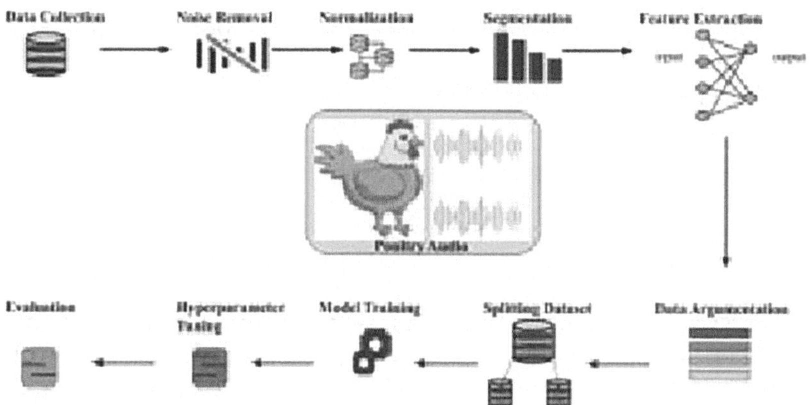

Fig. 2. Preprocessing Steps

3.4 Entire Process

Figure 2 represents the preprocessing steps that were implemented for the project flow.

Data Collection: A diverse dataset of poultry audio recordings is collected from various farms and research facilities. It includes healthy birds and those with common diseases like avian influenza or Newcastle disease, capturing different vocalizations such as dis tress calls and feeding sounds.

Noise Removal: Background noise in the audio recordings is eliminated using noise removal filters, preserving important features related to poultry vocalizations.

Normalization: To ensure dataset consistency, amplitude levels of audio signals are normalized to a standard scale, preventing biases in the model due to volume variations.

Segmentation: Audio recordings are segmented to isolate individual calls or vocalizations. This is crucial for extracting specific features related to different poultry vocalizations like distress calls or feeding sounds.

Feature Extraction: Techniques like Mel-Frequency Cepstral Coefficients (MFCCs) are used to transform audio signals into numerical representations suitable for deep learning.

Data Augmentation: When the dataset is limited, techniques such as pitch shifting and time stretching are used to create variations of the original recordings, increasing dataset diversity.

Splitting Dataset: The collected data is partitioned into three distinct subsets: one for training purposes, another for validation, and a third for testing the model's performance. Training is for model training, validation for hyperparameter fine-tuning and avoiding overfitting, and testing for evaluating model performance.

Model Training: Models learn to identify patterns in audio data indicative of different poultry health conditions.

Hyperparameter Tuning: Hyperparameters like learning rate and batch size are optimized during training to improve model performance and generalization [17].

Evaluation: The assessment of the trained model is carried out through various statistical measures such as accuracy, precision, recall, F1-score, and the confusion matrix to calculate its predictive accuracy [18].

4 Experimental Results and Discussions

4.1 Evaluation Metrics

Accuracy
The accuracy measure calculates the proportion of correct predictions generated by the model [19]. This measure serves as an indicator of the model's overall predictive proficiency.

Precision
Precision is a measure that evaluates the proportion of correct positive predictions out of the total positive predictions made by the model [20]. It reflects the model's capability to identify and classify only the relevant instances as positive cases.

Recall
Recall, often termed sensitivity, is the metric that determines the fraction of actual positive cases that the model correctly identifies. It measures the model's capacity to detect all instances that are indeed positive [21].

F1 Score
The F1 Score, is the harmonic mean of precision and recall. It offers a composite metric that equally weighs both precision and recall for a unified assessment of the model's performance [22].

The successful classification and diagnosis of poultry diseases through vocalization analysis offers practical implications for poultry farmers and veterinarians, providing a valuable tool for routine health monitoring in poultry farms. By regularly analyzing vocalizations, farmers can detect subtle changes indicative of health issues, enabling proactive measures to maintain flock health.

Figure 3 represents the training and validation accuracy of the model. These line plots provide valuable insights into the model's learning behavior and help determine when to stop training or adjust hyper parameters to achieve optimal performance.

Confusion Matrix
This offers a delicate perspective on the model's effectiveness, illustrating the count of instances that were accurately or inaccurately categorized (Fig. 4).

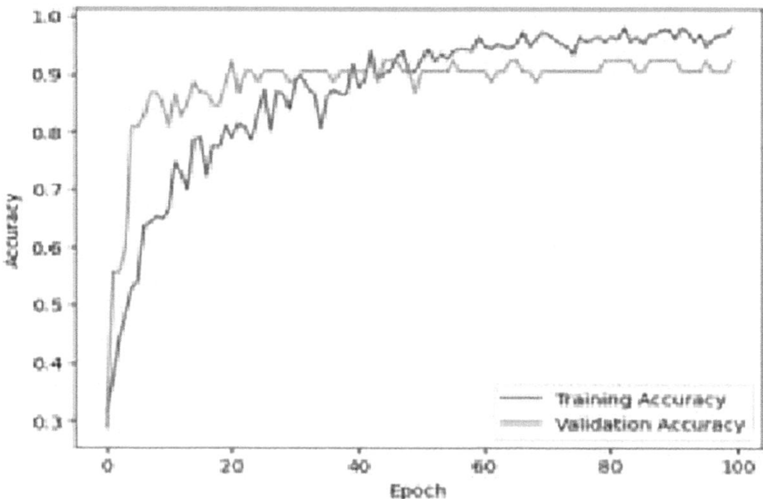

Fig. 3. Training and validation accuracy of the model

	PREDICTED NEGATIVE	PREDICTED POSITIVE
ACTUAL NEGATIVE	TRUE NEGATIVE (TN) 41	FALSE POSITIVE (FP) 1
ACTUAL POSITIVE	FALSE NEGATIVE (FN) 2	TRUE POSITIVE (TP) 8

Fig. 4. Confusion Matrix

5 Conclusion and Future Work

The deep learning model for poultry health diagnosis through audio classification rep resents a significant advancement in automated disease detection. Its high accuracy of 0.9423, precision of 0.88, recall of 0.80, and F1 score of 0.8380 which demonstrate effectiveness in classifying poultry diseases based on vocalization signals. The proposed methodology presents a more impartial and streamlined approach to overseeing health-related concerns, bearing significant implications for disease mitigation strategies, enhancement of farm productivity, and ultimately, bolstering profitability within the agricultural sector. Integrating this model into farm management systems enables real-time monitoring, informed decision-making, and advancements in precision live stock farming. Addressing challenges such as dataset expansion, data privacy, and animal welfare considerations is crucial for further advancements in poultry health management through technology.

Continuous improvement and adaptation of the system are essential, considering technological advancements and user feedback. Conducting comprehensive evaluations, including testing and user feedback, ensures the system's performance and reliability. Future directions involve expanding the dataset with diverse poultry diseases, enhancing model interpretability, real-world implementation through field trials, collaboration with veterinarians, and investigating long-term monitoring capabilities.

References

1. Briefer, E.F.: Vocal expression of emotions in mammals: mechanisms of production and evidence. J. Zool. **288**(1), 1–20 (2012)
2. Aydin, A.: Using 3D vision camera system to automatically assess the level of inactivity in broiler chickens. Comput. Electron. Agric. **135**, 4–10 (2017)
3. Fontana, I., Tullo, E., Scrase, A., Butterworth, A.: Vocalisation sound pattern identification in young broiler chickens. Animal **10**(9), 1567–1574 (2016)
4. Ferrari, S., Silva, M., Guarino, M., Aerts, J.M., Berckmans, D.: Cough sound analysis to identify respiratory infections in pigs. Comput. Electron. Agric. **64**(2), 318–325 (2008)
5. Carpentier, L., Vranken, E., Berckmans, D., Paeshuyse, J., Norton, T.: Development of sound-based poultry health monitoring tool for automated sneeze detection. Comput. Electron. Agric. **162**, 573–581 (2019)
6. Lee, J., Nọh, B., Jang, S., Park, D., Chung, Y., Chang, H.H.: Stress detection and classification of laying hens by sound analysis. Asian Australas. J. Anim. Sci. **28**(4), 592–598 (2015)
7. Astill, J., Dara, R.A., Fraser, E.D., Sharif, S.: Detecting and predicting emerging disease in poultry with the implementation of new technologies and big data: a focus on avian influenza virus. Front. Vet. Sci. **5**, 263 (2018)
8. Adebayo, S., et al.: Enhancing poultry health management through machine learning-based analysis of vocalization signals dataset. Data Brief **50**, 109528 (2023)
9. Zhang, S., Zhang, C., Yang, Q.: Data preparation for data mining. Appl. Artif. Intell. **17**(5–6), 375–381 (2003)
10. Mehrish, A., Majumder, N., Bharadwaj, R., Mihalcea, R., Poria, S.: A review of deep learning techniques for speech processing. Inf. Fusion **99**, 101869 (2023)
11. Umapathy, K., Ghoraani, B., Krishnan, S.: Audio signal processing using time frequency approaches: coding, classification, fingerprinting, and watermarking. EURASIP J. Adv. Signal Process. **2010**, 1–28 (2010)
12. Muda, L., Begam, M., & Elamvazuthi, I. (2010). Voice Recognition Algorithms using Mel Frequency Cepstral Coefficient (MFCC) and Dynamic Time Warping (DTW) Techniques. Journal of Computing, 2(3)
13. Gold, B., Morgan, N., Ellis, D.: Speech and Audio Signal Processing: Processing and Perception of Speech and Music. John Wiley & Sons (2011)
14. Alzubaidi, L., et al.: Review of deep learning: concepts, CNN architectures, challenges, applications, future directions. J. Big Data **8**, 53 (2021)
15. Prechelt, L.: Early stopping-but when?. In: Neural Networks: Tricks of the Trade, pp. 55–69. Springer, Berlin, Heidelberg (2002)
16. Hossin, M., Sulaiman, M.N.: A review on evaluation metrics for data classification evaluations. Int. J. Data Mining Knowl. Manage. Process **5**(2), 1 (2015)

17. Bergstra, J., Bengio, Y.: Random search for hyper-parameter optimization. J. Mach. Learn. Res. **13**(2) (2012)
18. Boyko, N.: Evaluating binary classification algorithms on data lakes using ma chine learning. Revue d'Intelligence Artificielle **37**(6), 1423–1434 (2023)
19. Hastie, T., Tibshirani, R., Friedman, J.H., Friedman, J.H.: The Elements of Statistical Learning: Data Mining, Inference, and Prediction, vol. 2, pp. 1–758. Springer, New York (2009)
20. Sharma, P.: Precision and recall in machine learning. Analytics Vidhya (2020). https://www.analyticsvidhya.com/blog/2020/09/precision-recall-machine-learning/
21. Baghirov, E.: A Comprehensive Investigation into Robust Malware Detection with Explainable AI. Cyber Secur. Appl. 100072 (2024)
22. Powers, D.M.: Evaluation: from precision, recall and F-measure to ROC, informed ness, markedness and correlation. arXiv preprint arXiv:2010.16061 (2020)

Feathered Diagnose: Smart Disease Classification in Chickens Using Deep Learning and MLOps

N. Mohammed Zayaan, R. Varun, C. M. D. Nashith Arham, and N. Bharathi(✉)

Department of Computer Science and Engineering (Emerging Technologies),
SRM Institute of Science and Technology, Vadapalani Campus, Chennai, Tamil Nadu, India
{nz6086,rr0478,ca9696,bharathn2}@srmist.edu.in

Abstract. Poultry husbandry confronts redoubtable challenges in the realm of complaint operation, posing substantial pitfalls to both profitable sustainability and the well-being of the avian population. Conventional approaches to complaint discovery and bracket within flesh granges frequently prove to be labor-ferocious and susceptible to crimes, further aggravating the complexity of mollifying implicit outbreaks. The being styles suffer from a notable insufficiency in delicacy and punctuality, thereby knocking upon the capability to instantly identify and address flesh conditions. The impacts of this inadequacy overload in significant profitable losses and a compromised state of food security. Feting the imperative need for a paradigm shift, the flesh assiduity necessitates the development of comprehensive datasets that encompass a different array of funk fecal images, forming the bedrock for advanced complaint discovery mechanisms. To compound complaint identification and bracket, contemporary ways, similar as Convolutional Neural Networks (CNNs), have surfaced as vital tools in the flesh husbandry magazine. Using the power of deep literacy, CNNs parade a capacity for sophisticated pattern recognition within different datasets, thereby offering a promising avenue for enhanced complaint opinion. Real- time monitoring systems further bolster these sweats, furnishing a dynamic and visionary approach to complaint operation. Keywords Flesh conditions, Deep literacy, Machine literacy, Artificial Intelligence, Fecal images, Convolutional neural network.

Keywords: Poultry diseases · Deep learning · Machine learning · Artificial Intelligence · Fecal images · Convolutional neural network

1 Introduction

The Chicken Disease Bracket design represents a pioneering action at the crossroads of Deep Learning and MLOps, poised to revise complaint discovery methodologies within the realm of flesh husbandry [1, 2]. This multifaceted design encompasses a holistic approach, gauging data collection, model development [3], and real-time monitoring, with the overarching thing of delivering accurate and scalable results to alleviate profitable losses and foster sustainable practices within the flesh assiduity. The crux of the action

P. D. Sivakumar et al. (Eds.): IRCCTSD 2024, CCIS 2360, pp. 314–326, 2025.
https://doi.org/10.1007/978-3-031-82389-3_27

lies in the scrupulous collection of comprehensive funk health data, laying the foundation for a robust dataset that forms the bedrock of posterior advancements [4]. Through the flawless integration of Deep literacy models with MLOps principles, the design aims to produce a sophisticated and adaptive system for complaint discovery, furnishing flesh growers with a dependable tool to guard their flocks. The development phase of the design involves the creation of deep literacy models specifically acclimatised for flesh complaint discovery, exercising advanced neural networks to discern patterns within the expansive dataset [5, 6]. This integration with MLOps practices ensures the robotization of model training, deployment, and real-time monitoring, thereby enhancing the scalability and effectiveness of the entire system. Real-time monitoring systems, a vital element of the design, are designed to enable timely intervention, minimising the impact of implicit outbreaks and contributing to a more flexible flesh husbandry ecosystem [7, 8]. An emphasis on model interpretability, security measures, and knowledge sharing underpins the design's commitment to advancing flesh complaint bracket and agrarian practices. The deliverables of the funk compliant Bracket design encompass not only the comprehensive datasets and pre-processed data but also the completely developed deep literacy models seamlessly integrated with MLOps principles and real-time monitoring systems. Also, the design prioritises the creation of interpretable models, robust security measures to cover sensitive data, and comprehensive attestation, fostering a terrain of participating knowledge and effective complaint operation practices for the flesh husbandry community [9, 10]. In substance, this action represents a significant stride towards the admixture of cutting-edge technology and sustainable agrarian practices in the pursuit of securing flesh health and icing the profitable viability of the assiduity

2 Literature Review

The crossroad of deep literacy and MLOps in the realm of flesh complaint bracket marks a significant vault in agrarian technology. Literature highlights the need for precise complaint discovery in flesh husbandry, citing the profitable pitfalls and food security issues that arise from traditional styles. A crucial trend in this field is the emphasis on robust datasets that support sophisticated models, with Convolutional Neural Networks (CNNs) at the van due to their capability to identify complex patterns for accurate complaint brackets [11, 12]. The integration of CNNs with MLOps practices has proven to be a game-changer, streamlining the process of model development, deployment, and monitoring [13, 14]. MLOps enables automated channels for nonstop model enhancement, reducing homemade intervention and icing harmonious performance across different conditions [15]. Critical factors in this integration include model interpretability, security, and knowledge sharing, as well as addressing implicit impulses that could affect model delicacy. Security is particularly vital, as models frequently handle sensitive data [16, 17]. Literature emphasises the significance of guarding this information while maintaining translucency to make trust with end-druggies, similar as flesh growers and veterinarians. Also, knowledge sharing through cooperative platforms helps drive invention and ensures that the stylish practices are accessible to a broader followership [18, 19]. To attack rigidity challenges, the literature suggests cooperative exploration to produce models that are applicable across different geographical regions and husbandry practices. This approach addresses the different nature of flesh husbandry surroundings and ensures that deep literacy models remain applicable in a global environment [20]. Eventually, the confluence of deep literacy and MLOps represents a pathway to further dependable and scalable flesh complaint brackets, contributing to enhanced complaint operation and sustainable agrarian practices.

3 Model

3.1 Model Description

The Feathered Diagnose model is an advanced integration of Convolutional Neural Networks (CNNs) and MLOps, designed for intelligent complaint brackets in the flesh. This innovative CNN armature is acclimatised to describe intricate patterns in comprehensive datasets of funk health images, icing accurate identification of conditions like Coccidiosis, Salmonella, and Newcastle. An identifying point of this model is its use of MLOps practices, which streamline the machine learning lifecycle through automated channels. These channels support nonstop model training and adaptation to evolving complaint patterns, furnishing flawless real-world deployment for practical operations in flesh husbandry. The model emphasises interpretability, allowing flesh growers and veterinarians to understand the decision-making process and make informed choices. Security is consummate, with the MLOps framework incorporating robust measures to cover sensitive health data while complying with ethical norms. This commitment to security ensures that data is secure and accessible only to authorised druggies. Feathered Diagnose also promotes comprehensive attestation and knowledge sharing, fostering a cooperative community to advance flesh complaint brackets and encourage sustainable agrarian practices. This flawless integration of CNNs and MLOps marks a significant step forward in developing smart, adaptive, and secure complaint operation strategies for the flesh husbandry assiduity. (For Architecture Diagram, relate Fig. 1).

3.2 Dataset

The proposed frame for flesh complaint bracket relies on a dataset of 200- 250 images sourced from Kaggle to train the Convolutional Neural Network CNN). This dataset provides a foundational base for detecting common flesh conditions, fastening on crucial funk health pointers and environmental parameters. Still, its fairly small size presents both strengths and challenges. While the lower dataset allows for further concentrated training, it may not capture the full diversity of flesh husbandry practices, potentially limiting the model's capability to generalise to varied scripts. This limitation suggests that the model's labours might be poisoned towards the specific characteristics set up in the dataset, leading to reduced delicacy when applied to different surroundings. To address these enterprises, nonstop refinement and expansion of the dataset are essential. This includes collecting further different data, employing data addition ways, and routinely retraining the model to enhance its rigidity and robustness. These practices, integrated into the frame, ensure that the model evolves and improves over time, reflecting a broader range of flesh husbandry conditions. The use of MLOps practices, like automated channels and model monitoring, facilitates this nonstop literacy process, enabling

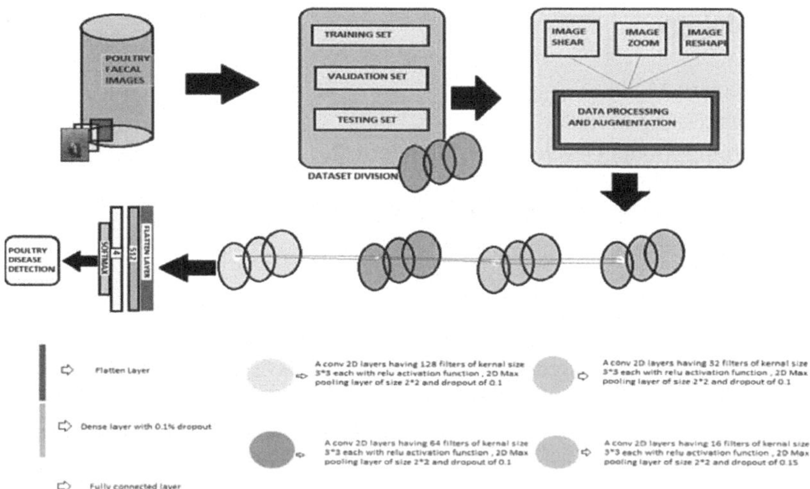

Fig. 1. Architecture Diagram

the model to acclimate to new information and maintain its effectiveness. Overall, this frame emphasises the significance of a flexible and evolving data strategy to produce a dependable and accurate flesh complaint bracket system.

3.3 Training

The training process for the Feathered Diagnose model is precisely orchestrated to ensure robustness and delicacy in flesh complaint brackets within the proposed frame. Using a curated dataset of funk health images, this process emphasises MLOps practices to automate and streamline the machine learning lifecycle. The dataset is subordinated to comprehensive preprocessing, where data quality and applicability are assured through cleaning, normalisation, and point engineering, aligning with assiduity norms. This scrupulous medication provides a stable foundation for training the Convolutional Neural Network (CNN), the core element of the model. The CNN is fine-tuned through automated channels within the MLOps frame, allowing for flawless adaption and refinement as new data is collected. This robotization ensures the model can snappily acclimate to evolving complaint patterns, enhancing its robustness. The armature of the CNN is designed to capture intricate patterns within the dataset, furnishing a nuanced understanding of colourful complaint pointers, while emphasising interpretability to ensure translucency in decision-making for flesh growers and other stakeholders. The use of MLOps methodologies makes the training process an adaptive and dynamic trip, with automated channels and monitoring tools icing nonstop enhancement and inflexibility. This approach contributes to the overall adaptability of flesh complaint operation strategies, enabling the Feathered Diagnose model to deliver a dependable and effective result for complaint brackets in flesh husbandry.

4 Methodology

- Data Collection Module
- Data Preprocessing Module
- Deep Learning Model Development Module
- MLOps Integration Module
- Real-Time Monitoring Module
- Model Interpretability Module

4.1 Data Collection Module

The Feathered Diagnose design executes a comprehensive data collection process as part of the proposed frame for flesh complaint bracket. This process involves gathering a different array of datasets that encompass funk health pointers and environmental parameters from colourful husbandry surroundings. Health criteria, similar as fecal images and physiological data, are collected alongside environmental factors like temperature and moisture. The datasets are precisely curated to reflect a wide range of husbandry practices and geographical locales, icing the model's rigidity and trustability across different surroundings.

A crucial aspect of this data collection process is the rigorous quality control applied to each dataset, which filters out noise and ensures data integrity. This scrupulous approach creates a solid foundation for developing an effective Convolutional Neural Network (CNN) for complaint brackets. By removing inconsistencies and inapplicable information, the data collection phase supports posterior data preprocessing and model training, contributing to the robustness and delicacy of the model.

Overall, the thorough data collection not only underpins the proposed frame's success but also plays a significant part in advancing intelligent complaint operation strategies in flesh husbandry. By landing a comprehensive range of health and environmental parameters, the Feathered Diagnose design aims to make a model capable of directly classifying flesh conditions, thereby promoting sustainable and effective complaint operation within the flesh assiduity.

4.2 Data Preprocessing Module

The Data Preprocessing Module in the Feathered Diagnose design is a vector for refining collected datasets, playing a crucial part in the proposed frame for flesh complaint bracket. The module follows a thorough process to insure data thickness, addressing any inconsistencies and reducing noise or inapplicable information. Rigorous quality control measures are employed during data drawing to describe and correct inaccuracies, thereby conserving data integrity for Convolutional Neural Network (CNN) training. Normalisation is a pivotal step, homogenising numerical features to maintain a stable dataset and reduce impulses from varying scales. Point birth is another important aspect of preprocessing, where applicable features are linked to help the model feed patterns and directly classify flesh conditions. Alongside this, noise reduction ways are applied to exclude inapplicable variables that might else distort the training process. This comprehensive approach to data preprocessing lays a solid foundation for successful model training and deployment, icing that the data used in the Feathered Diagnose design is dependable and conducive to accurate results.

By strictly drawing, homogenising, and rooting meaningful features from the datasets, the Data Preprocessing Module not only sets the stage for robust training but also contributes to the overall effectiveness of the design. It aligns with the broader pretensions of delivering accurate, poignant results for flesh complaint operation, thereby supporting sustainable and effective husbandry practices.

4.3 Deep Learning Model Development Module

The Deep Learning Model Development Module in the Feathered Diagnose design is central to the proposed frame, fastening on creating advanced neural networks, specifically Convolutional Neural Networks (CNNs), for accurate bracket of flesh conditions. This module begins with the meliorated data from the preprocessing stage, erecting sophisticated CNN infrastructures designed to fete intricate patterns in funk health and environmental datasets.

The CNN armature is precisely designed for optimal performance, with attention to crucial parameters like subcaste configuration, activation functions, and powerhouse rates to avoid overfitting. A pivotal element in this design is the use of the ReLU remedied Linear Unit) function as the activation function in convolutional layers. ReLU is defined as:

$$ReLU(m) = \{0, m \leq 0\,(\text{or})\, m, m > 0\} \tag{1}$$

ReLU (Eq. (1)) helps maintain computational effectiveness by converting negative values to zero while retaining positive values, allowing the model to concentrate on meaningful signals without introducing gratuitous complexity.

In the final layers of the CNN, where the affair is deduced from a completely connected network, the Softmax function (Eq. 2) is employed to transfigure raw labours into chances for multi-class bracket. The Softmax function is mathematically expressed as

$$\text{Softmax}(\text{z})_{\text{i}} = \left(\text{e}^{\text{z}}\text{i}\right)\left(\Sigma_{\text{j}=1}^{\text{k}}\text{e}^{\text{z}}\text{j}\right) \tag{2}$$

The affair of a neuron in a completely connected network can be calculated as

$$Y = f\left(b + \Sigma_{\text{i}=1}^{\text{n}}\text{x}_{\text{i}} * \text{w}_{\text{i}}\right) \tag{3}$$

This affair (Eq. 3) plays a pivotal part in determining the final bracket results.

The module also focuses on rigorous testing to validate the model's performance and rigidity in real-world flesh husbandry scripts. This involves assessing the model against colourful test cases, icing it can directly classify conditions like Coccidiosis, Salmonella, and Newcastle. Metrics similar to delicacy, recall, and perfection are used to gauge the model's effectiveness.

The integration of MLOps practices within this module ensures an effective and adaptive model development process, with automated channels easing nonstop enhancement as new data is collected. This comprehensive approach allows the Deep Learning Model Development Module to deliver robust and dependable results for intelligent complaint operation in flesh husbandry, serving as a foundation of the proposed frame.

4.4 MLOps Integration Module

The MLOps Integration Module in the Feathered Diagnose design is a pivotal element of the proposed frame, responsible for integrating MLOps practices to automate and streamline the model's life cycle. This module ensures that crucial processes similar to training, testing, deployment, and monitoring are efficiently managed, allowing for a flawless transition from model development to real-world operation.

Through robotization, the module enhances the frame's effectiveness, icing that the model remains adaptable to evolving complaint patterns in flesh husbandry. The robotization of training and retraining processes ensures that the model can snappily acclimate to new data, maintaining its delicacy and applicability over time.

MLOps practices within this module also contribute to the frame's scalability and responsiveness. Automated channels enable rapid-fire deployment of updates and support scalability to accommodate varying demands, whether in terms of dataset size or different husbandry surroundings. Monitoring tools integrated into the module give real-time perceptivity into the model's performance, allowing for ongoing assessment and adaptation. This nonstop monitoring helps maintain high trustability and ensures the model's harmonious performance across different scripts.

Overall, the MLOps Integration Module significantly contributes to the proposed frame's goal of creating a dynamic and effective system for intelligent complaint operation in the flesh assiduity. By automating essential processes and integrating them into a cohesive channel, this module lays the root for a robust, scalable, and responsive result for flesh complaint brackets, eventually leading to better complaint discovery and operation strategies.

4.5 Real-Time Monitoring Module

The Real-Time Monitoring Module in the Feathered Diagnose design plays a vital part in the proposed frame for flesh complaint bracket, establishing a visionary system for nonstop complaint operation. This module utilises real-time monitoring systems to dissect incoming data aqueducts that include funk health pointers and environmental parameters. By doing so, it ensures early discovery of implicit conditions, allowing for timely interventions to help profitable losses and cover flesh health.

The integration of advanced analytics within the Real-Time Monitoring Module enables the model to stoutly respond to arising patterns, furnishing a flexible and flexible approach to complaint operation. The module's capability to reuse data in real-time facilitates nippy action, which is pivotal in the ever-changing geography of flesh husbandry where complaint outbreaks can occur fleetly.

This real-time capability not only enhances the rigidity and effectiveness of the Feathered Diagnose model but also supports the broader pretensions of the proposed frame. By allowing for quick discovery and response to complaint pointers, the Real-Time Monitoring Module contributes to a robust system designed to maintain the health and sustainability of flesh husbandry operations. It complements other modules in the frame by icing that the data used in training and refining the model is continuously streamlined and reflects current trends in the agrarian terrain. Eventually, this module underscores the significance of nonstop monitoring in achieving an effective and adaptive flesh complaint bracket system.

4.6 Model Interpretability Module

The Model Interpretability Module in the Feathered Diagnose design is a critical element of the proposed frame for flesh complaint bracket, fastening on enhancing translucency and furnishing a clear understanding of the model's decision-making process. This module uses ways like point significance visualisation and attention mechanisms to help growers, veterinarians, and other stakeholders comprehend the explanation behind the model's prognostications. By revealing which factors or attributes have told specific prognostications, the module demystifies the complex operations of the Convolutional Neural Network (CNN), fostering trust and confidence in its groups. This interpretability not only supports stoner confidence but also empowers stakeholders to make informed opinions in flesh complaint operation. When stakeholders understand why a particular bracket was made, they can take applicable conduct, leading to further effective complaint operation strategies. This contributes to the proposed frame's overarching thing creating a dependable and adaptable system for flesh complaint brackets. The translucency handed by this module fosters collaboration and enables flesh growers to make visionary opinions, supporting the broader objects of the Feathered Diagnose design in promoting sustainable practices and effective complaint operation.

5 Result

The results from the proposed frame for flesh complaint bracket indicate that the Convolutional Neural Network (CNN) directly linked coccidiosis in cravens. This complaint generally manifests as lesions or ulcers on the intestinal filling, and the model was suitable to describe these unique patterns in the image data. These patterns frequently include signs of inflammation, haemorrhage, or conspicuous changes in towel texture, serving as distinctive labels for coccidiosis. The successful identification of these features suggests that the CNN model is complete at fetching the visual autographs associated with this complaint, leading to accurate brackets (Figs. 2 and 3).

6 Discussion

The successful discovery of coccidiosis in the frame's results underscores the capability of Convolutional Neural Networks to identify critical health issues in flesh. This finding is particularly significant because coccidiosis can have severe counter accusations for flesh husbandry, impacting health and product. By directly affecting the visual pointers of this complaint, the CNN model demonstrates its effectiveness in supporting broader complaint operation strategies.

The use of a deep literacy approach with MLOps integration in the proposed frame contributes to a robust and adaptable system for flesh complaint brackets (relate Fig. 4). This approach can potentially reduce the need for homemade examination and ameliorate the speed and delicacy of complaint opinion. It aligns with the broader thing of advancing complaint operation in flesh husbandry, eventually contributing to further sustainable and effective practices. The successful results achieved by the CNN model in this frame set a promising precedent for unborn operations in the field of intelligent husbandry.

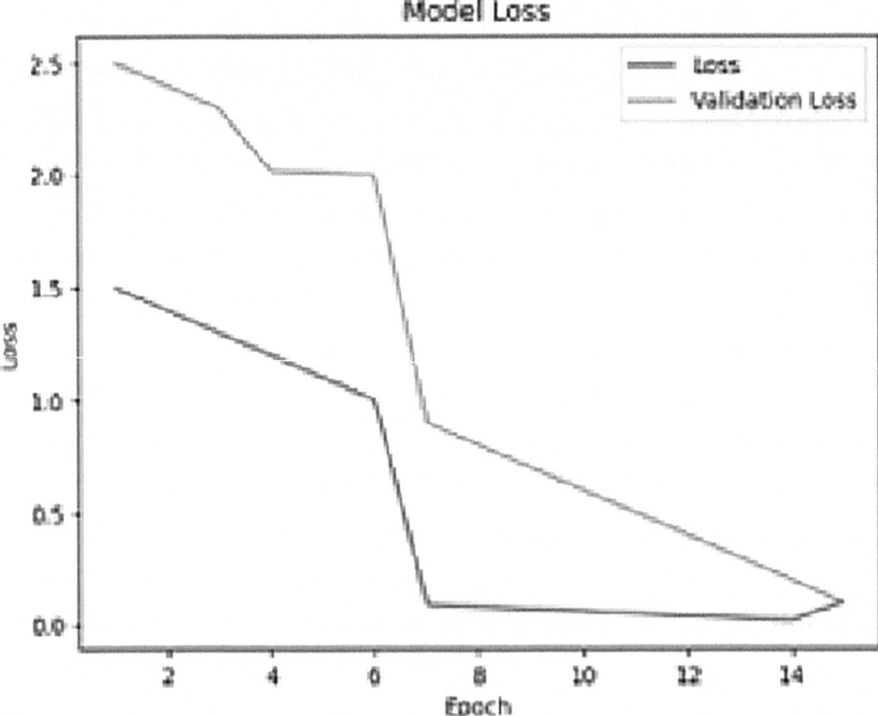

Fig. 2. Loss Graph

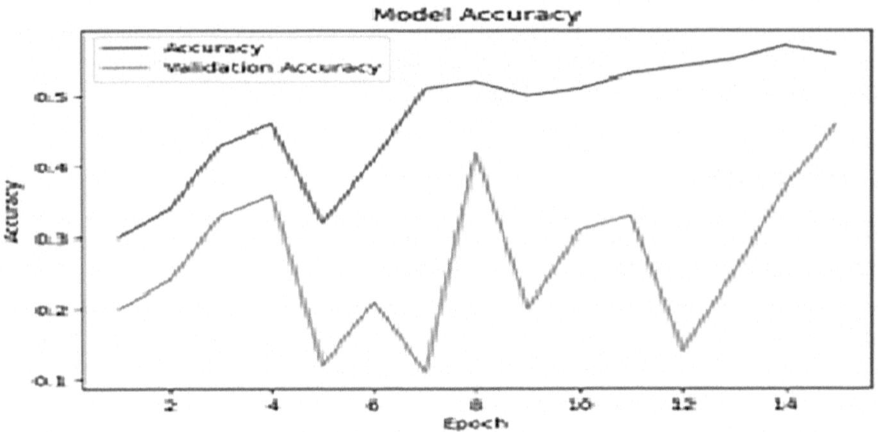

Fig. 3. Accuracy Graph

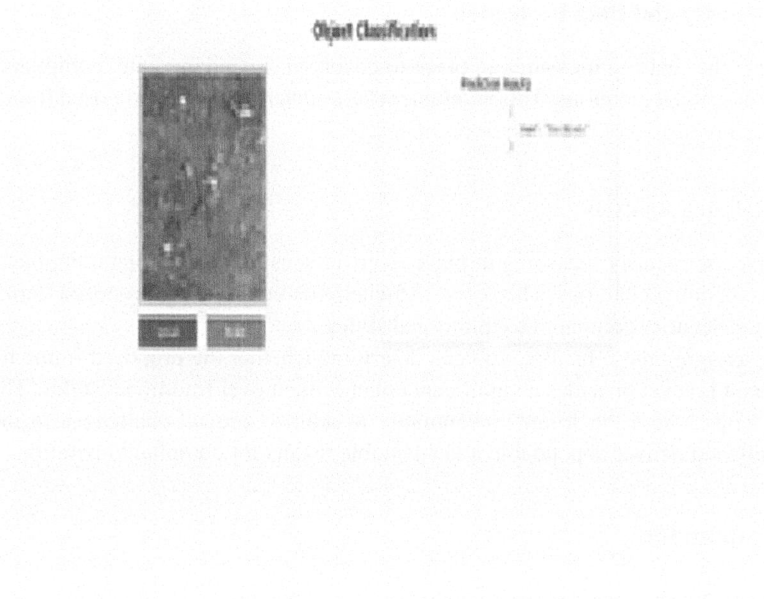

Fig. 4. Output Screen

7 Evaluation

Assessing the proposed frame for flesh complaint bracket involves several crucial criteria and a comparison with being styles:

7.1 Delicacy, Precision and Recall

To measure delicacy, perfection, and recall, compare the model's prognostications with a ground verity dataset containing labelled exemplifications of flesh conditions like coccidiosis and salmonella, as well as healthy cases. Delicacy is calculated as the proportion of correct prognostications to the total prognostications. Precision measures the rate of true cons to all prognosticate cons, while recall is the rate of true cons to the total factual cons.

7.2 Effectiveness and Scalability

Assess effectiveness by measuring the time needed to classify a set of images, comparing this with traditional styles or other models. Scalability is estimated by adding the dataset size and observing the impact on performance and resource operation.

7.3 Robustness

Test the robustness of the model by introducing noise or variations in the input data to pretend different conditions. A robust model should maintain delicacy despite these disturbances. Compare this with being styles to gauge the proposed frame's adaptability.

7.4 Security and Data Protection

Examine the security measures in place to cover sensitive data, icing compliance with data protection regulations. This assessment helps determine if the proposed frame offers enhanced security compared to traditional styles.

7.5 Relative Analysis

Examine the security measures in place to cover sensitive data, icing compliance with data protection regulations. This assessment helps determine if the proposed frame offers enhanced security compared to traditional styles.

By assaying these factors, you can determine whether the proposed frame for flesh complaint bracket provides a significant enhancement over traditional styles. This approach helps assess the frame's eventuality to address crucial challenges in the flesh assiduity and deliver dependable and adaptable results for complaint operation.

8 Conclusion

In conclusion, the proposed frame in the Feathered Diagnose design offers a comprehensive and innovative result to the challenges of complaint operation in flesh husbandry. By integrating Convolutional Neural Networks (CNNs) with MLOps principles, the frame significantly improves upon traditional styles of complaint discovery, furnishing increased delicacy, effectiveness, and scalability. This approach aims to reduce the profitable losses and food security pitfalls frequently associated with homemade labour and error-prone conventional practices.

The frame's objectification of real-time monitoring, model interpretability, and robust security measures establishes a dynamic and visionary system for managing flesh conditions. Its success in classifying conditions similar to coccidiosis demonstrates the eventuality of this technology to revise the flesh assiduity, offering intelligent strategies that can acclimatise to changing surroundings and evolving complaint patterns, eventually securing flesh health and icing profitable adaptability.

The Feathered Diagnose design demonstrates a strong commitment to advancing agrarian practices and promoting sustainable flesh husbandry. This holistic approach not only improves upon traditional styles but also encourages invention within the assiduity. With a focus on rigidity and nonstop enhancement, the design lays a solid foundation for the future of intelligent flesh complaint operation.

Looking ahead, unborn work on this design could explore several crucial areas. One avenue is expanding the dataset to include a broader range of flesh conditions, further enhancing the model's generalizability and robustness. Also, the frame could be extended to integrate other advanced technologies similar as edge computing and IoT bias for more expansive real-time monitoring and data collection. Another important aspect of unborn work is perfecting model interpretability, allowing druggies to understand the explanation behind the model's prognostications. This could involve developing more advanced visualisation tools or ways to give clearer perceptivity into the model's decision-making processes. Likewise, collaboration with the flesh husbandry community can play a significant part in enriching the frame, icing it meets the practical requirements of assiduity stakeholders. This collaboration could lead to the development of stoner-friendly operations that can be extensively espoused, promoting sustainable practices and better complaint operation.

Eventually, the Feathered Diagnose design sets a promising line for intelligent flesh complaint operation. By continuously evolving and conforming to new challenges, this frame can pave the way for a more secure, sustainable, and prosperous future for the flesh assiduity.

References

1. https://www.sciencedirect.com/science/article/pii/S2772375523000515
2. https://www.sciencedirect.com/science/article/pii/S0164121223000109#:~:text=This%20pipeline%20consists%20of%20four,AI%20to%20these%20four%20stages
3. https://proceedings.neurips.cc/paper_files/paper/2015/file/86df7dcfd896fcaf2674f757a2463eba-Paper.pdf
4. Kholil, M., Waspada, H.P., Akhsani, R.: Classification of infectious diseases in chickens based on feces images using deep learning. IEEE Access,
5. Ubbens, J., Cieslak, M., Prusinkiewicz, P., Stavness, I.: The use of plant models in deep learning: an application to leaf counting in rosette plants. Plant Methods **14**(1), 6 (2018). https://doi.org/10.1186/s13007-018-0273-z
6. Owomugisha, G., Quinn, J., Mwebaze, E., Lwasa, J.: Automated vision-based diagnosis of banana bacterial wilt disease and black sigatoka disease. In: 1st International Conference on the Use of Mobile ICT in Africa (2019)
7. Sadeghi, M., Banakar, A., Khazaee, M., Soleimani, M.: An intelligent procedure for the detection and classification of chickens infected by clostridium perfringens based on their vocalisation. RevistaBrasileira de Ciencia Avıcola **17**, 537–544 (2015). https://doi.org/10.1590/1516-635X1704537-544
8. Zhuang, X., Bi, M., Guo, J., Wu, S., Zhang, T.: Development of an early warning algorithm to detect sick broilers. Comput. Electron. Agric. **144**, 102–113 (2018). https://doi.org/10.1016/j.compag.2017.11.032
9. Hepworth, P.,Nefedov, A., Muchnik, I., Morgan, K.: Broiler chickens can benefit from machine learning: Support vector machine analysis of observational epidemiological data. J. R. Soc. Interface **9**, 1934–1942 (2012). https://doi.org/10.1098/rsif.2011.0852
10. Hemalatha, C., Muruganand, S., Maheswaran, R.: Recognition of poultry disease in real time using extreme learning machine. In: International Conference on Interdisciplinary Research in Engineering & Technology, pp. 44–50 (2014)
11. National Action Plan for Egg & Poultry-2022 For Doubling Farmers' Income by 2022 Department of Animal Husbandry, Dairying & Fisheries Ministry of Agriculture & Farmers Welfare, Government of India
12. Zhang, Z., Han, Y.: Detection of ovarian tumors in obstetric ultrasound imaging using logistic regression classifier with an advanced machine learning approach. IEEE Access **8**, 44 999–45 008 (2020). https://doi.org/10.1109/ACCESS.2020.2977962
13. Ashraf, R., et al.: Deep convolution neural network for big data medical image classification. IEEE Access **PP**, 1 (2020). https://doi.org/10.1109/ACCESS.2020.2998808
14. El-Kereamy, A., et al.: Deep learning for image-based cassava disease detection Front. Plant Sci. 1852 (2017). https://doi.org/10.3389/fpls.2017.01852
15. Ferentinos, K.: Deep learning models for plant disease detection and diagnosis. Comput. Electron. Agric. **145**, 311–318 (2018). https://doi.org/10.1016/j.compag.2018.01.009

16. Mbelwa, H., Mbelwa, J., Machuve, D.: Deep convolutional neural network for chicken diseases detection. Int. J. Adv. Comput. Sci. Appl. **12**(2), 759–765 (2021). https://doi.org/10.1016/j.compag.2018.01.009
17. Machuve, D., Nwankwo, E., Mduma, N., Mbelwa, H., Maguo, E., Munisi, C.: Machine learning dataset for poultry diseases diagnostics. Zenodo (2021). https://doi.org/10.5281/zenodo.4628934
18. LeCun, Y., Bottou, L., Bengio, Y., Haffner, P.: Gradient-based learning applied to document recognition. Proc. IEEE **86**(11), 2278–2324 (1998). https://doi.org/10.1109/5.726791
19. Nair, V., Hinton, G.E.: Rectified linear units improve restricted boltzmann machines. Proc. ICML. **27**, 807–814 (2010)
20. Bridle, J.: Training stochastic model recognition algorithms as networks can lead to maximum mutual information estimation of parameters. Adv. Neural. Inf. Process. Syst. **2**, 211–217 (1989)

Author Index

P. D. Sivakumar et al. (Eds.): IRCCTSD 2024, CCIS 2360, pp. 327–330, 2025.
https://doi.org/10.1007/978-3-031-82389-3

The manufacturer's authorised representative in the EU is Springer Nature Customer Service Centre GmbH, Europaplatz 3, 69115 Heidelberg, Germany. If you have any concerns regarding our products, please contact ProductSafety@springernature.com

Printed and bound by CPI Group (UK) Ltd, Croydon, CR0 4YY
15/07/2026
02167617-0005